Cell Synchrony

Studies in Biosynthetic Regulation

Contributors to This Volume

R. M. BOCK
WILLIAM F. BUSBY
I. L. CAMERON
J. R. COOK
KATSUMA DAN
JOHN W. DANIEL
L. H. ELROD
JOSEPH ENGELBERG
R. EPSTEIN
W. D. FISHER
H. O. HALVORSON
HENRY R. HIRSCH
T. W. JAMES
M. LA BERGE
KARL G. LARK
JOYCE C. LEWIN
ELLEN MATTINGLY
J. M. MITCHISON
G. M. PADILLA
POTU N. RAO
BERNHARD E. REIMANN
ELLIOT ROBBINS
MATTHEW SCHARFF
ROBERT R. SCHMIDT
NOBORU SUEOKA
P. TAURO
W. S. VINCENT
BENJAMIN E. VOLCANI
G. L. WHITSON
D. H. WILLIAMSON
B. W. WILSON

Cell Synchrony

Studies in Biosynthetic Regulation

Edited by

Ivan L. Cameron

Department of Anatomy
Upstate Medical Center, State University of New York
Syracuse, New York

George M. Padilla

Department of Physiology and Pharmacology
Duke University Medical Center
Durham, North Carolina
and
Wrightsville Marine Bio-Medical Laboratory
Wilmington, North Carolina

1966

ACADEMIC PRESS *New York and London*

ACADEMIC PRESS INC.
111 Fifth Avenue, New York, New York 10003

United Kingdom Edition published by
ACADEMIC PRESS INC. (LONDON) LTD.
Berkeley Square House, London W.1

LIBRARY OF CONGRESS CATALOG CARD NUMBER: 66-17288

PRINTED IN THE UNITED STATES OF AMERICA

List of Contributors

Numbers in parentheses indicate the pages on which the authors' contributions begin.

R. M. BOCK, Department of Biochemistry, University of Wisconsin, Madison, Wisconsin (102)

WILLIAM F. BUSBY, Scripps Institution of Oceanography, University of California, La Jolla, California (169)

I. L. CAMERON,* Biology Division, Oak Ridge National Laboratory, Oak Ridge, Tennessee (269)

J. R. COOK, Department of Zoology, University of Maine, Orono, Maine (153)

KATSUMA DAN, Biology Department, Tokyo Metropolitan University, Tokyo, Japan (307)

JOHN W. DANIEL, McArdle Laboratory for Cancer Research, University of Wisconsin, Madison, Wisconsin (117)

L. H. ELROD, Biology Division, Oak Ridge National Laboratory, Oak Ridge, Tennessee (269)

JOSEPH ENGELBERG, Department of Physiology and Biophysics, College of Medicine, University of Kentucky, Lexington, Kentucky (14, 332)

R. EPSTEIN, Department of Bacteriology, University of Wisconsin, Madison, Wisconsin (102)

W. D. FISHER, Biology Division, Oak Ridge National Laboratory, Oak Ridge, Tennessee (289)

H. O. HALVORSON, Department of Bacteriology, University of Wisconsin, Madison, Wisconsin (102)

HENRY R. HIRSCH, Department of Physiology and Biophysics, College of Medicine, University of Kentucky, Lexington, Kentucky (14)

T. W. JAMES, Department of Zoology, University of California, Los Angeles, California (1, 236)

* Present address: Department of Anatomy, Upstate Medical Center, State University of New York, Syracuse, New York.

M. LA BERGE, Department of Bacteriology, University of Wisconsin, Madison, Wisconsin (102)

KARL G. LARK, Department of Physics, Kansas State University, Manhattan, Kansas (54)

JOYCE C. LEWIN,* Scripps Institution of Oceanography, University of California, La Jolla, California (169)

ELLEN MATTINGLY, Biology Division, Oak Ridge National Laboratory, Oak Ridge, Tennessee (256)

J. M. MITCHISON, Department of Zoology, University of Edinburgh, Edinburgh, Scotland (328)

G. M. PADILLA,† Biology Division, Oak Ridge National Laboratory, Oak Ridge, Tennessee (269, 289)

POTU N. RAO, Department of Physiology and Biophysics, College of Medicine, University of Kentucky, Lexington, Kentucky (332)

BERNHARD E. REIMANN, Scripps Institution of Oceanography, University of California, La Jolla, California (169)

ELLIOTT ROBBINS, Department of Cell Biology, Microbiology and Medicine, Albert Einstein College of Medicine, Yeshiva University, New York, New York (353)

MATTHEW SCHARFF, Department of Cell Biology, Microbiology and Medicine, Albert Einstein College of Medicine, Yeshiva University, New York, New York (353)

ROBERT R. SCHMIDT, Department of Biochemistry and Nutrition, Virginia Polytechnic Institute, Blacksburg, Virginia (189)

NOBORU SUEOKA, Department of Biology, Princeton University, Princeton, New Jersey (38)

P. TAURO, Department of Bacteriology, University of Wisconsin, Madison, Wisconsin (102)

W. S. VINCENT, Department of Anatomy and Cell Biology, University of Pittsburgh, Pittsburgh, Pennsylvania (328)

BENJAMIN E. VOLCANI, Scripps Institution of Oceanography, University of California, La Jolla, California (169)

G. L. WHITSON, Biology Division, Oak Ridge National Laboratory, Oak Ridge, Tennessee (289)

D. H. WILLIAMSON, John Innes Institute, Bayfordbury, Hertford, England (81)

B. W. WILSON, Department of Poultry Husbandry, University of California, Davis, California (236).

* Present address: Department of Oceanography, University of Washington, Seattle, Washington.

† Present address: Department of Physiology and Pharmacology, Duke University Medical Center, Durham, North Carolina, and Wrightsville Marine Bio-Medical Laboratory, Wilmington, North Carolina.

Preface

As an experimental technique, cell synchrony was developed primarily for the amplification of time-limited events within the cell cycle. As such, cell synchrony is finding ever-increasing use with molecular biologists, cell biologists, and, even more so, with developmental biologists who seek to understand cell division and cell specialization. Although concerted efforts have been made only during the last twelve years to obtain synchronized cultures of cells, this particular approach has already opened new and important experimental and conceptual avenues to the study of cell regulation.

The idea of collating into a book experimental material on cell synchrony was first proposed to the editors largely as a result of a Cell Synchrony Conference held at the Biology Division of the Oak Ridge National Laboratory in October, 1964 [Padilla, G. M., Cameron, I. L., and Whitson, G. L. (1965), *Science* **147,** 175]. The book is directed to cell biologists and biochemists alike. We have attempted to widen the scope of the volume by including contributions from workers whose theories and methods employ procaryotic and eucaryotic cells. In this work the reader will find a broad coverage of many different cell systems, the substance of which contributes to the understanding of the evolutionary diversity between different types of cells. He will also be provided with clues for the classification of regulatory processes which operate in different cell types and in the same cell under different synchrony procedures.

We hope that the present volume will show a definite degree of differentiation within the school of cell synchronizers. Therefore we feel that this compilation will complement "Synchrony in Cell Division and Growth" [Zeuthen, E., Editor (1964), Wiley (Interscience), New York] and serve as a stimulus to further investigations in an area that demands intuitiveness, inventiveness, and patience. We hope that the suspicion that cell synchrony was at best an unnatural, artificial, and possible specious occupation for cell biologists will have been dissipated once and for all.

March, 1966

I. L. Cameron
G. M. Padilla

Contents

Chapter 3

SYNCHRONOUS REPLICATION OF THE CHROMOSOME IN *BACILLUS SUBTILIS*

Noboru Sueoka

Chapter 4

CHROMOSOME REPLICATION IN *ESCHERICHIA COLI*

Karl G. Lark

Chapter 5

NUCLEAR EVENTS IN SYNCHRONOUSLY DIVIDING YEAST CULTURES

D. H. Williamson

Chapter 6

PERIODIC ENZYME SYNTHESIS IN SYNCHRONOUS CULTURES OF YEAST

H. O. Halvorson, R. M. Bock, P. Tauro, R. Epstein, and M. La Berge

Chapter 7

LIGHT-INDUCED SYNCHRONOUS SPORULATION OF A MYXOMYCETE—THE RELATION OF INITIAL METABOLIC CHANGES TO THE ESTABLISHMENT OF A NEW CELL STATE

John W. Daniel

Chapter 8

STUDIES ON CHLOROPLAST REPLICATION IN SYNCHRONIZED *EUGLENA*

J. R. Cook

Chapter 9

SILICA SHELL FORMATION IN SYNCHRONOUSLY DIVIDING DIATOMS

Joyce C. Lewin, Bernhard E. Reimann, William F. Busby, and Benjamin E. Volcani

Chapter 10

INTRACELLULAR CONTROL OF ENZYME SYNTHESIS AND ACTIVITY DURING SYNCHRONOUS GROWTH OF *CHLORELLA*

Robert R. Schmidt

Chapter 11

ENERGETICS AND THE SYNCHRONIZED CELL CYCLE

B. W. Wilson and T. W. James

Chapter 12

SYNCHRONY OF CELL DIVISION IN ROOT MERISTEMS FOLLOWING TREATMENT WITH 5-AMINOURACIL

Ellen Mattingly

Chapter 13

THE PHYSIOLOGY OF REPETITIVELY SYNCHRONIZED *TETRAHYMENA*

G. M. Padilla, I. L. Cameron, and L. H. Elrod

Chapter 14

MORPHOGENETIC AND MACROMOLECULAR ASPECTS OF SYNCHRONIZED *TETRAHYMENA*

G. L. Whitson, G. M. Padilla, and W. D. Fisher

Chapter 15

BEHAVIOR OF SULFHYDRYL GROUPS IN SYNCHRONOUS DIVISION

Katsuma Dan

Chapter 16

A METHOD OF MAKING SYNCHRONOUS CELL CULTURES BY DENSITY GRADIENT CENTRIFUGATION

J. M. Mitchison and W. S. Vincent

Chapter 17

EFFECTS OF TEMPERATURE ON THE MITOTIC CYCLE OF NORMAL AND SYNCHRONIZED MAMMALIAN CELLS

Potu N. Rao and Joseph Engelberg

Chapter 18

SOME MACROMOLECULAR CHARACTERISTICS OF SYNCHRONIZED HELA CELLS

Elliott Robbins and Matthew Scharff

CHAPTER 1

Cell Synchrony, A Prologue to Discovery*

T. W. James

I. INTRODUCTION

If wisdom is a virtue that comes with age, there is good reason to approach the subject of cell synchrony historically and hope that a backward glance will pay dividends in future use. This is particularly so with a subject that has such a short history and displays such a variety of dimensions. Some might say that it should not be the subject at all and that our focus of attention ought to be the cell. Cell synchrony is a problem that touches principles whose foundations are the regulation of the cell cycle and involves methods which require theory and machinery to be solved. This volume is a recognition of these facts. The work on cell synchronization, like all endeavors aimed at solving real problems, demonstrated that complicated first methods always yielded to simplification. This has already been seen in the promising new methods, such as synchrony by selection, which have appeared and are beginning to be used.

* Supported by National Science Foundation grant 1643 awarded to T. W. James.

This is not to say that the old methods will have to be discarded or that they lack value. They have contributed and will continue to contribute to our understanding of the cell, and when these methods are reexamined, they may be used with greater understanding.

The development of techniques was of necessity the first direction synchrony took, but 2 other dimensions became superimposed upon it: the research objective of the investigator and the cell he selected. With the understanding that our view of the topic is not to be interpreted as a review, and our neglect of certain areas is necessitated by our desire to emphasize others, we will examine the subject along these 3 dimensions—method, objective, and cell species.

The first deliberate attempts to synchronize cell growth and division were not just accidental observations of synchrony. They were concerted efforts to produce simultaneous cell activities in mass populations imitating the division behavior seen in natural systems such as the simultaneous cleavages of marine eggs. Among the first cells to be artificially synchronized were *Chlorella* (Tamiya *et al.*, 1953), *Amoeba* (James, 1954, 1959), bacteria (Hotchkiss, 1954; Maaløe and Lark, 1954; Barner and Cohen, 1955), and *Tetrahymena* (Scherbaum and Zeuthen, 1954; Zeuthen and Scherbaum, 1954). Although the goals and the methodology have expanded and flourished in subsequent years, few would say that the sum of fundamental information obtained about the cell and the cell cycle offers little more than a prologue to future chapters in cell biology.

Discovering what controlled the onset of cell division was the obvious premium and also the driving force for the production of cultures of cells in synchronous division. Following the invention of techniques, measurements on the time course of DNA, RNA, and protein synthesis were initiated. Because of the belief that the effects of the synchronizing agents on the synthetic patterns of these constituents were less important than the sequence in which they occurred, they were temporarily ignored. As new organisms and new methods were introduced, this neglected aspect of the problem became more apparent. Unfortunately, some criticisms of these first methods gave the impression that one could synchronize a culture of cells without having to pay for the alignment by the introduction of distorting forces, but this critical point of view led to some new thoughts on the problem, i.e., a change from induction to selection methods for obtaining synchronous cultures. Happily, the nonacceptance of the early techniques did not preclude the acceptance of the idea of synchrony.

II. A CONCEPT OF CELL SYNCHRONY

Abbo and Pardee (1960) stressed the meaning of the word "synchronous" in contrast to the term "synchronized." The dictionary definition of synchronous is "... Concurrent in time, contemporaneous; simultaneous; or in its physical context, having the same period and phase...." In the past when the term was applied to the activity of cells, it was generally taken to be the property displayed as the simultaneous division of cells in a population, particularly the simultaneous cleavage of marine eggs or cells in an embryonic tissue. The term in this context suggests an intrinsic property of cells which does not require the intervention of an external agent other than some ontogenetic process in the development of the population. Cultures of cells in which divisions occur simultaneously are said to be synchronous. While the meaning of the term "synchronized" in its adjectival form can be taken to imply a system showing simultaneous divisions, in the broader verbal sense it can be equated to the production of a relationship between the inducing procedure and the concurrent division of a population of cells, i.e., a phase relationship between the synchronizing regime and the division phase of the population. The degree to which these ideas governed the early work on synchronization is conjecture, but it is clear that the procedures which have developed over the past 12 years reflect a considerable difference in investigators' concepts of the cell, the cell cycle, and synchronization. The distinction between synchronized and synchronous can now be taken to emphasize the difference between active production or induction of such cultures and the more passive role taken when selection methods are used to isolate synchronous subcultures from logarithmic parent cultures.

A. Division-Oriented versus Cycle-Oriented Synchrony

An acquaintance with the early work also leads one to appreciate the subtleties of concepts that have been important to the design of induction methods. The recognition of natural synchrony undoubtedly set the pattern for future thoughts on the nature of the cell cycle. For some time people had been repeating the rather hackneyed truism about the activity of the cell during the interphase of the cell cycle and the misnomer that has been perpetuated by calling it the "resting phase." Yet the examination of some early methods suggests that a vestige of the resting phase idea still remained. This was evident in the assumption that to make a population of cells divide simultaneously, interphase activities of the cell could be ignored. We would call this approach to synchronization "division-oriented" because the primary purpose is the production of simultaneous cell divisions irrespective of the kind and duration of treatment that is necessary to bring the result about. The significance of the interphase to this concept of synchronization is played down since the duration of treatment does not reflect the cellular generation time (Zeuthen and Scherbaum, 1954). Excellent simultaneity of cell divisions was and is obtained by this method, but at the sacrifice of producing oversized cells. Despite these distortions, the cells produced by such methods are highly viable and maintain their potential for future divisions. Another sacrifice is the distortion of the DNA synthetic period by such treatments relative to the DNA synthetic patterns over the cell cycle obtained by other methods (Scherbaum *et al.*, 1959; Padilla and Cameron, 1964; and Prescott, 1960).

Cycle-oriented induced synchrony, as distinct from the division-oriented method, was developed early (James, 1954, 1959, 1960; Padilla, 1960; Padilla and James, 1960; Cook, 1960; Tamiya *et al.*, 1953; and Lorenzen, 1957). These are typified by techniques that seek to fit or entrain the treatment to the generation time of the cell. Shifts in both temperature and light have been used in this manner, and their use is characterized by a repeating cycle such that the inducing cycle is in phase with the cell cycle. Single shifts were also used, but in many of these no attempt was made to integrate them with the generation time of the cells. There is a sacrifice with these methods, also, i.e., the method introduces biochemical transients whenever shifts are instituted, and the synchronous fission period shows a greater spread in time than that found using division-oriented techniques. The cell size is not obviously distorted, and, since the shifts are not severe, changes in metabolism are more accessible to analysis. For example, the DNA *S* period may be displaced relative to the time of fission but it is of the same character as found in unsynchronized cells. Since both division-oriented and cycle-oriented systems cannot be considered ergodic because they do not fit Engleberg's criteria (see Chapter 2), namely the constancy of the environment, comparisons between cycle-oriented and division-oriented systems have only limited meaning.

The distinction between division-oriented and cycle-oriented synchrony is interesting in terms of the historical development of methods and in terms of how strongly some of these attitudes govern the production of new synchronous systems. It can be argued that "division synchrony" can result from "cycle synchrony" and vice versa, and the position that is taken will depend on how strongly one assumes cell growth and cell division to be coupled.

B. General Comments on the Specificity of Inducing Agents

The specificity of the synchronizing agents has and should be of major concern. Shifts in temperature and light were among the first used and are perhaps the least specific. The most specific method, undoubtedly, was that utilizing thymine to control the division of the thymineless mutant of *Escherichia coli* (Barner and Cohen, 1955). But obviously, if one asks what event in the cell cycle is inhibited or triggered by either changes in temperature, light, nutrition, respiratory gases, or inhibitors, agreement would be impossible to find, mainly because the specificity of any agent must be evaluated against the cell type that is being employed. The procaryote-eucaryote spectrum of cellular complexity is a useful guide and a conceptual necessity in making comparisons of this kind (Stanier and van Niel, 1962). We shall emphasize this aspect of the problem later. Obviously, the more complicated the cell type the greater the opportunity for interaction between different mechanisms in the cell and the agent being used. Thus, some synchronizaing agents may be specific for bacteria and other procaryotic cells but may not work effectively on eucaryotic systems. For example, if one confines this question to cycle-oriented systems in which eucaryotic cells are synchronized, moderately low temperatures tend to permit interphase activity, while warm temperatures favor mitotic activity (Padilla, 1960). In the case of light cycles, light favors interphase growth and darkness favors division processes (Cook, 1960). But these are hardly specific, and the explanation of the effects are not this simple. (See Rao and Engleberg, Chapter 17.) In most temperature-induced systems the DNA *S* period is inhibited or delayed by the cold phase and is reinitiated by the warm phase, but the temperature changes cannot be interpreted as specific agents in that one cannot expect their effects to be confined to one or a few such components. Both light cycle- and temperature cycle-induced synchrony display elements that might be called nutritional. In the case of light control, the relationship of photosynthesis to nutrition is obvious; while in the nonphotosynthetic forms the effects of temperature on respiratory and diffusion rates are tantamount to a nutritional control. The nonspecificity of these agents is further compounded by the possibility that differential synthesis of energy-providing elements may be occurring, i.e., synthesis of chloroplasts and mitochondria (James, 1965). And, of course, appraising the specificity of an agent should also include the effects of temperature on pathway selection mechanisms.

Furthermore, depending on the degree of specificity, the agent may initiate the appearance of specific biochemical entities independent of the cell cycle. This is particularly so wherever temperature is used and rate phenomena are being studied. To eliminate these it is necessary to show, at least in the comparative sense, that the magnitude of the activity or rate is inherent to the cell at that stage irrespective of the temperature used in the determination (James, 1961).

TABLE I

Methods for Obtaining Synchronous Cultures

Category	Principle	Representative references
INDUCED SYNCHRONY	Synchronous cultures are produced by treatment of the culture	
Temperature methods	Temperature changes are used as the treatment	
Cycles	Culture is subjected to a cold-warm cycle equal to one generation. Cycle is repeated with renewal of medium with each cycle	James (1954, 1959, 1965); Padilla and James (1960); Padilla and Cameron (1964)
Shifts	Temperature is changed from a low to high level. Divisions occur in warm period	Maaløe and Lark (1954); Hotchkiss (1954)
Shocks	Temperature is raised and lowered in rapid succession in a series of pulses. Pulses range from optimal or suboptimal to supraoptimal. Divisions occur following a period at the lower temperature	Scherbaum and Zeuthen (1954); Zeuthen and Scherbaum (1954)
Light methods	Light changes are cue for the entrainment of cell cycle	
Cycles	Light and dark cycles are used to entrain cultures of photosynthetic cells such that divisions occur one each cycle. Strength and duration of light period must be such as to give balanced growth	Tamiya *et al.* (1953); Lorenzen (1957); Cook (1960); Schmidt (Chapter 10)
Nutritional methods	Growth on basal medium is followed by addition of enriched medium called a shift-up, or vice versa, a shift-down. Shifts from starvation level to enriched medium are also used	
Auxotrophic control	Uses a change in a single requirement holding other requirements in excess. Genetic strains with specific requirements are most effective	Barner and Cohen (1955); Burns (1959, 1961); Lewin (Chapter 9); Stevenson (1961)
Heterotrophic control	Makes use of changes on broad spectrum requirements, i.e., carbon source, nitrogen source. Divisions occur after a lag following enrichment	Williamson and Scopes (1960); Maaløe and Kurland (1963)
Gasometric methods		
Aerobic-anaerobic	Uses single or multiple cycles of bubbling nitrogen then air through the culture in a fixed pattern in time to induce synchrony. Controlled gas tensions may also be a mode of control	R. J. Neff and Neff (1964)
Inhibitory methods	Employs an inhibitor of cell division or a stage in the cell cycle and a washout procedure to accumulate cells at a specific stage of cycle. Excess of natural metabolites are most promising, i.e., thymidine	Xeros (1962); E. C. Anderson and Petersen (1964)
SELECTION SYNCHRONY	Synchronous cultures obtained by isolation of a specific stage	
Sizing methods	Depend on correlation between cell size and cell age. Selects small fraction from a logarithmic culture to start synchronous subculture	
Filtration	Small size bacteria are obtained by expressing slurry of cells through a filter pad. Care is taken to maintain other conditions constant. Small cells are subcultured	Maruyama and Yanagita (1956); Abbo and Pardee (1960); P. A. Anderson and Pettijohn (1960)
Sedimentation velocity	Depends on difference on sedimentation rate with cell size. Small cells are removed from upper portion of a density gradient to form synchronous subculture	Mitcheson and Vincent (Chapter 16)

TABLE I—*Continued*

Category	Principle	Representative references
Grow-off methods	Adsorb cells onto filter pack. Parent cell attaches, daughter cell falls off at fission	Helmstetter and Cummings (1963)
Substrate attachment	Tissue culture cells and amoeboid forms attach to the substrate during interphase of the cell cycle. At time of division they tend to round up and detach. Collection of detached forms provides subculture of synchronous cells	Terasima and Tolmack (1963)

C. A Summary of Synchronizing Methods

Table I is a series of brief statements of synchronizing methods indicating various agents or various methods of selection. It is presented to provide a perspective from which some comparative insight may be gained. The 2 major categories, synchrony by induction and synchrony by selection, are not just a classification convenience but represent an important transition in methodology.

As can be seen from the few cases of selection synchrony cited above, this approach to the production of synchronous cultures is still in its infancy. One has only to think of this problem in a general form to realize that mechanical separation on the basis of any one of several biological properties that change over the cell cycle can be used to obtain synchronous subcultures. Loss of locomotive apparatus, such as flagella (James, 1963), that occurs during the fission period might also be exploited to obtain synchronous subcultures. Phototropism (Pittendrigh and Bruce, 1959), galvanotropism, or geotropism may also vary over the cell cycle in a manner that could be used for this same end.

Obtaining a synchronous culture of cells by the use of a selection technique is theoretically the most satisfactory method. If properly done, the cell population should be free from the distortions that result with changes in temperature, light, nutrition, or other factors. That this should prove to be so, in practice, requires that the selection process itself not introduce the same undesirable distortions. Since the method depends on the momentary existence at some time in the cycle of a specific physical characteristic upon which an isolation technique can operate, it is obvious that the percentage yield from a logarithmically growing culture will be small. The efficiency with which this can be done and the purity of the resulting sample can be determined only from a thorough knowledge of the system. The introduction of automated methods, to effect mechanical separation of cells at a given stage, may provide the experimenter with a continuous culture of cells at whatever age he desires and thus alleviate the problem of low yield.

III. SYNCHRONY AND THE EVOLUTION OF CELL TYPES

The general pattern of cell growth and division that is the foundation of cell theory has tended to emphasize the similarities rather than the differences between cell types. While this tendency to point to the unity rather than the diversity between cells has many pedagogical virtues, it has the disadvantage of leading one into making false analogies. One of the most sobering views of the diversity among cells is the dichotomy between the procaryotic and the eucaryotic types (Stanier and van Niel, 1962). The value of this classification is that it provides 2 alternative interpretations, one being that these are distinct and easily separable categories and the other is that these categories are in reality poles of cell evolution.

An examination of the distinguishing characteristics between the procaryotic cells, namely, bacteria and blue-green algae, and the eucaryotes, namely, higher protists, plant and animal tissue cells, etc., gives rise to a spectrum of questions that can be asked at the level of cell growth and the division cycle. It is a simple matter to ask about the meaning of these different degrees of complexity and consider what the use of synchronization techniques could add to their elucidation.

One example is apparent, i.e., the synthesis of DNA over the cell cycle. In many procaryotic cells that have been examined the general conclusion is that DNA synthesis is essentially linear over the generation time except for a short period at the time of fission (Hanawalt *et al.*, 1961; Abbo and Pardee, 1960; Schaechter *et al.*, 1959). Other patterns have been found but only when environmental conditions such as temperature, nutrition, etc., are out of the normal growth range, or synchrony has been induced by unknown changes in the environment. Even in these cases the synthetic phase often appears to be linear. Linear synthesis is consistent with the present findings that the bacterial chromosome is a single, circular, long molecule of DNA (Cairns, 1963a). The functional relationship between the replicon, initiator, and mesosome in bacteria is being explored in several laboratories (Jacob *et al.*, 1963; Lark, 1963; Maaløe and Hanawalt, 1961). The work of Lark presented in a later chapter illustrates the importance of these concepts to the duplication of DNA and the cell cycle. The refined synchronization techniques using the highly specific thymineless mutant of *Escherichia coli* (Barner and Cohen, 1955) were not only the foundation of these investigations but are essential to experiments aimed at studying the onset and control of the bacterial cell cycle. To bridge the conceptual gap between the procaryotic chromosome and those found in eucaryotic cells, one must ask and answer many questions. For example, is an individual chromosome of the eucaryotes a single DNA molecule? If so, according to

Cairns (1963b) it would be about 20,000 μ long. Is it circular? Is the centromere comparable to the mesosome in bacteria? Does the histone control duplication and transcription as well as structurally stabilize the chromosome? And, finally, how could these questions be answered by the use of synchronized cells?

The synthesis of DNA in eucaryotic cells is now known to be discontinuous in time and has been well documented by labeling techniques (Howard and Pelc, 1953) which have made G_1, S, and G_2 periods of cell cycle part of the cytological literature. Work on synchronous cultures of cells have supported and added to this data, but just how the synthetic period reflects the synthesis of individual chromosomes is only beginning to be explored. The relationship between the synthetic period and the structure of the chromosomes may not be much different than that seen in procaryotes. At this juncture, it is interesting to note that Cairns' estimate of the time required to synthesize 20,000 μ of DNA (one human chromosome), assuming rates similar to those in bacteria, is 6½ hours. This is in good agreement with other observations on the duration of the S period of mammalian cells (Cameron and Greulich, 1963).

One reason that the discovery of the length of the bacterial chromosome proved surprising is that it seems to be a mechanical enigma. Apparently its fragility was more assumed than real. It was the preconceived assumption of mechanical fragility that made the length of the molecule so incredibly unbelievable. That this preconception should also operate in our thinking about eucaryote chromosomes is perhaps the reason for believing that they, too, could not be single molecules. The fallacy in this assumption is that there is no mechanical basis to justify a belief that an increase in length from 1000 μ to 20,000 μ will alter the fragility within the cell, since at this length distant portions of the molecule can have little mechanical influence on other regions.

Taylor's work (1963) on the duplication of *Crepis* chromosomes, in a sense, anticipated the finding of replication polarity in bacterial chromosomes, and his model for DNA structure in chromosomes recognizes the necessity of having a linear tandem array of molecules running the length of the unit. Inoue's (1964) polarization studies on the chromosomes in the sperm of cave crickets would also support the idea of the chromosome as either a single molecule or a tandemly linked set of molecules. If there is a structural correspondence between procaryotic and eucaryotic chromosomes at the DNA level, then one must ask if circularity also is to be found in the latter. The dissynchrony of chromosomal synthesis (Taylor, 1963) led to the recognition of the nonuniform labeling of certain late chromosomes. Labeling appeared to occur first at the centromere and then one arm and finally the other arm. Would details of this pattern support the conclusion that the chromosomal DNA is a collapsed circle which, when labeled from one point (centromere) and in one direction, would yield Taylor's observed pattern?

It is obvious that some analysis for changes in several enzymes through the *S* period of a synchronized eucaryotic cell might be an alternate means of seeing if a mappable sequence occurs in the eucaryotic chromosomes which can be related to time in the *S* period. Unfortunately, the temporal spread of the synchronized *S* period results from the inclusion of several variables, namely, the duration of DNA synthesis in the individual cell and the variation of the time of onset of DNA synthesis between cells in the populations. The dissynchrony at the chromosomal level that was noted by Taylor (1963) further suggests that although the *S* period may be a composite of the synthesis of several chromosomes occurring simultaneously or nearly so, some chromosomes may be out of phase in terms of the time of their onset or completion. Therefore, the observed duration of the *S* period and the shape of the DNA increase curve in synchronized cultures can be attributed to the interplay of all of these variables. The total synchronized *S* period would be the sum of the time required to complete the longest chromosome plus the time required to complete the out-of-phase portions of the others, as well as the dispersions that are contributed by dissynchrony between cells in the population.

In other eucaryotic cells, in particular the colored and colorless Euglenids (Cook, 1961; Blum and Padilla, 1962) and *Tetrahymena* (Padilla and Cameron, 1964), the shape of the DNA increase curve (*S* period) in synchronized cells is varied, being essentially linear in some cases while a steep step in others. These were measured on temperature- and light-induced synchronous systems and therefore they may be displaced in time or otherwise distorted. Nevertheless, they indicate that any one of several patterns in the G_1, *S*, and G_2 period arrangement can exist without affecting the subsequent division of the cell. The key to this problem, i.e., to what degree DNA, RNA, and protein synthesis control the onset of division, is less apparent from the mass of experiments that have been done on higher cell types than from the work on bacteria and yeasts (see Lark, Chapter 4 and Halvorson, Chapter 6) which indicates that some phase of protein synthesis is essential to the initiation of the bacterial cycle (Lark, 1963; Maaløe and Hanawalt, 1961). Furthermore, studies on the time of enzyme synthesis in synchronized bacteria (Abbo and Pardee, 1960; Sueoka, Chapter 3) go far to suggest that the DNA chromosome is both the linear map and the temporal map which is readout in terms of the stepwise increase (perhaps doubling) of each enzyme over the cell cycle. In cells that have only one linkage group, synchronous cultures offer a method of mapping the position of cistrons as an adjunct to genetic and transfer mapping procedures. Halvorson's work on yeast (Chapter 6) is wonderful proof that this also occurs in higher cell types and in those that contain several chromosomes. These data again support the idea that the readout of the chromosome has a definite polarity. If transcription happens in the *S* period of the bacterial and the yeast cell cycle, which is to say approximately over its entire cycle, does it also occur in the *S* period of higher cell forms where there are well-formed nuclear membranes, numerous differentiated chromosomes, and pauses in DNA synthesis during the G_1 and G_2 and the mitotic periods? As stated previously, some hint of duplication polarity in higher chromosomes has been found by Taylor (1963) in *Crepis* and hamster and by Prescott and Bender (1962) in hamster chromosomes. Furthermore, the time-linked and polar synthesis of DNA and RNA along the nucleus of Euplodes (Prescott and Kimball, 1961) adds to the probability that this scheme may be correct. The temporal association of nucleic acid synthesis as shown in synchronized cultures of other cells (Blum and Padilla, 1962; Iwamura, 1960) can also be taken as examples that imitate to some degree the *S* period coupled synthesis in bacteria and yeast.

The study of RNA and protein synthesis in synchronous cultures is still fallow ground, which to be properly examined requires special attention to some highly specific enzymes and their mRNA's. We shall not discuss this aspect of the problem of cell type and synchronization but will concern ourselves with another topic which we feel is pertinent to the synchronizability of cells and their evolutionary position.

In the evolutionary sense, the derivation of higher cell types can be postulated to occur by several routes (Lederberg, 1952; Stanier and van Niel, 1962). One is by the evolution of an extensive and complex genome through duplication and mutation followed by the retention of the new genome in the cell, i.e., a gradual transition from a unichromosomal to multichromosomal cell, and the other is by the summation of several independent genomes to form a symbiotic union in a single cell, each contributing its characters to the whole. Plastids and other loci of extranuclear DNA are strong arguments for the latter hypothesis. Thus, problems related to the integration of the activity of the nuclear with the extranuclear genetic material over the cell cycle becomes an area where the use of synchronization techniques are warranted. In the evolutionary sense, how do each of the DNA synthetic periods of the cytoplasmic units become meshed in the final form? Extensive work in areas related to this question is lacking. First, the evaluation of morphogenetic changes in subcellular structure over the cycle is seriously needed in a large number of different eucaryotic cell types. The reasons for this lack hinge on the difficulty of assigning cell age to cells seen in electron micrographs unless there is some internal age marker. Although synchronized populations of cells provide the best opportunity for finding cells in rare stages, the population's statistical composition is such that there is no means of knowing the age precisely.

The functional activity of the extranuclear DNA and its associated structures also needs study over the cycle. One is tempted to say that this can be done only through the use of synchrony except that such certainty is always subject to refutation by the success of other techniques. Nevertheless, the mode of synthesis of the DNA in extranuclear structures as a function of the cycle raises many challenging questions.

How and when are mitochondria, chloroplasts, basal bodies, kinetosomes, Golgi apparatus, lysosomes, etc., synthesized over the cycle? How many of these contain DNA? What is it coded for? Is it sufficient in amount to be the code for the production of the structure and the function? How is its synthetic function controlled relative to the nuclear DNA? What is the source of the ribosomes that are important to such extranuclear synthesis? Do these ribosomes come from the nucleolus? A list of questions of this kind could be interminable, but discovery of answers to any would be a delight. If progress in cell biology is linked to our capacity to control our experimental material, the use of synchronous cultures will help charge the future with discovery.

REFERENCES

Abbo, F. E., and Pardee, A. B. (1960). *Biochim. Biophys. Acta* **39,** 478.
Anderson, E. C., and Petersen, D. F. (1964). *Exptl. Cell Res.* **36,** 423.
Anderson, P. A., and Pettijohn, D. A. (1960). *Science* **131,** 1098.
Barner, H. D., and Cohen, S. S. (1955). *Federation Proc.* **14,** 177.
Blum, J. J., and Padilla, G. M. (1962). *Exptl. Cell Res.* **28,** 512.
Burns, V. W. (1959). *Science* **129,** 566.
Burns, V. W. (1961). *Exptl. Cell Res.* **23,** 582.
Cairns, J. (1963a). *Cold Spring Harbor Symp. Quant. Biol.* **28,** 43.
Cairns, J. (1963b). *Endeavour* **22,** 141.
Cameron, I., and Greulich, R. (1963). *J. Cell Biol.* **18,** 31.
Cook, J. R. (1960). Ph.D. Thesis, University of California, Los Angeles.
Cook, J. R. (1961). *Biol. Bull.* **121,** 277.
Hanawalt, P. C., Maaløe, O., Cummings, D. J., and Schaechter, M. (1961). *J. Mol. Biol.* **3,** 156.
Helmstetter, C. E., and Cummings, D. J. (1963). *Proc. Natl. Acad. Sci. U.S.* **50,** 767.
Hotchkiss, R. D. (1954). *Proc. Natl. Acad. Sci. U.S.* **40,** 49.
Howard, A., and Pelc, S. R. (1953). *Heredity* **6,** Suppl., 261.
Inoue, S. (1964). Seminar presented at University of California, Los Angeles.
Iwamura, T. (1960). *Biochim. Biophys. Acta* **42,** 161.

Jacob, F., Brenner, S., and Cuzin, F. (1963). *Cold Spring Harbor Symp. Quant. Biol.* **28,** 329.

James, T. W. (1954). Ph.D. Thesis, University of California, Berkeley.

James, T. W. (1959). *Ann. N.Y. Acad. Sci.* **78,** 501.

James, T. W. (1960). *Ann. N.Y. Acad. Sci.* **90,** 550.

James, T. W. (1961). *Pathol. Biol.* **9,** 510.

James, T. W. (1963). *In* "Cell Growth and Cell Division" (R. J. C. Harris, ed.), pp. 9–26. Academic Press, New York.

James, T. W. (1965). *Exptl. Cell Res.* **38,** 439.

Lark, K. G. (1963). *In* "Molecular Genetics" (J. H. Taylor, ed.), p. 153. Academic Press, New York.

Lederberg, E. M. (1952). *Physiol. Rev.* **32,** 403.

Lorenzen, H. (1957). *Flora* **144,** 473.

Maaløe, O., and Hanawalt, P. C. (1961). *J. Mol. Biol.* **3,** 144.

Maaløe, O., and Kurland, C. G. (1963). *In* "Cell Growth and Cell Division" (R. J. C. Harris, ed.), p. 93. Academic Press, New York.

Maaløe, O., and Lark, K. G. (1954). *In* "Recent Developments in Cell Physiology" (J. A. Kitching, ed.), p. 159. Butterworth, London and Washington, D. C.

Maruyama, Y., and Yanagita, T. (1956). *J. Bacteriol.* **71,** 542.

Neff, R. J., and Neff, R. H. (1964). *In* "Synchrony in Cell Division and Growth" (E. Zeuthen, ed.), pp. 213–246. Wiley (Interscience), New York.

Padilla, G. M. (1960). Ph.D. Thesis, University of California, Los Angeles.

Padilla, G. M., and Cameron, I. L. (1964). *J. Cellular Comp. Physiol.* **64,** 303.

Padilla, G. M., and James, T. W. (1960). *Exptl. Cell Res.* **20,** 401.

Pittendrigh, C. S., and Bruce, V. G. (1959). *In* "Photoperiodism and Related Phenomena in Plants and Animals," Publ. No. 55, p. 475. Am. Assoc. Advance Sci., Washington, D. C.

Prescott, D. M. (1960). *Exptl. Cell Res.* **19,** 228.

Prescott, D. M., and Bender, M. A. (1962). *Exptl. Cell Res.* **26,** 260.

Prescott, D. M., and Kimball, R. F. (1961). *Proc. Natl. Acad. Sci. U.S.* **47,** 686.

Schaechter, M., Bentzon, M. W., and Maaløe, O. (1959). *Nature* **183,** 1207.

Scherbaum, O. H., and Zeuthen, E. (1954). *Exptl. Cell Res.* **6,** 221.

Scherbaum, O. H., Louderback, A. L., and Jahn, T. L. (1959). *Exptl. Cell Res.* **18,** 150.

Stanier, R. Y., and van Niel, C. B. (1962). *Arch. Mikrobiol.* **42,** 17.

Stevenson, I. (1961). *Can. J. Microbiol.* **1,** 661.

Tamiya, H., Iwamura, T., Shibata, K., Hase, E., and Nihei, T. (1953). *Biochim. Biophys. Acta* **12,** 23.

Taylor, J. H. (1963). *In* "Cell Growth and Cell Division" (R. J. C. Harris, ed.), pp. 161–177. Academic Press, New York.

Terasima, T., and Tolmack, L. J. (1963). *Biophys. J.* **3,** 11.

Williamson, D. H., and Scopes, A. W. (1960). *Exptl. Cell Res.* **20,** 338.

Xeros, N. (1962). *Nature* **194,** 682.

Zeuthen, E., and Scherbaum, O. H. (1954). *In* "Recent Developments in Cell Physiology" (J. A. Kitching, ed.), pp. 141–156. Butterworth, London and Washington, D. C.

CHAPTER 2

On the Theory of Synchronous Cultures

Joseph Engelberg and Henry R. Hirsch

I. INTRODUCTION

As suggested by theory (Engelberg, 1961, 1964b; Hirsch and Engelberg, 1965a,b,c) and shown by experiment (Scherbaum and Zeuthen, 1955; Anderson and Petersen, 1964), the synchronization of a real culture decreases with time and the culture approaches steady-state (exponential) growth. A number of questions having theoretical significance and practical implications suggest themselves. (*a*) How can the statistical distribution of cell doubling-times be inferred from data on the decay of synchronization?

(*b*) Given the doubling-time distribution how can the future history of a culture be predicted? (*c*) How can the instantaneous state of a synchronous culture be mathematically described? (*d*) How can the age distribution of cells be calculated at any time?

In this study answers to the preceding questions will be sought in relation to an ergodic culture, a culture having particularly simple mathematical properties. The first part of this report will be devoted to definitions and discussions of terms; the next part will be concerned with the development of methods for the determination of doubling-time distributions, age distributions, and state functions from data on the decay of synchronization of ergodic cultures; the last part will consist of a brief analysis of a special nonergodic culture, the cyclic culture. The methods used in this study are statistical in nature and are applicable to cultures containing large numbers of cells.

II. FINITE DIFFERENCE CONVENTIONS

The approach of finite difference mathematics lends itself to the type of analysis with which we will be dealing. The time axis is subdivided by a series of equidistant points. The points are sequentially labeled by integral numbers . . . $-2, -1, 0, 1, 2, \ldots (t-1), t, (t+1) \ldots$. The difference in time between two adjacent points will be chosen as the unit of time. Thus, at point t, t units of time have elapsed since zero time.

Let n_t be the number of cells in a culture at time t. Then the number of cells, Δn_t, added to the culture *between* t and $(t+1)$ is defined to be

$$\Delta n_t = n_{t+1} - n_t \tag{1}$$

It is a rather simple matter to go from equations written in finite difference notation to the forms of differential and integral calculus. Thus,

$$F_a = \sum_{T=1}^{T=a} f_T \rightarrow F(a) = \int_0^a f(T)\, dT$$

$$\Delta n_{t-a-1} \rightarrow dn(t-a)/dt$$

It is to be noted that in the limit of continuous functions a single unit of time is negligible and $(t-a-1) \rightarrow (t-a)$.

III. DOUBLING-TIME DISTRIBUTIONS

The doubling-time, T, of a cell may be defined as the time from one anaphase to the next. The doubling-time is a probabilistic quantity which varies from division to division. For a given cell in anaphase one specifies the probability of a daughter of this cell reaching the next anaphase a definite time, T, later; one does not attempt to predict the exact doubling-time of the daughter.

The pattern of doubling-times in a given culture can be described by a statistical distribution function. The *double-time distribution* function, f, is a function such that the *component* f_T of this function is the fraction of cells having a doubling-time of T time units. Of the cells which are born in the time interval between t and $t + 1$, for example, the fraction f_3 will divide again in the time interval between $t + 3$ and $t + 4$. It will be assumed that $f_0 = 0$, i.e., no cell can pass through its mitotic cycle in zero time.

A representative doubling-time distribution is shown in Fig. 1. The distribution has been truncated at $T = w$ and $T = z$, implying that $f_T = 0$ for $T < w$ and $T > z$. The postulated cut-off at w may be justified on physicochemical grounds—there will be some minimum time needed for a cell to synthesize a cell mass equal to itself. This minimum time is assured by the finite time needed for metabolites to diffuse into the cell and from the finite nature of chemical reaction rates.

The truncation at $T = z$ is an approximation motivated by the fact that in proliferating cultures the pattern of cellular proliferation is dominated by cells having short doubling-times. This can be seen by means of a simple example. Consider the 2-point, doubling-time distribution, $f_{10} = 0.98$, $f_{20} = 0.02$. Thus, 2% of the cells dividing at a given time require roughly twice the average doubling-time for their next division. By use of techniques described in other sections of this paper it can be shown that when this culture is in exponential growth the 2% slow dividers contribute only 1% of the cells dividing at any given time. Thus, by neglecting 2% of the population one makes an error of only 1% in the growth curve. It follows that there will always exist a point on the doubling-time distribution where the distribution can be truncated. The location of this point will

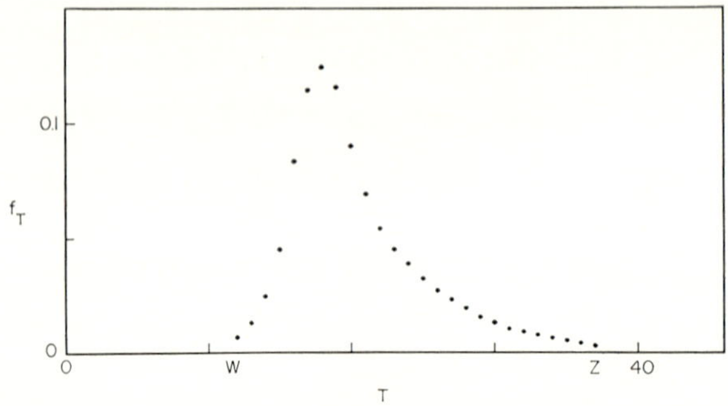

FIG. 1. Doubling-time distribution function, f. The probability of a cell which has passed through anaphase passing through its next anaphase a time T later is f_T. It is assumed that no cell has a doubling-time, T, shorter than w, and that a negligible fraction of the population has a doubling-time larger than z.

depend upon what error one is willing to tolerate. However, the resulting error is always smaller than the fractional area under the f curve which is neglected. This comes about, of course, because the fast dividers divide more often and thus contribute more cells to the culture.

Since the components of f are fractions of a population it follows that

$$\sum_{T=1}^{\infty} f_T = 1 \qquad \text{or} \qquad \sum_{T=w}^{T=z} f_T = 1 \tag{2}$$

Nondividing cells (cf. Engelberg, 1964c) are represented by the component f_∞. The presence of nondividing cells, however, will be neglected in this study.

For use in calculations of age distributions it is convenient to define the *cumulative doubling-time distribution*, F_a, by

$$F_a = \sum_{T=1}^{T=a} f_T \tag{3}$$

IV. ERGODIC CULTURES

The laws which govern the pattern of cellular proliferation in cell cultures are often of a complex nature. To make our subject amenable to mathematical analysis the discussion will be confined to a rather simple model of cellular proliferation. This model bears a relationship to real cultures which is similar to that which exists between the ideal gas model and real gases. For systems which depart appreciably from ideal behavior correction factors are introduced into the equations making these then suitable for the analysis of more complex situations.

In the subject under study the model analogous to the ideal gas is the *ergodic culture*. An ergodic culture is defined to be *a culture in which the doubling time distribution, f_T, does not vary with time*. The ergodic property implies a constancy of the cellular environment as well as of the cellular gene pool. Constancy of the cellular environment results when the chemical composition and physical parameters, such as the temperature of the growth medium, remain constant with time. Constancy of the gene pool implies that for the period of experimental observation genetic changes which alter f_T do not take place. A culture would not be expected to be ergodic when nutrients are depleted and metabolic products accumulate in the growth medium, or when mutants which alter the doubling-time distribution arise. A culture which has been maintained for many generation periods under conditions where fresh medium is introduced at a rate sufficient to maintain constancy of the cellular environment may be expected to be a good approximation to an ergodic culture.

What presently used synchronous cell systems are likely to be ergodic? To establish the ergodic nature of a culture one needs to determine the doubling-time distribution function as a function of time. In certain situations, however, there are *a priori* considerations which would suggest ergodicity, as in the following examples. (*a*) Helmstetter and Cummings (1963): Bacteria are adsorbed on a membrane filter. The filter is washed continuously with growth medium. It is found that only bacteria in the process of dividing are eluted from the surface of the filter. (*b*) Terasima and Tolmach (1963): These workers take advantage of the fact that mammalian cells grown attached to glass in tissue culture round up at the time of division. The dividing cells are washed off the glass surface and constitute a synchronous cell sample. In both cases if the cellular environment is kept reasonably constant during the process of washing or elution, the resulting culture may be expected to be approximately ergodic.

V. GROWTH CURVE PREDICTION

When a portion of the growth rate curve (Δn_t versus t) of an ergodic culture is known over a time interval greater than z, then the entire future history of the growth pattern can be predicted. Let us compute the number of cells, Δn_t, which divide at time t (i.e., between t and $t+1$). Consider the Δn_{t-j} cells which divided at some previous time $t-j$. Immediately after this division there were $2\Delta n_{t-j}$ cells of age zero. The fraction of these cells dividing again at time t is f_j. Hence, of the cells born at time $t-j$, $2f_j\,\Delta n_{t-j}$ cells divide again at time t. To find the total number of cells, Δn_t, dividing at t we need only sum over such contributions; hence

$$\Delta n_t = 2(f_1\,\Delta n_{t-1} + f_2\,\Delta n_{t-2} + \cdots + f_j\,\Delta n_{t-j} + \cdots + f_z\,\Delta n_{t-z})$$

$$\Delta n_t = 2\sum_{j=1}^{j=z} f_j\,\Delta n_{t-j} \qquad (4)^*$$

* The connection between the integral form and the finite difference form of the growth rate equation may be shown as follows. In the integral form the growth rate equation (Engelberg, 1964b) reads

$$dn(t) = \left[2\int_{t'=-\infty}^{t'=t} f(t-t')\,dn(t')\right] dt$$

In finite difference notation, when the fundamental interval of time is one time unit, this equation becomes

$$\Delta n_t = 2\sum_{t'=-\infty}^{t'=t} f_{t-t'}\,\Delta n_{t'}$$

Letting $t - t' = j$,

$$\Delta n_t = 2\sum_{j=\infty}^{j=0} f_j\,\Delta n_{t-j}$$

Since $f_j = 0$ for $j > z$ and $f_0 = 0$, the equation can be written

$$\Delta n_t = 2\sum_{j=1}^{j=z} f_j\,\Delta n_{t-j}$$

Once Δn_t is known, Δn_{t+1} can be computed, and so on. It follows that when the growth curve of an ergodic culture is known over a time interval equal to or exceeding z time units, that the entire future history of the growth curve can be predicted.

VI. DETERMINATION OF THE CELL DOUBLING-TIME DISTRIBUTION FUNCTION FROM THE DECAY OF SYNCHRONIZATION

Equation (4) permits one to compute the number of cells in an ergodic culture for all future times, (*a*) when the growth pattern of the culture is known for some initial period and (*b*) when the doubling-time distribution function is known. In this section we will be concerned with the inverse problem of finding the doubling-time distribution function when the growth pattern during the decay of synchronization is known.

When the growth pattern of an ergodic culture is known over a time interval greater than or equal to $2z - 1$, Eq. (4) can be written z times to yield the following set of simultaneous, linear equations:

$$\begin{aligned} \tfrac{1}{2}\Delta n_t &= \sum_{j=1}^{j=z} f_j \, \Delta n_{t-j} \\ \tfrac{1}{2}\Delta n_{t+1} &= \sum_{j=1}^{j=z} f_j \, \Delta n_{t+1-j} \\ &\vdots \\ \tfrac{1}{2}\Delta n_{t+z-1} &= \sum_{j=1}^{j=z} f_j \, \Delta n_{t+z-1-j} \end{aligned} \tag{5}$$

The Δn values in these equations are known from the growth data. The f_j values are the unknowns. This set of z simultaneous equations can be solved for the z unknown f_j values. Thus, given $2z - 1$ successive Δn values for an ergodic culture one can, in principle, calculate the doubling-time distribution of the cells.

Computational Problems

It has been shown above that, in principle, it is possible to compute the doubling-time distribution, f, when (*a*) the growth curve of a culture is known over a sufficiently long period of time and, (*b*) *the culture is ergodic* over this period of time. There are a number of practical problems associated with these computations.

What is z? This number is not known, yet it is needed to set up the required equations. One assumes a reasonable trial number z_1 for z. The assumption can be guided by the following considerations: When the growth curve (n versus t) of a culture is known, an R plot (Engelberg, 1964a) may be drawn. The elapsed time between 2 adjacent peaks on this plot is very close to the mean doubling-time of the cells. From the decrease with time of the oscillations of the R plot the standard deviation of the doubling-time distribution can be estimated (Engelberg, 1964b*). The first estimate, z_1, for z might be a time interval equivalent to the mean doubling-time, $\bar{T}$, plus 2 standard deviations. In the ideal case, where the growth data are known with unlimited accuracy, it can be shown that the determinant of the coefficients of the right side of Eq. (5) vanishes when $z_1 > z$. This comes about because the equations in the set are then no longer independent—the $(z + 1)$th equation is a linear combination of the first z equations. Thus, the behavior of the determinant of Eq. (5) for different trial values of z may be used as a guide to the selection of the optimal value of z.

The necessarily finite accuracy of the growth data restricts the number of simultaneous equations that can be successfully solved for the f_T values. If the number of equations set up exceeds the number of equations warranted either by the accuracy of the data or by the number of digits carried in the calculation (rounding off errors), spurious values will be obtained for the f_T. The number of equations needed can be reduced by noting that f_T is of necessity zero for small T. A first estimate for w (Fig. 1) can be obtained by taking w equal to the mean doubling-time minus 2 standard deviations. This reduces the number of equations in the set of simultaneous equations (Eq. 5) by w equations. It is evident that in practical applications these calculations need be performed on a computer and involve trial and error procedures.

The calculation of f_T from the growth curve of a culture cannot be carried out when the culture is in steady-state growth. In the steady-state the growth pattern is independent of the details of the doubling-time distribution. Thus, the growth curve needs be analyzed during the period of the decay of synchronization of cell division; the greater the degree of synchronization of the culture, the greater the accuracy with which the doubling-time distribution can be calculated from the growth data.

VII. THE INITIAL STATE OF A CULTURE

Two time intervals of significance in experimental procedures used to synchronize cells are the induction period and the harvest period. During the *induction period* the cells are subjected to a perturbation. At the end of this period the cells are returned to their normal culture conditions. There now follows the *harvest period* during which the cell number is observed as a function of time and during which the cells can be harvested at different times for study. In the following discussion, time is measured from the beginning of the harvest period. The question arises, how can one mathematically characterize the state of an ergodic culture at zero time?

* See also Hirsch and Engelberg, 1965b.

Consider the beginning of the harvest period with t smaller than z. The summation of Eq. (4) can be split into two parts:

$$\begin{aligned}\Delta n_t &= 2 \sum_{j=1}^{j=z} f_j \, \Delta n_{t-j} \\ &= 2 \sum_{j=1}^{j=t} f_j \, \Delta n_{t-j} + 2 \sum_{j=t+1}^{j=z} f_j \, \Delta n_{t-j} \qquad (t < z) \end{aligned} \tag{6}$$

Δn_t is the number of cells dividing at time t. The first sum is the number of cells dividing at time t which last divided at some time greater than zero. The second sum gives the number of cells dividing at time t which last divided before zero time. Let n_0 equal the number of cells present in the culture at time zero, and let

$$S_t = \left(2 \sum_{j=t+1}^{j=z} f_j \, \Delta n_{t-j}\right) \Big/ n_0 \qquad (t < z) \tag{7a}$$

The quantity S_t is the fraction of cells present in the culture at time zero which divide for the first time at time t. To each unit of time in the interval $0 \leqslant t < z$ there corresponds a number S_t. For any given culture the set of z numbers

$$S = \{S_0, S_1, \cdots, S_t, \cdots, S_{z-1}\} \tag{7b}$$

is defined to be the *initial state function** (Hirsch and Engelberg, 1965a). S_t is said to be the tth *component* of the initial state function. Since each component is a fraction of the initial cell number

$$\sum_{t=0}^{t=z-1} S_t = 1 \tag{7c}$$

The initial state function is essentially an initial condition on the integral or difference equations which describe the proliferative pattern of an ergodic culture. Given the initial state function, S, and the doubling-time distribution, f, the complete growth pattern of an ergodic culture can be predicted. This can be seen by rewriting Eq. (6) with the help of Eq. (7a) as

$$\Delta n_t = 2 \sum_{j=1}^{j=t} f_j \, \Delta n_{t-j} + n_0 S_t \tag{8}$$

* The initial state functions and other state functions used in this study are normalized functions, i.e., the sum of the components of each function equals one [see Eq. (7c)]. This was not the case in previous studies (Hirsch and Engelberg, 1965a,b,c) where the initial state function was used in an unnormalized form and the components added up to n_0.

From this equation the growth rate curve of an ergodic culture can be calculated; thus,

$$\Delta n_0 = n_0 S_0 ,$$

$$\Delta n_1 = 2f_1 \Delta n_0 + n_0 S_1 = 2f_1 n_0 S_0 + n_0 S_1 , \text{ etc.}$$

The ability of the initial state function to act as an initial condition follows from the fact, that once it is known at what time a group of cells divide for the first time, the doubling-time distribution function can be used to calculate the fraction of these cells which divide again at given subsequent times. When a culture contains a large number of cells this is equivalent to being able to predict the growth pattern of the culture.

How can the initial state function actually be determined? Two methods suggest themselves. The first is time-lapse cinemicrography where the time at which each cell divides in the harvest period is recorded by direct observation. The second can be used in the case of a highly synchronized culture such as that obtained as described above by the method of Terasima and Tolmach (1963). Here, clearly, the pattern of cell division during the very first part of the harvest period results only from the first division of the synchronized cells, since the time interval over which most of the cells divide is smaller than w. Thus, the pattern of Δn_t versus t at the beginning of the harvest period of such a culture can be used in itself to compute the initial state function. If the period of time over which these highly synchronized cells first divide is relatively small, then, to a good approximation, one can even ignore the details of the growth pattern during this period and to assume that the cells divide simultaneously. To this approximation the initial state function is an impulse function. The growth pattern of cultures for which the initial state function is an impulse function has been studied by numerical and analytic means and reported in previous publications (Engelberg, 1964b; Hirsch and Engelberg, 1965a,b). Given the specification of the initial state of an ergodic culture and the doubling-time distribution function one can predict the future history of the culture. From this point of view it is clear that the initial state function is indeed a specification of the initial state of a synchronized culture. It follows that the initial state function of an ergodic culture has implicit in it the degree to which a cell population has been synchronized by an experimental procedure.

VIII. THE STATE FUNCTION

The initial state function may be easily generalized. What we are after is a description of the instantaneous state of an ergodic culture at *any* time t such that, given this description plus the doubling-time distribution, the future history of the culture can be predicted. The *state function*,

$S(t)$, which will be defined below, is such a description. In general this *function* differs for different values of t. For example, $S(0)$ is the initial state function previously denoted by S.

The reasoning behind the definition of the state function, $S(t)$, is analogous to that used in the case of the initial state function. As we are interested in a function for time t, t ceases to be a variable in the definition of this function. Instead, we define a new time variable, k. The k axis has its origin at time t (Fig. 2). Beginning with Eq. (4) we have

$$\Delta n_t = 2 \sum_{j=1}^{j=z} f_j \, \Delta n_{t-j}$$

so that

$$\begin{aligned} \Delta n_{t+k} &= 2 \sum_{j=1}^{j=z} f_j \, \Delta n_{t+k-j} \\ &= 2 \sum_{j=1}^{j=k} f_j \, \Delta n_{t+k-j} + 2 \sum_{j=k+1}^{j=z} f_j \, \Delta n_{t+k-j} \qquad (k \geq 0) \end{aligned} \tag{9a}$$

The second sum on the right gives the number of cells dividing at time $t + k$ which last divided previous to time t. Let $S_k(t)$ be the fraction of the n_t cells present in the culture at time t which divide for the first time subsequent to t at time $t + k$. Then,

$$\Delta n_{t+k} = 2 \sum_{j=1}^{j=k} f_j \, \Delta n_{t+k-j} + n_t S_k(t) \tag{9b}$$

The state function, $S(t)$, of the culture at time t is the set of components $S_k(t)$:

$$S(t) = \{S_0(t), S_1(t), \cdots, S_k(t), \cdots S_{z-1}(t)\} \tag{9c}$$

Given f and $S(t)$ for an ergodic culture the proliferative pattern for the culture subsequent to time t can be predicted. Conversely, given f and the proliferative pattern subsequent to t the state function for time t can be calculated from Eq. (9b) as follows:

$$S_k(t) = \left(\Delta n_{t+k} - 2 \sum_{j=1}^{j=k} f_j \, \Delta n_{t+k-j}\right) / n_t \tag{9d}$$

An application of this equation in the case of a simple example will be given below in Section XII, 5.

A sample state function of an ergodic culture, having the doubling-time distribution $f_1 = 0$, $f_2 = 0.2$, $f_3 = 0.8$, $f_j = 0$ for $j > 3$, is shown in Fig. 2. The growth data (n_t versus t) and growth rate data (Δn_t versus t) are shown at the bottom of the figure. At $t = 0$ the cells in the culture

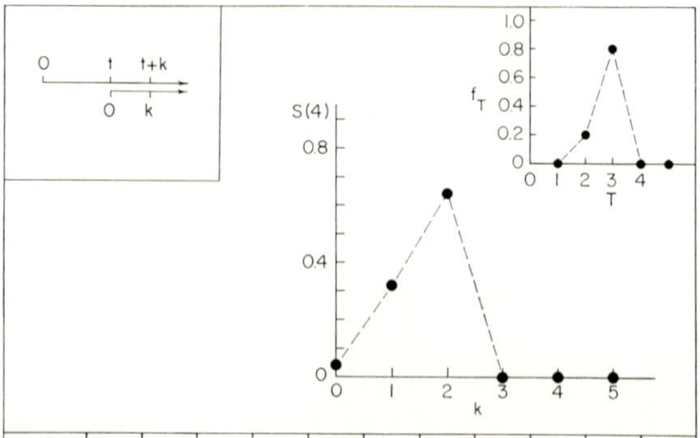

t	0	1	2	3	4	5	6	7	8	9	10
Δn_t	1	0	0.40	1.60	0.16	1.28	2.62	0.77	3.10	4.51	2.47
n_t	1	2	2	2.4	4.0	4.16	5.44	8.06	8.83	11.93	16.44

FIG. 2. State function, $S(t)$, of an ergodic culture at time $t = 4$. The component $S_1(4)$ equals 0.32 indicating that of the 4×10^5 cells present in the culture at time $t = 4$, 32% of the cells divide within the time interval between $t = 5$ and $t = 6$. n_t is the number of cells ($\times 10^{-5}$) present in the culture at time t. Δn_t is the number of cells ($\times 10^{-5}$) which are added to the culture by cell division within the time interval between t and $t + 1$. The doubling-time distribution, f, of the culture is shown in the upper right-hand corner. The relationship between the time variables t and k is shown in the upper left-hand corner.

divide with perfect synchrony. At $t = 4$ the state function is (Section XII, 5)

$$S(4) = [0.04, 0.32, 0.64]$$

The component $S_1(4) = 0.32$ indicates that of the 4×10^5 cells present in the culture at $t = 4$, 32% of the cells divide in the time interval between $t = 5$ and $t = 6$.

When an ergodic culture is in the steady-state, the state function is independent of time, that is, it is the same for all t. A stationary state function, one which does not vary with time, therefore, corresponds to steady-state (exponential) growth. The state function of a synchronous culture (Fig. 2) (*a*) differs from the limiting state function for steady-state growth and (*b*) varies with time, becoming in the limit of large time the state function, $S(U)$, for steady-state growth. The convergence of the state function $S(t)$ of a synchronous culture to $S(U)$ suggests that differences between $S(t)$ and $S(U)$ be used as a measure of synchrony. This possibility will be explored below.

A. Vector Space Representation of State Function

The preceding methods allow one to give a unique description to the instantaneous state of an ergodic culture by specifying the state function.

It would be useful to have a formalism for visualizing the difference between two arbitrary states of an ergodic culture as well as the course taken by a synchronous culture as it approaches steady-state growth. This may be realized by imagining the representation of the state function in a multidimensional vector space. A given state function is a set of z numbers. We imagine a space in which there are z mutually perpendicular axes. To represent a given state function in this space $S_0(t)$ is plotted along the first axis, $S_1(t)$ along the second, . . . and $S_{z-1}(t)$ along the zth axis. As would be the case in ordinary three-dimensional space, this set of z components defines a single and unique point in the z-dimensional space. Thus, a given state of an ergodic culture is represented by a single point (or vector) in its state (vector) space.

The components of the state function are positive numbers lying in the range from 0 to 1. Thus, state points lie only in the first (positive) quadrant of state space. The sum of the components of the state function always equals one [Eq. (7c)]; hence, the state points are confined to the hyperplane which cuts each state space coordinate axis at the point 1. It follows that *to every possible state of an ergodic culture there corresponds a unique point in the state plane, and conversely, every point in the state plane* (within the first quadrant) *corresponds to a possible state of an ergodic culture.*

There exists a point U in the state plane which corresponds to steady-state growth. Since all cultures which are not in the steady-state approach in time, but in principle never reach the steady-state, it is evident that the state point of any culture which does not lie at U will move with time towards U along a trajectory in the state plane. This trajectory spirals around the point U, but never reaches it. The location of the point U in state space is a function of, and is determined by, the doubling-time distribution.

In Fig. 3 such a trajectory is shown for an especially simple ergodic culture where the state function has only 3 components. When $z = 3$ the state points all fall in a 2-dimensional plane in ordinary 3-dimensional space. Because the number of components of this state function is finite, the trajectory of the state point in the state plane is discontinuous. The state point hops from one point to another with each unit of time. In the limit where the state function becomes a continuous function, the trajectory of the state point in the state plane becomes continuous. The dotted lines between points in Fig. 3 are drawn to aid the eye and are not part of the state point trajectory. Point 0 represents the state, $S(0) = [1,0,0]$, of the culture at time zero. At this time all the cells in the culture divide with perfect synchrony. With time the state point is seen to "spiral" about the

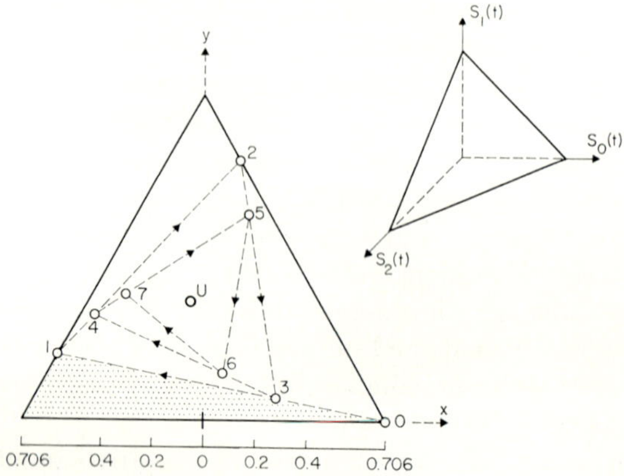

FIG. 3. Vector space representation for the state function, $S(t)$, corresponding to the culture of Fig. 2. The space is defined by three mutually perpendicular axes $S_0(t)$, $S_1(t)$, $S_2(t)$. The state points of the culture, however, lie only in the first quadrant of the space and in the plane which crosses the coordinate axes at the points $S_0(t) = 1$, $S_1(t) = 1$, $S_2(t) = 1$. The motion of the state point in this plane is indicated by the dotted lines and arrows. State point No. 4, for example, defines the state function for $t = 4$. The state point for steady-state growth is shown at U. All the state points within the stippled region are terminal points. The scale of the state space is shown at the bottom. The origin of the x, y coordinate system is indicated by the cross mark on the x axis.

steady-state growth point, U, and to approach it. The distance from the state point to U, however, does not decrease monotonically with time.

Equation (4) guarantees that the trajectory between a given state point and U is uniquely determined. It follows that any 2 trajectories chosen among the family of all possible trajectories never cross. For were the trajectories to cross at a state point then the point of crossing would have two possible trajectories corresponding to it, rather than one unique trajectory. In interpreting this condition it is to be noted that in the case of a discontinuous trajectory, the trajectory consists only of the state points and not of the dotted lines which may be drawn between them.

B. Terminal Points

We have so far discussed how, given the state of an ergodic culture at time t and its doubling-time distribution, one can predict the future proliferative pattern of the culture. From a given state point, P, one can usually also move backwards in time and calculate the trajectory which could have led up to state P. A numerical example of such a backward projection is given in Section XII, 8. There will exist in every state plane, however, a set of state points, T, from which it is possible to project a normal forward trajectory towards U though it is not possible to project a backward trajectory. Such points will be called terminal points. These can be detected from the fact that when the state function of the state point which precedes T along the backward trajectory is calculated, it is found to have one or more negative components. It should be remembered that each component of the state function represents a certain fraction of cells which enter the culture. Thus, it is not possible in a culture from which cell lysis is excluded to have a negative cell increment corresponding to a negative component. Terminal points represent synchronous cultures which can be prepared by an experimenter; however, they do not represent states which would arise as a consequence of the spontaneous proliferative pattern of an ergodic culture. In the illustrative example of Fig. 3, the stippled region contains the terminal points. This is shown below in Section XII, part 8. Trajectories may originate in this region; however, backward trajectories starting from points within this region do not exist.

A simple but important consequence of this line of reasoning is that partially synchronized cultures do not generally project back to the state of perfect synchrony. For an ergodic culture there exists only a single, unique trajectory which leads from perfect synchrony to the steady-state.

IX. DEGREE OF SYNCHRONIZATION

For the experimenter concerned with the development and maintenance of synchrony of cell division, as well as with the use of synchronous cultures over periods of time over which the synchrony decays, a numerical index which expresses how close a given culture is to perfect synchrony and how far it is from steady-state (exponential) growth would be of value. A number of proposals have been made in recent years for such an index (reviewed in Engelberg, 1964a). All of these proposals have merit and do, indeed, describe some aspect of what is meant by a synchronous culture. A difficulty with these proposals has been that each has attempted to define an index that would be appropriate for the description of all possible synchronous cultures. The variety of methods for synchronizing cell cultures, however, makes it doubtful that a single, universal synchronization index can be defined. The 4 classes of experimental technique, (1) environmental perturbation procedures (temperature shock, nutrient variation, light-dark variation), (2) selection procedures (harvesting during or immediately following mitosis), (3) egg fertilization and (4) procedures yielding cyclic cultures (cyclic temperature variation, light-dark cycles), may be incommensurable from the point of view of a single synchronization index unless one is satisfied with comparing a single aspect of the cultures.

As is evident from the concept of the state function, the characterization of the state of a culture at a given time, requires the specification of a large set of numbers, namely, the components of the state function. When one speaks of a synchronization index, however, one speaks of a single number. It is clear that a single number cannot, in general, replace a set of numbers without a loss of information. The question then arises, is it possible to devise a universal criterion which will allow one to associate a single number with the idea "degree of synchronization" for any culture no matter how synchronized? Such a criterion is indeed possible when practicality rather than depth of insight is the aim. For example, an experimenter may wish to isolate a substance from cells which are in the process of undergoing mitosis. Here the aim is to obtain a culture with as many as possible of the cells in mitosis. For this purpose mitotic index is a good "synchronization index." As soon as, however, the heterogeneity of the mitotic cells becomes of importance, this index loses its usefulness. It is, therefore, our feeling that the formulation of a synchronization index which would be applicable to any synchronous culture, no matter how prepared and no matter for what purpose, will be difficult if not impossible. For special classes of synchronous cultures, however, such a formulation is possible. In what follows we will discuss the problems associated with the formulation of a synchronization index for the class of ergodic cultures.

In the light of vector space representation of the state of a culture it would appear reasonable to define the degree of synchronization of an ergodic culture at time t in terms of the distance, s, between its state point at time t and the state point U for steady-state growth. This distance squared is given by

$$s^2 = (S_0(t) - S_0(U))^2 + (S_1(t) - S_1(U))^2 + \cdots + (S_{z-1}(t) - S_{z-1}(U))^2$$

Thus, the geometrical distance between a state point P and the point U is a measure of the deviation of the components of P from those of U.

Unfortunately, s can not in itself be used as a measure of the degree of synchronization since along a given trajectory the decrease in s is not, in general, monotonic. As the trajectory spirals around U, s becomes larger and smaller during each revolution though on the average it becomes smaller for successive revolutions. Hence, it is the average value of s over one revolution, i.e., a period of z time units, which might be used as a measure of the degree of synchronization of the culture.

This again points to the fact that if one wishes to express the deviation of a culture from steady-state growth by a single number, this number needs be derived from the analysis of the proliferative pattern over about one population doubling-time (Engelberg, 1964a). The requirement probably derives from the type of statistics implicit in f where one considers only the time at which a cell divides while saying nothing about what happens to the cell between divisions.

The method described here for specifying the degree of synchronization of an ergodic culture while of theoretical interest is not likely to be of practical importance; the required calculations are too complicated. It is interesting that a more practical, semiempirical method proposed by one of the authors (Engelberg, 1961) bears some resemblance to the above method.

X. AGE DISTRIBUTION

The age, a, of a cell may be defined to be the time which has elapsed since the last anaphase. When the ages of all the cells in a given culture are known at a given time t, an age distribution

$$A(t) = [A_0(t), A_1(t), \cdots, A_a(t), \cdots A_{z-1}(t)]$$

can be constructed. $A_a(t)$ is the fraction of cells of age a in the culture at time t. In general, the age distribution changes with time. When an ergodic culture is in the steady-state, the age distribution, $A(U)$, does not change with time.

Consider an ergodic population of cells in which the number of cells at time t is n_t. At time t the number of cells of age a in the population is $n_t A_a(t)$. Given (1) the growth curve (n_t versus t, or, Δn_t versus t) of this culture over the time interval from $t - z$ to t and (2) the doubling-time distribution of the cells, the components $A_a(t)$ of the age distribution at time t can be calculated as follows. All cells of age a at time t were born in the time interval between $t - a - 1$ and $t - a$, i.e., they were of age zero at time $t - a$. The number of cells of age zero at time $t - a$ is $2\Delta n_{t-a-1}$ (see Section II). However, though all the cells which are of age a at time t were born at $t - a - 1$, not all of the cells which were born at time $t - a - 1$ will be of age a at time t. This is because some of these cells may have divided at some time between $t - a - 1$ and t. Thus, the number of cells of age a at time t is $2\Delta n_{t-a-1}$ minus the number of cells which divided again, and

$$A_0(t) = 2\Delta n_{t-1}/n_t$$

$$A_1(t) = (2\Delta n_{t-2} - 2\Delta n_{t-2}f_1)/n_t = 2(1 - f_1)\,\Delta n_{t-2}/n_t$$

$$\vdots$$

$$A_a(t) = 2\Delta n_{t-a-1}(1 - f_1 - f_2 - \cdots - f_a)/n_t$$

$$= 2\Delta n_{t-a-1}\left(1 - \sum_{T=1}^{T=a} f_T\right)/n_t$$

$$A_a(t) = 2\Delta n_{t-a-1}(1 - F_a)/n_t \qquad (10)$$

where F_a is the cumulative doubling-time distribution function.

The age distribution $A_a(U)$, for the steady-state can be obtained as follows. In the steady-state the cell number increases exponentially; thus, we assume (Hildebrand, 1952)

$$\Delta n_t = \beta^t \tag{11}$$

where β is a constant. Since $\Delta n_t = n_{t+1} - n_t$, $n_t = \beta^t/(\beta - 1)$.* Substituting in Eq. (10) we obtain the age distribution for the steady-state

$$A_a(U) = 2(\beta - 1)\,\beta^{-(a+1)}(1 - F_a) \tag{12}$$

Thus, given the cumulative doubling-time distribution, F_t, of a culture, the age distribution of the culture in the steady-state can be computed from Eq. (12) (see also Powell, 1956; Scherbaum and Rasch, 1957). The number β is found as follows. Substituting (11) into the growth Eq. (4) we find

$$\beta^t = 2\sum_{j=1}^{j=z} f_j\beta^{t-j} \tag{13}$$

Rearranging this equation

$$2\sum_{j=1}^{j=z} \beta^{-j} f_j = 1 \tag{14}$$

The numerical value of β can be readily found from Eq. (14) by trial and error. The use of these equations will be illustrated in a subsequent section.

XI. STEADY-STATE GROWTH

Every synchronized ergodic culture which has a doubling-time distribution of nonzero variance approaches a state of pure exponential growth with time. In this state all of the statistical distributions which characterize the culture are invariant with time. The culture, in this limit, may be said to be in the *steady-state*. In the steady-state, for example, the state function and age distribution do not vary with time. During the decay of synchronization, on the other hand, these distributions change with time and approach the steady-state distributions.

It is customary to call cultures which are in the steady-state "exponential

* Assume that $\Delta n_t = \beta^t$ and $n_t = A\beta^t$, where A is a constant. Then,

$$\Delta n_t = n_{t+1} - n_t$$

$$\beta^t = A\beta^{t+1} - A\beta^t$$

$$A = 1/(\beta - 1)$$

and

$$n_t = \beta^t/(\beta - 1)$$

cultures" because the cell number increases exponentially. This nomenclature has led to some unfortunate consequences. Though all proliferating cultures which are in the steady-state are in exponential growth, nonergodic cultures which appear to be in exponential growth often are not in the steady-state. When fresh nutrient medium is not added to a bacterial culture, for example, the culture enters a stationary period of growth during which cell division virtually ceases. When fresh medium is subsequently added cell division commences after a lag period. To within the experimental error of measurement of cell number the culture soon attains a state of exponential growth. It is then, however, not as yet in the steady-state as shown by biochemical assays carried out on the culture at different times in the "exponential" growth period. It would, therefore, appear to be desirable that the word "exponential" not be used as a synonym for "steady-state."

For every ergodic culture there exists a mathematical transformation which, given one state point, generates the next (or previous) state point. Brouwer's fixed point theorem in topology states that "in a given region of space there is always at least one point unchanged by a given transformation if this transformation can never move a point in the region to outside the region" (Wei and Prater, 1962). The steady-state point U in the state space is precisely such a point.

XII. A NUMERICAL EXAMPLE

The use of the preceding equations can be illustrated by the following simple example involving an ergodic population having a doubling-time distribution with $z = 3$. Growth-rate data and the doubling-time distribution are given in Fig. 2.

1. *Doubling-time distribution, f.* $f_1 = 0, f_2 = 0.2, f_3 = 0.8$.
2. *Cumulative doubling-time distribution.* $F_1 = f_1 = 0, F_2 = f_1 + f_2 = 0.2, F_3 = f_1 + f_2 + f_3 = 1.0$.
3. *Growth equation.*

$$\Delta n_t = 2 \sum_{j=1}^{j=3} \Delta n_{t-j} f_j$$

$$= 2(\Delta n_{t-1} f_1 + \Delta n_{t-2} f_2 + \Delta n_{t-3} f_3)$$

$$\Delta n_t = 0.4 \Delta n_{t-2} + 1.6 \Delta n_{t-3} \tag{15}$$

Assume that the culture initially contains 10^5 cells and that these cells divide synchronously at time zero. Then $n_0 = 1 \times 10^5$, $\Delta n_0 = 1 \times 10^5$; $n_1 = n_0 + \Delta n_0 = 2 \times 10^5$, $\Delta n_1 = 0$; $n_2 = 2 \times 10^5$, $\Delta n_2 = 2(0.2)(1 \times 10^5) = 0.4 \times 10^5$; $n_3 = n_2 + \Delta n_2 = 2.4 \times 10^5$. Continuing by means of Eq. (15)

$$\Delta n_3 = 0.4\Delta n_1 + 1.6\Delta n_0$$
$$= (0.4)\ (0) + 1.6\ (1 \times 10^5) = 1.6 \times 10^5, \text{ etc.}$$

4. *Calculation of the doubling-time distribution from growth data.* The growth equation [Eq. (4)] is written three times.

$$\tfrac{1}{2}\Delta n_3 = \Delta n_2 f_1 + \Delta n_1 f_2 + \Delta n_0 f_3$$
$$\tfrac{1}{2}\Delta n_4 = \Delta n_3 f_1 + \Delta n_2 f_2 + \Delta n_1 f_3$$
$$\tfrac{1}{2}\Delta n_5 = \Delta n_4 f_1 + \Delta n_3 f_2 + \Delta n_2 f_3$$

Substituting the growth data from Fig. 2 for the Δn

$$\tfrac{1}{2}(1.60) = (0.40)f_1 + (0)f_2 + (1)f_3$$
$$\tfrac{1}{2}(0.16) = (1.60)f_1 + (0.40)f_2 + (0)f_3$$
$$\tfrac{1}{2}(1.28) = (0.16)f_1 + (1.60)f_2 + (0.40)f_3$$

Solving this set of three simultaneous equations for the three unknowns f_1, f_2, and f_3 we find $f_1 = 0$, $f_2 = 0.2$ and $f_3 = 0.8$. The reader can readily convince himself that if four simultaneous equations had been written, the determinant of the matrix of coefficients would have vanished and solutions for f_1, f_2, f_3, and f_4 could not have been obtained.

5. *The state function at $t = 4$.* The components of this function can be computed from Eq. (9d).

$$S_0(4) = \Delta n_4/n_4 = 0.16/4 = 0.04$$
$$S_1(4) = (\Delta n_5 - 2f_1\,\Delta n_4)/n_4 = [1.28 - 2(0)\ (0.16)]/4 = 0.32$$
$$S_2(4) = (\Delta n_6 - 2f_1\,\Delta n_5 - 2f_2\,\Delta n_4)/n_4$$
$$= [2.62 - 2(0)\ (1.28) - 2(0.2)\ (0.16)]/4 = 0.64$$

Thus, $S(4) = [0.04, 0.32, 0.64]$.

6. *Vector space representation of state function.* Given a state function $S(t) = [S_0(t), S_1(t), S_2(t)]$ we should like to find the pair of numbers $x(t)$ and $y(t)$ which locate the state point in the state plane. The coordinate axes for x and y are chosen as shown in Fig. 3. By application of the methods of analytic geometry (Lehmann, 1942) the equations of transformation which take S_0, S_1, and S_2 into x and y can be shown to be

$$x = 0.706[S_0(t) - S_2(t)]$$
$$y = 1.224S_1(t)$$

For state point No. 4 in Fig. 3 the state function is, as shown above, $S(4) = [0.04, 0.32, 0.64]$. In the state plane the coordinates of this point are

$$x = 0.706\ (0.04 - 0.64) = -0.42$$

$$y = 1.224\ (0.32) = 0.39$$

7. *Coordinates of the steady-state point U*. The growth equation

$$\Delta n_t = 0.4\Delta n_{t-2} + 1.6\Delta n_{t-3}$$

is a linear equation with constant coefficients. This equation can be solved in the steady-state limit by assuming a solution of the form (Hildebrand, 1952)

$$\Delta n_t = \beta^t$$

where β is a constant. Substituting this expression into the growth equation one obtains the equation

$$\beta^3 - 0.4\beta - 1.6 = 0$$

which upon solution by trial and error yields the value

$$\beta = 1.285$$

so that

$$\Delta n_t = (1.285)^t$$

Using this relationship the state function for the steady-state can be found (see 5 above) to be

$$S(U) = [0.283, 0.364, 0.352]$$

8. *Terminal Points*. By use of the equations in 5 (above) and the growth equation it can be shown that if

$$S(t) = [S_0, S_1, S_2]$$

then

$$S(t - 1) = [S_2/(1 - S_2),\ 1.6\ S_0/(1 - S_2),\ 1.6(S_1 - S_2/4)/(1 - S_2)]$$

If $S_1 < S_2/4$ then the third component of $S(t - 1)$ is a negative number. It follows that the point representing $S(t)$ is a terminal point whenever $S_2(t) > 4\ S_1(t)$. The stippled region in Fig. 3 is a region within which every point is a terminal point.

9. *Age distribution at* $t = 4$.

$$A_a(t) = 2\Delta n_{t-a-1}(1 - F_a)/n_t$$

$$\begin{aligned} A_0(4) &= 2\Delta n_3(1 - F_0)/n_4 \\ &= 2(1.60)\ (1 - 0)/4 \\ &= 0.80 \end{aligned}$$

$$\begin{aligned} A_1(4) &= 2\Delta n_2(1 - F_1)/n_4 \\ &= 2(0.4)\ (1 - 0)/4 \\ &= 0.20 \end{aligned}$$

$$\begin{aligned} A_2(4) &= 2\Delta n_1(1 - F_2)/n_4 \\ &= 2(0)\ (1 - 0.2)/4 \\ &= 0 \end{aligned}$$

Thus,

$$A(4) = [0.80, 0.20, 0]$$

XIII. NONERGODIC CULTURES: CYCLIC SYNCHRONOUS CULTURES

Cyclic synchronous cultures are important examples of nonergodic cultures. In these cultures a degree of synchrony is permanently preserved by subjecting the cells to periodic perturbations (Padilla and Cook, 1964) such as light-dark cycles (Lorenzen, 1964), and low-high temperature cycles (James, 1960).

It is easy to show, that with exception of a trivial case, cyclic cultures are not ergodic. Consider a typical growth rate curve for such a culture (Fig. 4). During the first part of each cycle there is no cell division; hence, the growth rate curve is a horizontal straight line. Toward the end of the cycle there is a burst of cell division. At the end of the cycle cells are re-

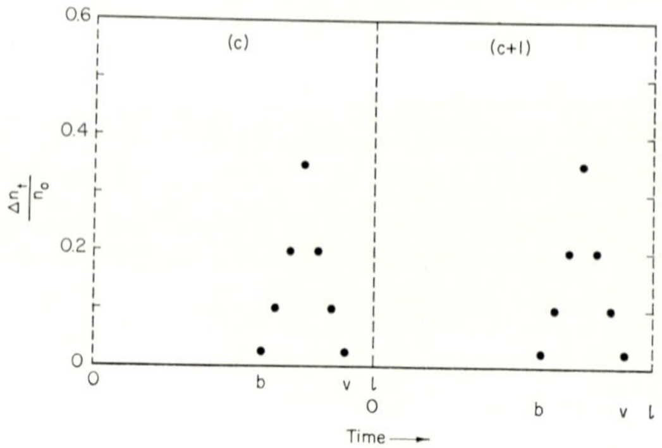

FIG. 4. Growth rate curve of a cyclic (permanently synchronized) culture for cycles c and $c + 1$. Each cycle is of duration l time units; the variable t assumes values from 0 to l. Δn_t is the number of cells added to the culture by cell division at time t. n_0 is the number of cells in the culture at $t = 0$. At $t = l$ the culture is diluted so that the number of cells at corresponding points in each cycle remain the same. There is no cell division from $t = 0$ to $t = b - 1$. The synchronous burst of cell division takes place from $t = b$ to $t = v$.

moved from the culture to establish the same cell number as was present at the beginning of the cycle.

We will assume that the culture has reached a steady-state in its periodic environment so that the growth rate curves of successive cycles are precisely alike. Each cycle has a length of l time units. The time t is measured from the beginning of each cycle; thus $0 \leqq t < l$. In Fig. 4 there is no cell division in the interval $0 \leqq t < b$. The burst of cell division takes place in the time interval $b \leqq t \leqq v$.

Consider the cells which double at a time t. The fraction of these cells having a doubling-time T is given by a distribution function $f_T(t)$. The time parameter, t, is shown explicitly to include the possibility that the doubling-time distribution varies over the cycle. To simplify the discussion we will assume that cells born in one cycle divide again during the next cycle, i.e., they do not next divide in the same cycle nor in cycles beyond the next cycle. If $f_T(t)$ has a nonzero variance, i.e., when all cells do not have exactly the same doubling-time, $f_T(t)$ cannot be the same function for all values of t. This can be seen from the fact that when the *same* doubling-time distribution f is applied during one cycle to the cells dividing at $t = b$ and $t = v$, this leads to cell divisions in the next cycle outside the time interval $b \leqq t \leqq v$ unless f has zero variance. It follows that a cyclic cell culture is nonergodic—the doubling-time distribution varies over the cycle. Only in the trivial case (f has zero variance) where all cells have a doubling-time precisely equal to the period of the cycle would a cyclic culture be ergodic.

We have concluded that cells dividing at different points along the synchronization cycle will, in general, have different doubling-time distributions. What relationships must exist among these distributions such that the growth curve remains invariant from cycle to cycle? Of the cells which divide at time $t = j$ during one cycle, some will divide during the next cycle at time $t = j$ others will divide at times other than j. For convenience let us define a set of distribution functions g_j^r where g_j^r is the fraction of the cells dividing at time $t = j$ in one cycle which divide again at time $t = r$ in the next cycle. The g_j^r functions are related to the doubling-time distributions $f_T(t)$ by

$$g_j^r = f_{T=l-j+r}(t = j) \tag{16}$$

The number of cells dividing at time r is given by

$$\Delta n_r = \sum_{j=b}^{j=v} \Delta n_j g_j^r \tag{17}$$

Since $\sum_r g_j^r = 1$ for each j, $(v - b)$ values of g_j^r need be specified for a given r. The variable r can assume $(v - b + 1)$ different values; hence,

$(v - b + 1)(v - b)$ different values of g_j^r need be given to specify the complete set of distribution functions. Equation (17) can be written $(v - b + 1)$ times. Thus, for a given growth rate curve, i.e., a given set of numbers Δn_t, only $(v - b + 1)$ linear, independent equations can be written but $(v - b + 1)(v - b)$ components of the distribution functions need be specified. It follows that there does not correspond a unique set of doubling-time distributions to a given growth rate curve. The particular set of distributions which corresponds to a given cyclic culture, however, is subject to the constraints of the $(v - b + 1)$ equations (17).

It is to be observed that the cyclic culture is a special case of a general biological phenomenon encountered in the study of biological rhythms. This is the entrainment of a biological oscillation (in this case the cell division cycle) by an external Zeitgeber (in this case the periodically varying environmental conditions).

These brief considerations may be summarized as follows: (*a*) Cyclic cultures are not ergodic. (*b*) To each cyclic culture there corresponds a set of doubling-time distributions, $\{f_T(t)\}$, (or, equivalently, the $\{g_j^r\}$) rather than a single doubling-time distribution. As a step in the direction of the clarification of the mechanism of cyclic proliferation it would be desirable to experimentally determine this set. In principle, this can be done by the analysis of time-lapse cinemicrography records. (*c*) The explanation of cyclic synchronization involves the explanation of the genesis of these distribution functions. (*d*) A more detailed analysis of Eqs. (17) would show that a cell which divides early in one cycle will tend to divide late in the next cycle and vice versa.

XIV. RESUMÉ

The discussion in this report has been restricted, for the most part, to the case of ergodic cultures, cultures in which the cell doubling-time distribution does not change with time. The instantaneous state of such cultures is described by means of a *state function*. Given the state function at a time t and the doubling-time distribution, the entire growth rate curve of an ergodic culture for times greater than t can be predicted. The representation of the state function in a multidimensional *state space* permits us to visualize, in geometric terms, the path of the culture in time. When the culture is in the steady-state the state function does not vary with time and the culture rests at a certain point, U, in the state space. By contrast, when the culture is initially perfectly or partially synchronized, the state point moves in time along a trajectory in state space which spirals about and approaches the steady-state point U.

For any given state other than U one can also calculate what states previous in time could have given rise to the given state. In calculating this "backward" trajectory in state space a terminal state is invariably reached. This state of the culture is characterized by the fact that it could not have arisen from a spontaneous ergodic process acting on previous states. The state point corresponding to such a state is a *terminal point.*

These studies point to the fact that cell doubling-time distribution functions are not scientific curiosities. They are fundamental descriptions of the statistical properties of cells in cultures and are indispensable to the prediction and interpretation of the proliferative pattern of such cultures. In principle, precise knowledge of a distribution implies knowledge of a large, if not infinite, set of numbers representing the different ordinates of the distribution. From a practical point of view it is entirely satisfactory to approximate the distribution by a smaller set of numbers. The minimum set consists of two numbers, essentially, the mean and the standard deviation of the distribution. If more information is available the accuracy of the analysis can be increased by increasing the number of descriptive parameters.

ACKNOWLEDGMENTS

We thank Professor Louis L. Boyarsky for acquainting us with the existence of Brouwer's fixed point theorem and for suggesting its application to the state space of an ergodic culture. This investigation was aided by Public Health Service Grant NIH-CA-06835 from the National Cancer Institute.

REFERENCES

Anderson, E. C., and Petersen, D. F. (1964). *Exptl. Cell Res.* **36**, 423.

Engelberg, J. (1961). *Exptl. Cell Res.* **23**, 218.

Engelberg, J. (1964a). *In* "Synchrony in Cell Division and Growth" (E. Zeuthen, ed.), p. 497. Wiley (Interscience), New York.

Engelberg, J. (1964b). *Exptl. Cell Res.* **36**, 647.

Engelberg, J. (1964c). *Exptl. Cell Res.* **34**, 111.

Helmstetter, C. E., and Cummings, D. J. (1963). *Proc. Natl. Acad. Sci. U.S.* **50**, p. 767.

Hildebrand, F. B. (1952). "Methods of Applied Mathematics," p. 237. Prentice-Hall, Englewood Cliffs, New Jersey.

Hirsch, H. R., and Engelberg, J. (1965a). *Bull. Math. Biophys.* Submitted for publication.

Hirsch, H. R., and Engelberg, J. (1965b). *J. Theoret. Biol.* **9**, 297.

Hirsch, H. R., and Engelberg, J. (1965c). *J. Theoret. Biol.* **9**, 303.

James, T. W. (1960). *Ann. N.Y. Acad. Sci.* **90**, 550.

Lehmann, C. H. (1942). "Analytic Geometry," pp. 282 and 290. Wiley, New York.

Lorenzen, H. (1964). *In* "Synchrony in Cell Division and Growth" (E. Zeuthen, ed.), p. 571. Wiley (Interscience), New York.

Padilla, G. M., and Cook, J. R. (1964). *In* "Synchrony in Cell Division and Growth" (E. Zeuthen, ed.), p. 521. Wiley (Interscience), New York.

Powell, E. O. (1956). *J. Gen. Microbiol.* **15**, 492.

Sherbaum, O., and Rasch, G. (1957). *Acta Path. Microbiol. Scand.* **41**, 161.

Scherbaum, O., and Zeuthen, E. (1955). *Exptl. Cell Res.* Suppl. 3, 312.

Terasima, T., and Tolmach, L. J. (1963). *Exptl. Cell Res.* **30**, 344.

Wei, J., and Prater, C. D. (1962). *Advan. Catalysis* **13**, 344.

CHAPTER 3

Synchronous Replication of the Chromosome in *Bacillus subtilis*

Noboru Sueoka

I. INTRODUCTION

The mode of replication of the bacterial chromosome has recently become much clearer in 2 systems: *Bacillus subtilis* (Yoshikawa and Sueoka, 1963a,b; Yoshikawa *et al.*, 1964; Oishi *et al.*, 1964) and *Escherichia coli* (Cairns, 1963; Nagata, 1963; Bonhoeffer and Gierer, 1963; K. G. Lark *et al.*, 1963). The current picture is that the bacterial chromosome starts replicating from a particular position on the chromosome (origin) and replication proceeds along the chromosome in one direction. The synchrony of chromosome replication further provides the basis of our experimentation on mapping genes on the chromosome, the regulation of chromosome replication, the sequence of gene function and the locations of hitherto unmapped genes such as rRNA, sRNA, and RNA and DNA polymerases loci.

A partial synchrony of chromosome replication was observed in *B. subtilis,* when the cells (strain W23) of stationary phase were transferred to fresh medium (Yoshikawa and Sueoka, 1963b). The synchrony was shown to be much better during spore germination (Oishi *et al.*, 1964). In this chapter our work on *B. subtilis* will be summarized, with special emphasis on chromosome replication synchrony, and more recent results discussed in the light of future problems.

II. NONSYNCHRONOUS REPLICATION

A. Theoretical Analyses

The working model of chromosome replication on which our work has been based is shown in Fig. 1. The theoretical analyses of the following problems based on the above model of chromosome replication have been the basis of our experimentation, the essence of which will be presented here. The analysis is applicable to the nonsynchronously dividing cell population. For details of the theory, see Sueoka and Yoshikawa (1965).

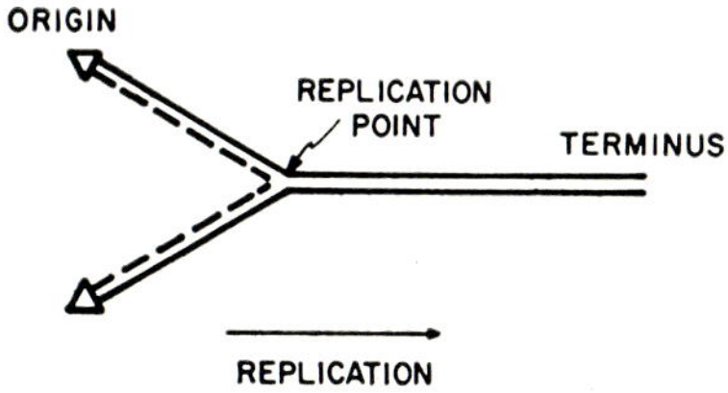

Fig. 1. A replication model of the chromosome. The solid line represents the old strand and the broken one—the newly synthesized strand. This model does not differ, in principle, from a circular chromosome model in which replication starts at a fixed point in one direction.

1. *Age Distribution of Chromosomes in Exponential, Nonsynchronous Population*

The total length of the chromosome is represented as 1, thus assigning position 0 to the origin and position 1 to the terminus. The position of the replication point (x), therefore, may represent the "age" of the chromosome. In an exponentially growing (nonsynchronous) cell population, the *age distribution function of chromosomes,* $f(x)$, should take

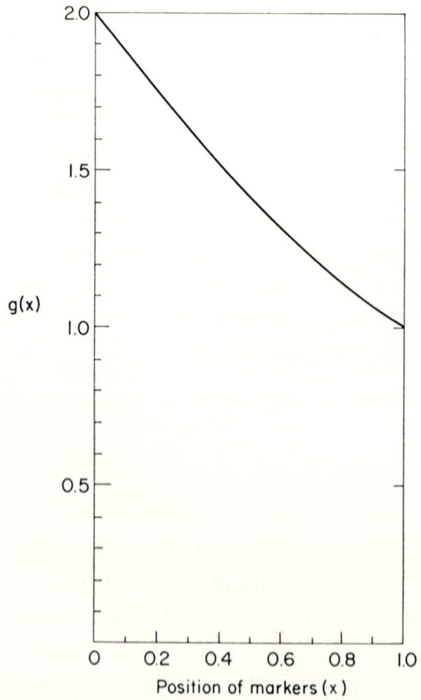

FIG. 2. Graphic representation of relative frequencies of genetic markers at different positions on one chromosome in an exponentially growing population. The marker frequency function (or mapping function) $g(X)$ is not normalized; the integration of $g(X)$ from $X = 0$ to $X = 1$ gives $1/\ln 2$.

the form of an age distribution function for binary dividing cells (Powell, 1956; C. Neuman, personal communication, 1962), and can be given

$$f(x) = (\ln 2)2^{1-x} \tag{1}$$

The derivation assumes that there is no resting period, each chromosome has one replication point, and the replication proceeds at a constant speed. The constant (ln 2) is given to normalize the function.

2. *Marker Frequency in Exponential, Nonsynchronous Population*

The frequency $g(X)$ of a marker (X) located at X on the chromosome in the exponential population can be obtained from the equation

$$g(X) = 2^{1-X} \tag{2}$$

This equation is called the *marker frequency* equation. A graphic representation of the equation is shown in Fig. 2. The function provides the basis of gene mapping by relative frequency of genetic markers in the

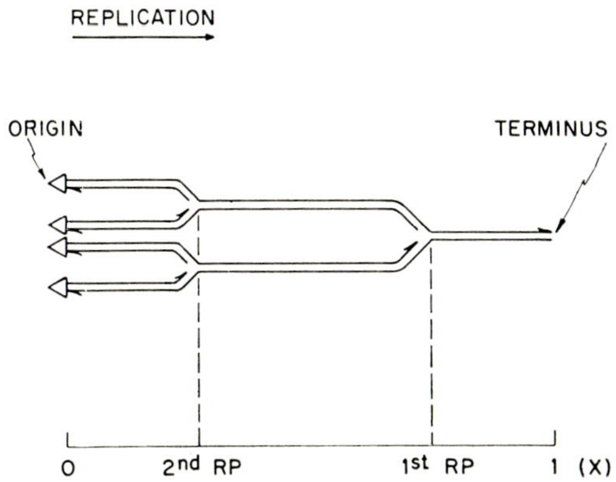

FIG. 3. A schematic representation of dichotomous replication of the chromosome. Before the first replication position (RP) reaches the terminus, the second RP starts from the origin. In an exponentially growing population, the simplest steady-state can be set if the distance between the first and second RP's occupies half the chromosome and each replication point proceeds at a constant rate. In reality the 2 replication points at the second RP may not necessarily be at the same position.

population. For technical problems associated with mapping see Yoshikawa and Sueoka (1963a) and Sueoka and Yoshikawa (1965).

The marker frequency equation has been generalized to include situations where there is more than one replication point per chromosome or where a part of one cell generation time constitutes a resting period, namely, without chromosome replication. The generalized equation, called the *equation of chromosome replication* (Sueoka and Yoshikawa, 1965), is shown below:

$$g_n(X) = 2^{n(1-X)} \tag{3}$$

In the equation, n represents the average number of replication positions per chromosome and can theoretically assume any positive values. The n-value of less than 1 indicates the presence of a resting period and that of more than 1, the multifork (or dichotomous) replication of the chromosome. A picture of dichotomous chromosome is shown in Fig. 3 ($n - 2$).

B. Genetic Map and Replication Order of Markers

The analysis of marker frequency in an exponential (nonsynchronous) population leads to the construction of a genetic map of *Bacillus subtilis* shown in Fig. 4. Note that the map also indicates the order of chromosome replication. The detailed description of the map construction will be given elsewhere (Yoshikawa and co-workers, 1966b).

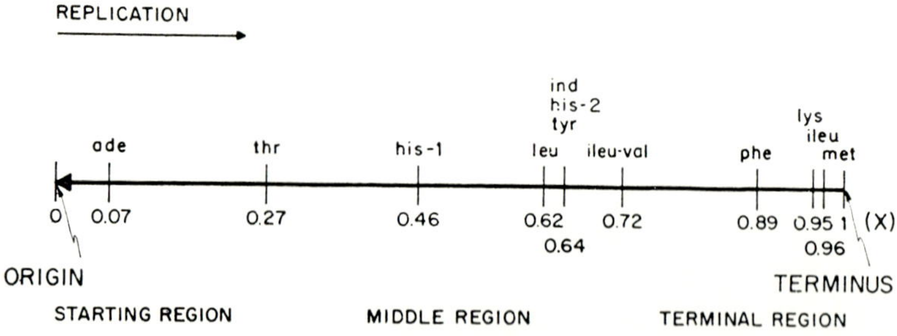

FIG. 4. A genetic map of *B. subtilis*. The position of each marker is simply based on the average of the results so far obtained by marked frequency analysis. Minor adjustments are expected in the future. For example, the *met* marker is assumed to have a position 1.00, which makes the *ade* position 0.07. Equally possible are the situations in which *ade* is located at 0.00 and *met* 0.93, or more generally, *ade* is in 0.00–0.07 and *met* in 0.93–1.00. The distance between *ileu* and *met* may be closer than indicated in the map because a linkage has been found for them by joint transformation. In our previous communication (Yoshikawa and Sueoka, 1963a), the *his* 2-*ind-tyr* group was mislocated, for *his* 1 and *his* 2 were unjustifiably assumed to be identical. A detailed account of the map will be given elsewhere.

III. SYNCHRONY OF CHROMOSOME REPLICATION

A. Criteria of Synchrony

The synchronization of chromosome replication has to satisfy the following criteria: (*a*) Replication should start simultaneously from a fixed position on the chromosome. (*b*) Replication of the chromosome should follow a definite sequence with regard to loci.

To satisfy these criteria, the replicated part of the chromosome should be identified and the extent of synchrony should be measured. In the *Bacillus subtilis* system, these points can be fulfilled by examining the replicated part of the chromosome by density transfer and genetic transformation.

B. Experimental Evidence

1. *Growth from the Stationary State*

Chromosomes of the W23 strain of *Bacillus subtilis* in the stationary growth phase are in nonreplicating form (Yoshikawa and Sueoka, 1963a). The release of growth from the stationary phase, therefore, should provide synchronous replication of the chromosome from the origin. This has been shown by transferring strain W23, brought to the stationary phase in a D_2O medium, to a fresh H_2O medium, and by examining the distribution of various markers in parental and repli-

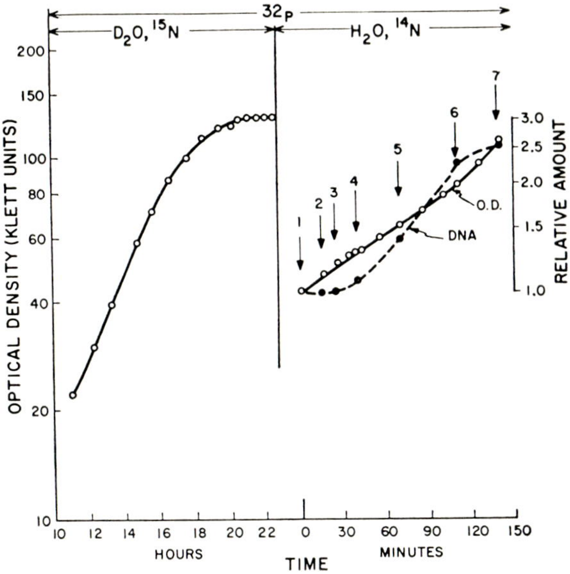

FIG. 5. Isotopic transfer of stationary cells. The DNA donor strain (W23) of *B. subtilis* previously conditioned to grow in 99.7% D_2O-N^{15}-P^{32} medium was brought to the stationary phase in 20 ml D_2O-N^{15}-P^{32} medium. Cells were collected by centrifugation and resuspended in 60 ml of a fresh H_2O-N^{14}-P^{32} medium. Specific radioactivity of P^{32} is the same in the D_2O-N^{15} and the H_2O-N^{14} media. Seven samples of 5–10 ml were taken at different times after the transfer. Cells were centrifuged, suspended in a versene saline solution, and their lysates prepared immediately after sampling. Increase of DNA after transfer was calculated from the relative amount of hybrid DNA observed in the profiles obtained by the CsCl density gradient centrifugation technique. (From Yoshikawa and Sueoka, 1963b.)

cated (hybrid) molecules (Yoshikawa and Sueoka, 1963b). The results are shown in Figs. 5 and 6.

The synchrony of the chromosome is clearly shown by the fact that transfer of markers is not uniform. However, the extent of synchrony is only fair in this experiment. The reason for this partial synchrony became clear when the ratio *ade/met* at the time of transfer was found to be not 1 but 1.13. This indicates that about 20% of the chromosomes were not in the completed form. In other words, the main reason for the partial synchronization is that in the D_2O medium the growth was slow and some of the stationary chromosomes were not complete at the time of transfer.

C. SPORE GERMINATION

Germination of bacterial spores provides an unusual opportunity for analyzing the process of development of a metabolically active state of

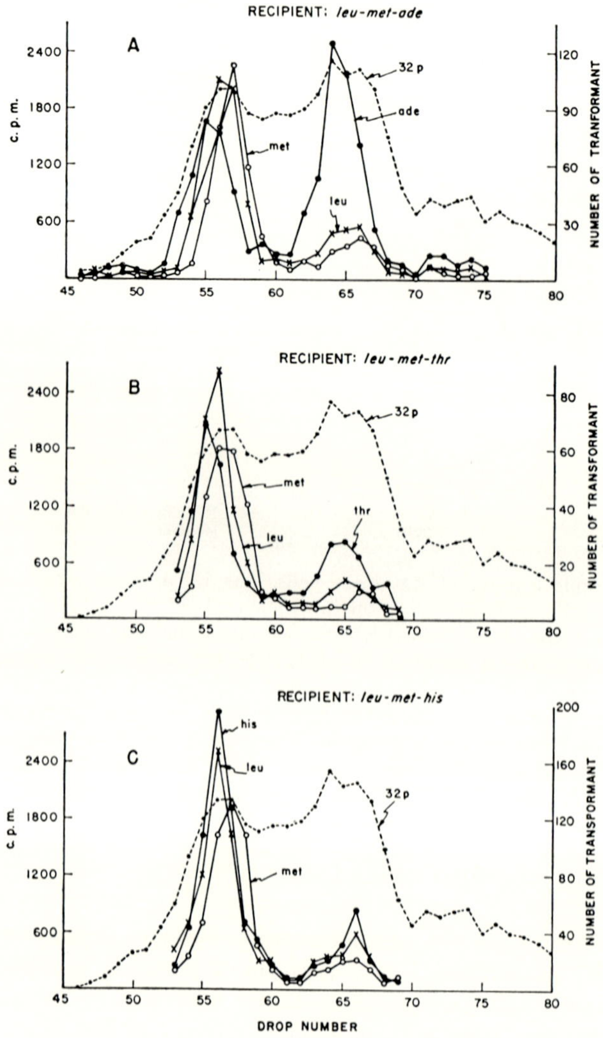

FIG. 6. Analyses of sample 5 of Fig. 5 for transforming activities. The base line ratio activity was subtracted from the P^{32} radioactive profile. The recipient strains used are 3 triple mutants. Mu8u5u6 (*leu-met-ade*), Mu8u5u5 (*leu-met-thr*) and Mu8u5u2 (*leu-met-his*). The markers *leu* and *met* come from a common origin (Mu8u5). The amounts of DNA used in all fractions were within a range of linear transforming activity. When the *leu-met-ade* mutant was used as the recipient, 100 μg/ml of adenine was added to the second growth medium for transformation. (From Yoshikawa and Sueoka, 1963b.)

the cell from an inert or dormant state (Woese, 1961). The mode of chromosome replication during spore germination, therefore, should provide the basis for the analysis of the regulation of gene function. First evidence of chromosome synchrony during spore germination was reported by Wake (1963) in which the ratio of *indole* to *methionine* markers changed as the germination proceeded. Our results on chromosome replication during spore germination provided the following information: (*a*) The chromosome in the spore is in completed form; (*b*) on germination, the genes replicate in definite sequential order from the origin to the terminus of the chromosome as in vegetative cells; (*c*) in an enriched medium, a multifork replication occurs, as in the exponential phase in the same medium (Yoshikawa *et al.*, 1964).

An excellent synchrony of chromosome replication has been observed during the spore germination both by marker ratio analysis (Fig. 7) (Yoshikawa *et al.*, 1964) and by density transfer experiments (Figs. 8, 9) (Oishi *et al.*, 1964). In the latter experiment, multifork replication was also unequivocally demonstrated. It is noted that the extent of the synchrony is excellent in the early part of the replication and somewhat degenerates later (Fig. 9). Improvement of the synchrony in the later stage deserves further investigation.

IV. APPLICATIONS OF CHROMOSOME SYNCHRONY

The synchrony of chromosome replication during spore germination (Oishi *et al.*, 1964) opens a number of new areas of investigation. Some of them are discussed below.

A. Mapping of the Genetic Markers

The location of the marker on the chromosome can now be obtained directly by the density transfer experiment during spore germination. Dichotomy further helps to locate markers in the middle and terminal regions of the chromosome. Moreover, the mapping by this method is more direct and has higher resolution than that of the marker frequency technique, especially for markers located in the starting region of the chromosome.

Two methods of density transfer experiments have been used for spore germination: (*a*) transferring D_2O spores to H_2O medium (Oishi *et al.*, 1964); (*b*) transferring spores of a thymine-requiring mutant (Farmer and Rothman, 1965) to a 5-bromouracil containing medium (Oishi and Sueoka, 1965). By using the system, we were able to subdivide adenine-requiring mutants into 2 groups which were known to be close to the origin by marker frequency analysis (Fig. 10) (O'Sullivan *et al.*,

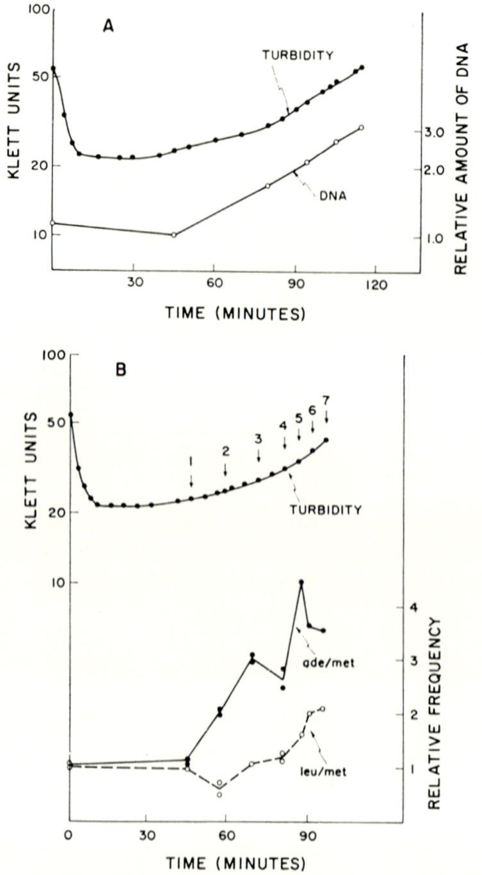

FIG. 7. Change of marker frequencies during germination of spores of strain W23. The germination mixture consisted of 1.9 ml purified spore suspension in H_2O (OD_{650} = 12.17), 19 ml H_2O and 40 ml Difco Penassay medium (1.75%). The mixture was incubated at 37°C in a growth flask with constant shaking. (A) DNA was assayed by the indole method (Keck, 1956). Five-ml samples from the culture were heated at 60°C for 10 minutes and centrifuged. Cells were suspended in 1 ml cold 5% trichloroacetic (TCA), kept in an ice bath for 30 minutes, centrifuged down, resuspended in 1 ml 5% TCA, incubated at 100°C for 15 minutes, and centrifuged. The supernatant (1 ml in 5% TCA) was mixed with 0.5 ml 12 *N* HCL and 0.5 ml 0.06% indole, incubated at 100°C for 10 minutes, and assayed for DNA. (B) Eight-ml samples were taken at various times, and DNA was isolated. Marker ratios were obtained relative to those of DNA from the stationary culture of strain W23. (From Yoshikawa *et al.*, 1964.)

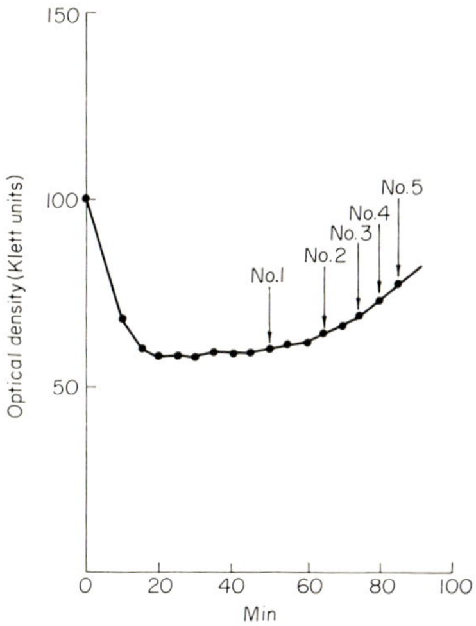

FIG. 8. Germination of *B. subtilis* D_2O spores. Spores were germinated in a Penassay medium at 37°C with constant shaking. Samples were taken at the time indicated and immediately heated at 60°C for 10 minutes to stop further DNA replication. The volumes of samples taken were 6 ml for Nos. 1, 2, and 3 samples, and 5 ml for Nos. 4 and 5 samples. These samples were then lysed and analyzed by CsCl density gradient centrifugation. (From Oishi *et al.*, 1964.)

1965). Interestingly enough, the density of DNA carrying group 1 *ade* markers is smaller than that of group 2 (Fig. 10).

The system also enabled us to examine the locations of rRNA's and sRNA's. Thus, the map positions of rRNA loci have been analyzed by germinating spores of a thymine-requiring mutant in the presence of 5-bromouracil (Oishi and Sueoka, 1965). Cell lysates from various stages of germination were centrifuged in a CsCl solution and DNA profiles were obtained. The hybridization pattern between the DNA and ribosomal RNA indicates that the ribosomal RNA loci are between the *ade* and *thr* markers (see Fig. 4). The similarity of 16 S and 23 S profiles should mean that the 2 loci are close together (Oishi and Sueoka, 1965). Such results are shown in Fig. 11. Theoretical treatment of the mapping by density transfer experiment in general will be shown elsewhere (Sueoka and Oishi, 1965).

B. ISOLATION OF SPECIFIC FRAGMENTS OF THE CHROMOSOME

Fractionation of DNA molecules originated from a genome has been attempted in various systems (Rolfe and Ephrussi-Taylor, 1961; Ganesan and Lederberg, 1964; Mindich and Hotchkiss, 1964). The chromosome synchronization and transformation in *Bacillus subtilis* during

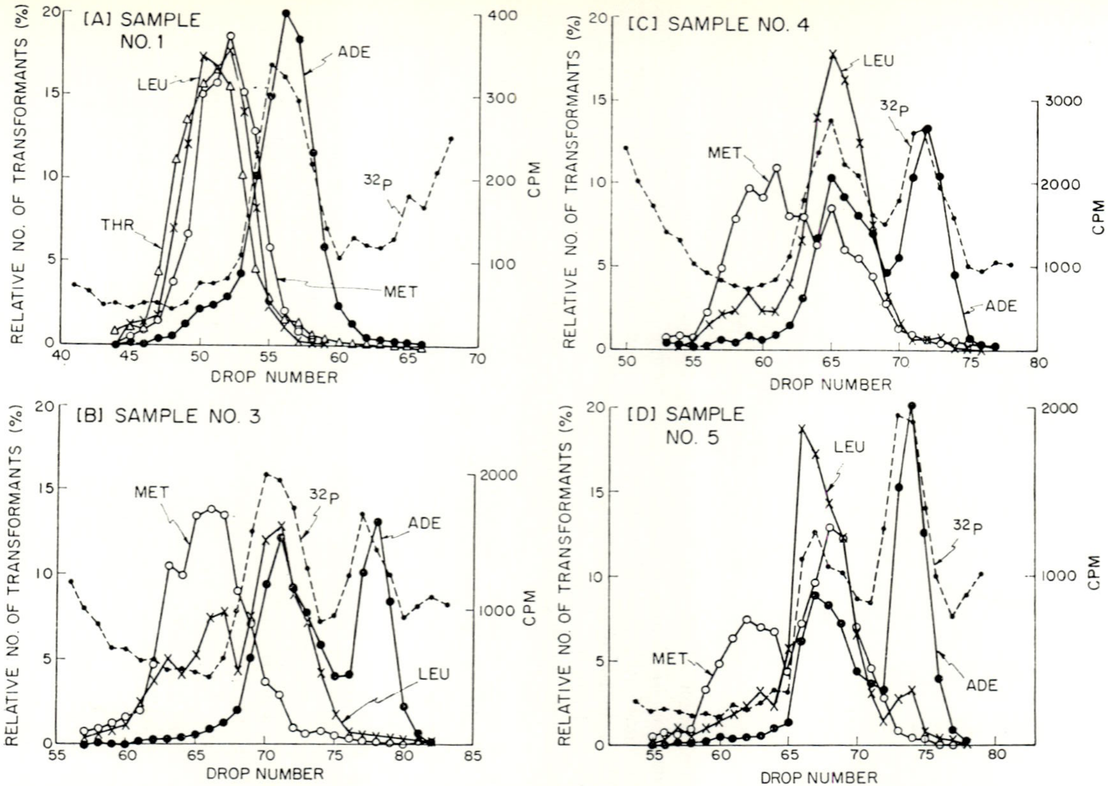

FIG. 9. Distribution of DNA from various samples indicated in Fig. 5 in CsCl density gradient centrifugation and analysis of transforming activities. Lysates of germinated spores prepared from samples in Fig. 8 were mixed with CsCl, centrifuged in a Spinco SW39 rotor, and fractionated. Forty per cent of each fractionation was used for assay of P^{32}-radioactivity and the rest was analyzed for transforming activity of various markers. The recipient strains used were 2 triple mutants: Mu8u5u16 (*leu-met-ade*) for Nos. 1, 3, 4, and 5 samples for leucine, methionine and adenine markers and Mu8u5u5 (*leu-met-thr*) for No. 1 samples for threonine marker. The relative number of transformants (ordinate) is the relative value (%) of the transformants of each fraction to the total number of transformants

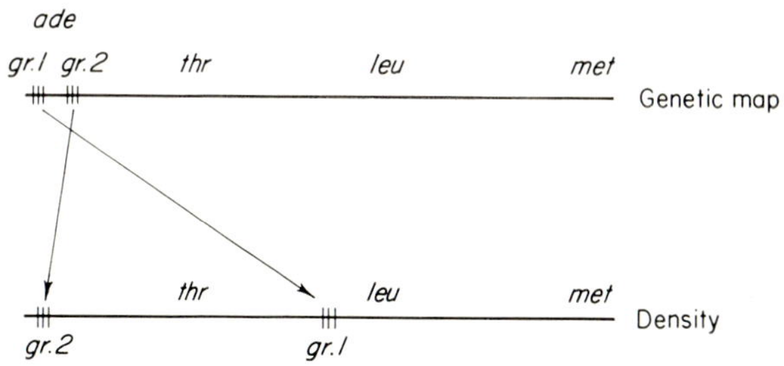

FIG. 10. Two groups of *ade* markers of *B. subtilis* revealed by transfer experiments during spore germination and by buoyant density of DNA. A clear distinction between *ade* groups 1 and 2 was observed by analyzing DNA of germinating spores in CsCl density gradient. Some of the *ade* markers (group 1) replicated earlier than the other (group 2), although both groups of *ade* markers were located close to the origin by marker frequency mapping. Densities of DNA carrying various markers were examined by CsCl density gradient centrifugation. The figure shows one situation representing the results of several experiments using different samples of DNA (strain W23) prepared from various stages of growth and by different methods of purification. (From O'Sullivan *et al.*, 1965.)

spore germination should theoretically, at least, allow us to isolate any part of the chromosome. Since, in practice, the highest extent of synchrony is required to obtain a homogenous sample of a small part of the genome, at present only the starting region can be isolated in a rather homogenous form. The physicochemical and genetic characterization of the tip of the chromosome should, therefore, be possible.

C. Regulation of the Chromosome Replication

Regulation of the chromosome replication has been the subject of active discussion (Maaløe, 1961; Jacob *et al.*, 1963; Sueoka and Yoshikawa, 1963).

The basic features of the regulation of chromosome replication are: (*a*) In normal growth conditions, chromosome replication is coordinated with cell division. (*b*) Initiation of the chromosome replication at the origin is repressed most of the time. (*c*) In order to initiate chromosome replication, protein synthesis is necessary (Maaløe and Hanawalt, 1961; C. Lark and Lark, 1964; Yoshikawa, 1965). (*d*) The replicating point seems to be associated with the so-called "membrane fraction" (Jacob *et al.*, 1963; C. Lark and Lark, 1964; Ganesan and Lederberg, 1965).

The chromosome synchronization is almost essential to analyze these problems further. Such an approach has already been initiated in the spore germination system. New protein synthesis is necessary for the first initiation of the DNA replication, despite the fact that some DNA polymerase exists in the spore (Yoshikawa, 1965). The elucidation

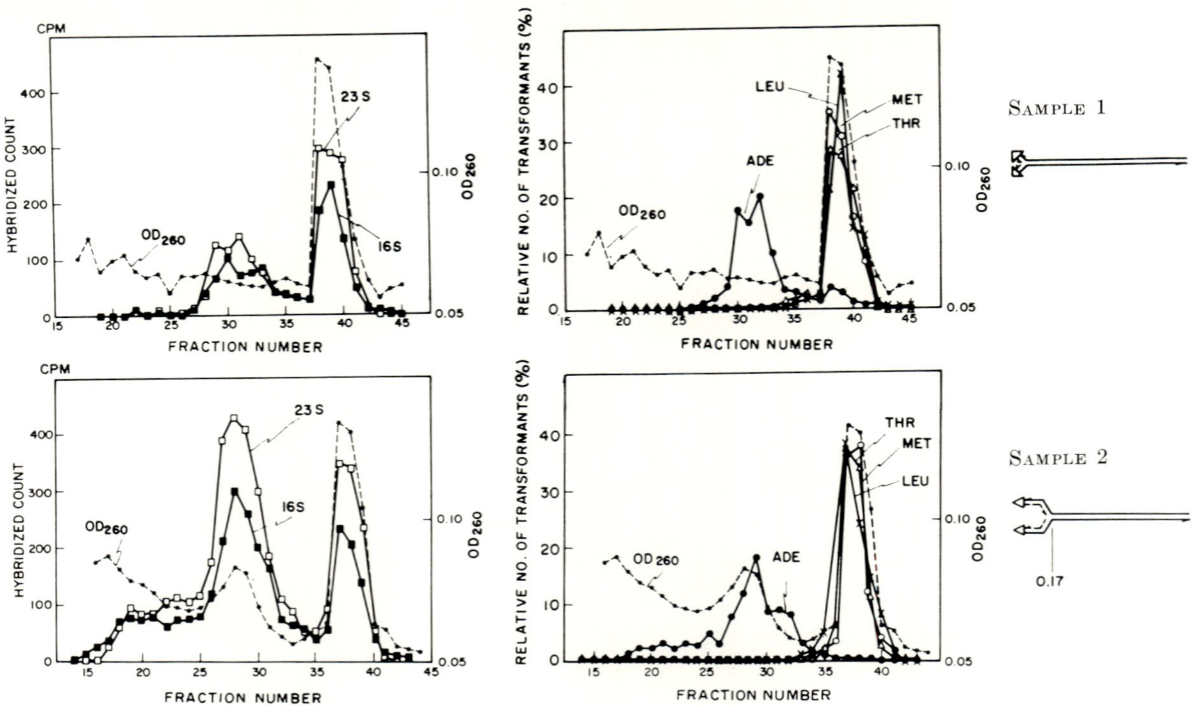

Fig. 11. Fractionation of various DNA samples during spore germination by CsCl density gradient centrifugation and analyses of transforming activities and hybridizability with H^3-labeled 16 S and 23 S RNA. Lysates of germinated spores prepared from the samples taken at various times were mixed with CsCl solution, centrifuged in Spinco SW39 rotor. Each 2 drops collected were diluted with tris buffer (0.01 *M*, pH 7.4) to adjust CsCl concentration. After measuring optical density at 260 mμ, 1.10 ml of each fraction was used for hybridization (0.55 ml each for 16 S and 23 S RNA). Radioactivity profiles are shown in the figure after subtracting the background value (80 cpm for 16 S RNA, 30 cpm for 23 S RNA). 0.04 ml of each fraction was analyzed for transforming activities of 4 markers. The recipient strains used were 2 triple mutants: Mu8u5u16 (*ade-leu-met*) for *ade* and *leu* markers and Mu8u5u5 (*thr-leu-met*) for *thr* and *met* markers.

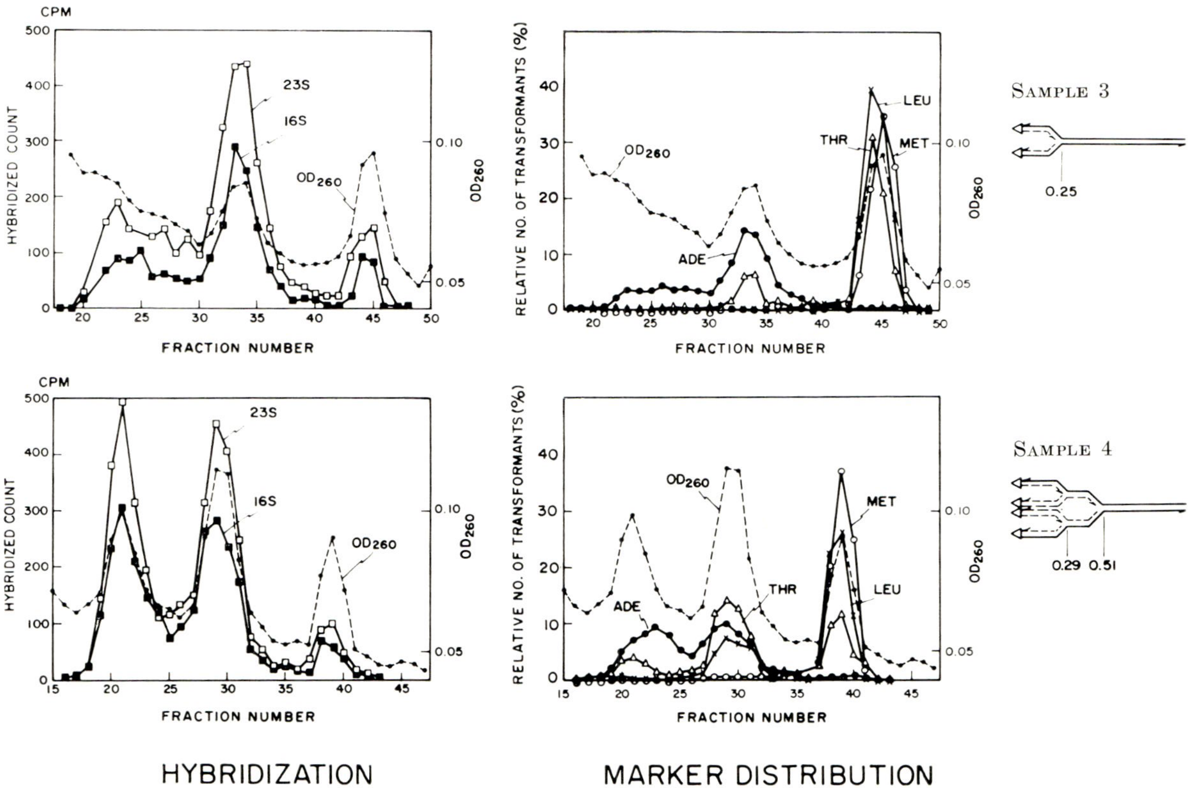

FIG. 11 (*legend cont.*). A relatively high background of optical density comes from ultraviolet absorbing low molecular weight material in lysates. The chromosome diagrams in the figure represent the status of chromosome replication in each sample calculated from the amount of DNA in parental, hybrid, and heavy peaks. (From Oishi and Sueoka, 1965.)

of the structure of the replication origin of the chromosome should be most instructive for our understanding of the regulation of the chromosome replication. The difference between the polar strain (W23) and the apparent nonpolar strain (W168), the conditions for multifork replication (Yoshikawa *et al.*, 1964), and the participation of the membrane fraction for the replication (Ganesan and Lederberg, 1965) should be some of the most instructive clues for our understanding of the initiation mechanism.

D. Regulation between Replication and Function of Genes

It has been reported in *E. coli* as well as *B. subtilis* that the capacity of the enzyme synthesis doubles when the corresponding gene duplicates (Masters *et al.*, 1964; Masters and Pardee, 1965). Gorman *et al.* (1964) report that bursts of enzyme occur at particular periods in one generation. These facts themselves do not indicate the order of the transcription, and, therefore, it should be examined further.

An interesting possibility is that during spore germination a sequential functioning of genes might be required for developing metabolically active cells from the dormant state (spores). If the transcription (RNA synthesis) follows the order of the genes on the chromosome during the spore germination, the order of the gene on the chromosome should be the basis of the regulation. This particular point is under investigation. In this connection, of special interest is that ribosomal RNA loci seem to lie in the starting region of the chromosome (Oishi and Sueoka, 1965).

V. FUTURE PROBLEMS

The structure of *Bacillus subtilis* chromosome has to be studied further to elucidate the following points: (*a*) Has the chromosome one continuous structure? (*b*) Is the chromosome circular? (*c*) Do episomes exist in *B. subtilis?* (*d*) How does the chromosome attach to the membrane?

These questions have been partly answered, as described before. However, the direct solutions are not yet available.

The structure of the origin of the chromosome should also be studied thoroughly in connection with the mechanism of the initiation of the replication. Loci for hitherto unmapped genes (e.g., sRNA, DNA, and RNA polymerases, aminoacyl-sRNA synthetases) should be identified on the chromosome.

Further questions on regulation of replication and gene function should become clear by studying detailed features of the DNA, RNA, and protein synthesis during the spore germination in which chromosome replicate synchronously. A possibility exists that the order of the genes on the chromosome has some meaning for the regulation of protein synthesis during the spore germination in which a rapid development of a metabolically active cell from a dormant cell (spore) occurs.

REFERENCES

Bonhoeffer, F., and Gierer, A. (1963). *J. Mol. Biol.* **7,** 534.

Cairns, J. (1963). *Cold Spring Harbor Symp. Quant. Biol.* **28,** 43.

Farmer, J. L., and Rothman, F. (1965). *J. Bacteriol.* **89,** 262.

Ganesan, A. T., and Lederberg, J. (1964). *J. Mol. Biol.* **9,** 683.

Ganesan, A. T., and Lederberg, J. (1965). *Biochem. Biophys. Res. Commun.* **18,** 824.

Gorman, J., Tamo, B., and Halvorson, H. O. (1964). *Biochem. Biophys. Res. Commun.* **15,** 43.

Jacob, F., Brenner, S., and Cuzin, F. (1963). *Cold Spring Harbor Symp. Quant. Biol.* **28,** 329.

Keck, K. G. (1956). *Arch. Biochem. Biophys.* **63,** 446.

Lark, C., and Lark, K. G. (1964). *J. Mol. Biol.* **10,** 120.

Lark, K. G., Repko, T., and Hoffman, E. J. (1963). *Biochim. Biophys. Acta* **76,** 9.

Maaløe, O. (1961). *Cold Spring Harbor Symp. Quant. Biol.* **26,** 45.

Maaløe, O., and Hanawalt, P. C. (1961). *J. Mol. Biol.* **3,** 144.

Masters, M., and Pardee, A. B. (1965). *Proc. Natl. Acad. Sci. U.S.* **54,** 64.

Masters, M., Kuempel, P. L., and Pardee, A. B. (1964). *Biochem. Biophys. Res. Commun.* **15,** 38.

Mindich, L., and Hotchkiss, R. D. (1964). *Biochim. Biophys. Acta* **80,** 73.

Nagata, T. (1963). *Cold Spring Harbor Symp. Quant. Biol.* **28,** 55.

Oishi, M., and Sueoka, N. (1965). *Proc. Natl. Acad. Sci. U.S.* **54,** 481.

Oishi, M., Yoshikawa, H., and Sueoka, N. (1964). *Nature* **204,** 1069.

O'Sullivan, A., Yoshikawa, H., and Sueoka, N. (1965) (in press).

Powell, E. O. (1956). *J. Gen. Microbiol.* **15,** 492.

Rolfe, R., and Ephrussi-Taylor, H. (1961). *Proc. Natl. Acad. Sci. U.S.* **49,** 1450.

Sueoka, N., and Oishi, M. (1965). In preparation.

Sueoka, N., and Yoshikawa, H. (1963). *Cold Spring Harbor Symp. Quant. Biol.* **28,** 47.

Sueoka, N., and Yoshikawa, H. (1965). *Genetics* **52,** 747.

Wake, R. G. (1963). *Biochem. Biophys. Res. Commun.* **13,** 67.

Woese, C. (1961). *In* "Spores" (H. O. Halvorson, ed.), Vol. 2, pp. 59–69. Burgess, Minneapolis, Minnesota.

Yoshikawa, H. (1965). *Proc. Natl. Acad. Sci. U.S.* **53,** 1476.

Yoshikawa, H., and Sueoka, N. (1963a). *Proc. Natl. Acad. Sci. U.S.* **49,** 559.

Yoshikawa, H., and Sueoka, N. (1963b). *Proc Natl. Acad. Sci. U.S.* **49,** 806.

Yoshikawa, H., O'Sullivan, A., and Sueoka, N. (1964). *Proc. Natl. Acad. Sci. U.S.* **52,** 973.

Yoshikawa, H., Hayashi, M. N., and Sueoka, N. (1966a). In preparation.

Yoshikawa, H., Neuman, R., and Sueoka, N. (1966b). In preparation.

CHAPTER 4

Chromosome Replication in *Escherichia coli*

Karl G. Lark

I. INTRODUCTION

In 1962 I summarized the state of our knowledge of DNA replication in bacterial, plant, and animal systems (K. G. Lark, 1963). At that time, little was known about the *in vivo* replication of DNA beyond a description of its synthesis in various systems and the measurements of the production and utilization of precursors.

A critical evaluation of the available information led to the conclusion that the precise regulation of DNA biosynthesis and chromosome replication must depend on factors other than the production and use of DNA precursors. It seemed necessary, therefore, to look to the primer and to

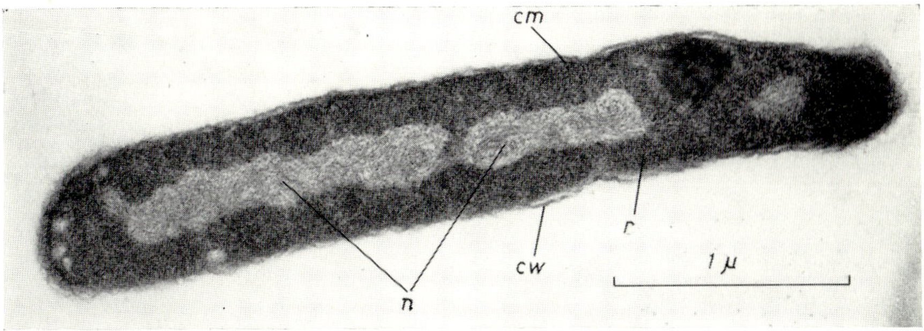

FIG. 1. Ultrathin section of *E. coli* after starvation in NaCl. (From Kellenberger and Ryter, 1958.) cw = cell wall; cm = cell membranes; r = ribosomes in cytoplasm; n = nucleus.

the template activity of DNA itself as the controlling factor in replication.

Since 1962 a number of experiments have been carried out which have determined whether replication of DNA is physically ordered with respect to the macromolecule itself. These studies have been, in turn, related to the regulation of DNA synthesis within the cell. Two systems have been used in these studies, *Escherichia coli* and *Bacillus subtilis*. The studies with *B. subtilis* are reported by N. Sueoka (Chapter 3). In discussing the experiments with *E. coli*, I have attempted to concentrate on those general aspects which may be of importance to experimentation in other biological systems. A more detailed review of the work in our laboratory will be published elsewhere (K. G. Lark, 1966).

II. THE *Escherichia coli* SYSTEMS

Escherichia coli (Fig. 1) is a gram negative bacterium that can be grown in synthetic medium. It will grow on a variety of nutrients with a growth rate which is about 20 minutes per division in an enriched nutrient broth, 40 minutes in a glucose minimal salts medium, and 4.5 hours in a medium in which acetate is the sole energy source (C. Lark, 1966; also Schaechter *et al.*, 1958). It divides by constriction of the cell wall. The cell wall of *Escherichia coli* is a complex structure consisting of several layers which possess different chemical and structural properties (Weidel *et al.*, 1960). Within this cell wall, a cell membrane may be discerned. The nuclear material is contained in central regions of the cell (Kellenberger, 1960) which also appear to be the site of RNA synthesis (Caro and Forro, 1961). In rapidly growing cells, at least 2 such regions (nuclei) may be discerned by cytological techniques. The average number of these regions has been correlated with the number of separable genetic units within the

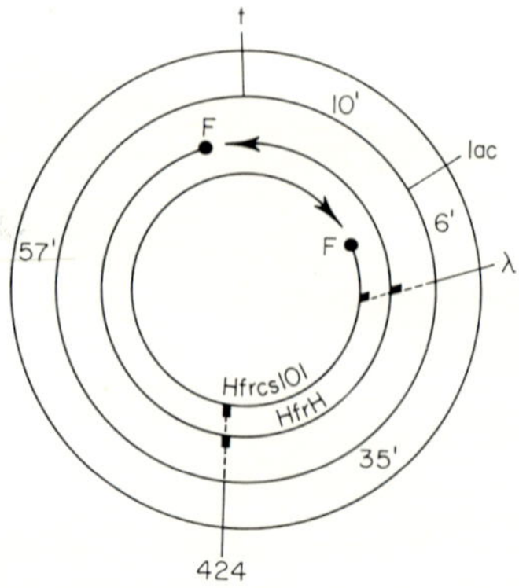

FIG. 2. Linkage map of *E. coli* K-12. (From Nagata, 1963b.) Only a few markers are shown. The distance between any 2 markers is expressed in terms of minutes of transfer of the Hfr chromosome during conjugation. t = threonine; lac = lactose; λ = prophage λ; 424 = prophage 424; F = the integrated F factor. A more complete map is given by Hayes (1964, pp. 568–569).

cell by studies of sectored colonies arising from cells which have undergone mutation (Witkin, 1951).

Not all strains of *E. coli* are amenable to genetic study. However, opposite mating types of the K-12 strains will transfer genetic material in a process known as conjugation (Lederberg, 1947; Hayes, 1952; Jacob and Wollman, 1961; Hayes, 1964). Thus, genetic material from certain donor (Hfr) strains is transferred to recipient cells (F^-) as a result of cell pairing. It has been shown that DNA is actually transferred from the donor to the recipient as the result of a mating act (Skaar and Garen, 1955; Silver, 1963). Transfer of genetic markers normally occurs in a sequence and can be interrupted at any time by separation of mating partners. This sequence is the same as that obtained from studies of genetic recombination and linkage, i.e., the genetic map. Different Hfr strains may begin the transfer of genetic material at different genetic regions of the chromosome (designated as their origin) and the transfer may proceed in either direction along the chromosome, the direction being specific for a given Hfr strain. However, the sequence of genetic markers transferred is always that of the genetic map (Jacob and Wollman, 1961; Hayes, 1964). This map represents a single circular linkage group (Fig. 2).

In addition to the genetic material contained in the chromosome, there exist fertility factors F, (or sex factors) which can be shown to have the properties of DNA (Jacob and Wollman, 1958, 1961; Marmur *et al.*, 1961) which can be transferred readily to recipient cells. These F factors belong to a class of extra chromosomal genetic material known as episomes. Frequently F factors may be obtained with a small portion of bacterial genetic material attached which duplicates information already present on the chromosome. Cells containing episomal as well as chromosomal information are thus diploid in certain characters and maintain 2 sets of genetic information. The existing data are compatible with the hypothesis that Hfr strains arise by intergration of the F episome into the normal *E. coli* chromosome (Jacob and Wollman, 1961; Campbell, 1962).

III. SEQUENTIAL REPLICATION OF THE *Escherichia coli* CHROMOSOME

In their original studies demonstrating the semiconservative nature of DNA replication, Meselson and Stahl (1958) also demonstrated that chromosome replication was sequential. Upon transfer of cells in the exponential phase of growth from "heavy" (N^{15}) to "light" (N^{14}) medium, all of the newly synthesized DNA was hybrid in density until at the end of one generation, only hybrid material was observed. At this time, when all of the original (heavy) DNA had disappeared, they began to observe the formation of light, or twice replicated, material (Fig. 3). This could only mean that all portions of the chromosome had replicated at least once before any individual portion began a second round of replication.

Their observations have since been supported (K. G. Lark *et al.*, 1963) by an experiment in which a portion of the chromosome of a thymine requiring strain, *E. coli* 15T, was labeled by brief exposure to thymine-H^3 after which the exponentially dividing culture was immediately transferred to medium containing a density label, 5-bromouracil. Although synthesis of hybrid density material commenced immediately, none of this was radioactive.

This is shown in Fig. 4. On the abscissa we have plotted the percent of the total DNA which is found in the hybrid form at any time; this is related to the amount of original DNA replicated by the equation:

$$\text{Fraction of DNA replicated} = \%\text{hybrid DNA}/(200 - \%\text{hybrid DNA})$$

The abscissa, then, is a function of the total amount of DNA synthesized or of the length of chromosome replicated.

The amount of radioactive DNA which has replicated at any time is plotted on the ordinate. This is simply the fraction of the total radioactive material appearing in the hybrid band. Radioactive hybrid material appeared only after about 50% of the chromosome had replicated and then only in very small amounts. More than 50% of the radioactive material was replicated after 80% of the nonradioactive portions of the chromosome had completed replication.

Experiments such as those of Meselson and Stahl and that shown in

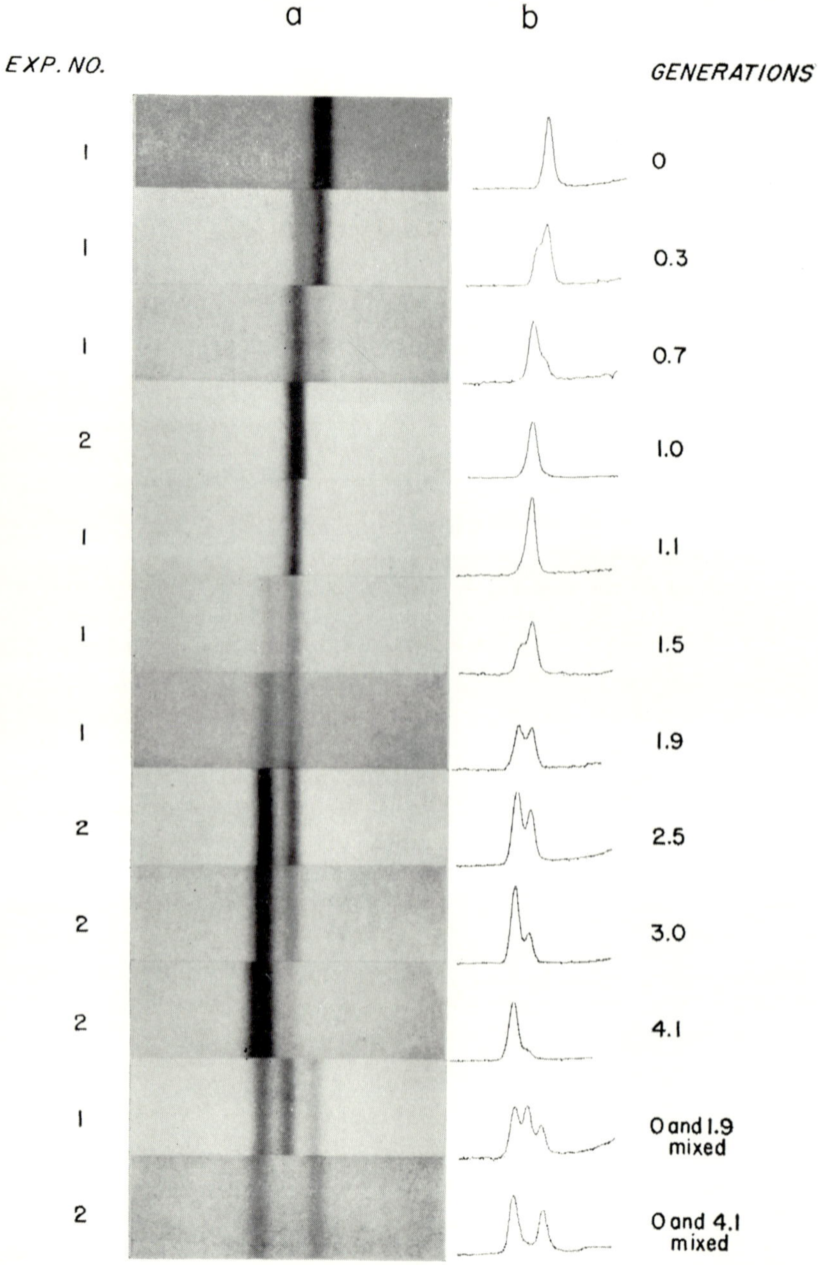

FIG. 3. Ultraviolet photographs showing DNA bands resulting from a density gradient centrifugation of lysates of *E. coli* obtained at intervals after transfer of a culture in exponential growth from N^{15} to N^{14} medium. (From Meselson and Stahl, 1958.) The time elapsing after the transfer is shown on the right as the fraction of the generation period. The density increases to the right of the photograph.

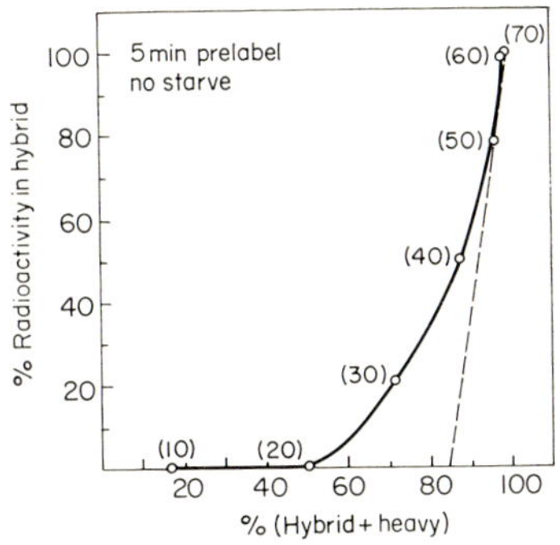

FIG. 4. Pattern of replication of a pulse of thymine-H^3 following immediate transfer to 5-bromouracil (5 BU) medium. (From K. G. Lark *et al.*, 1963.) *E. coli* $15T^-$ was labeled with thymine-H^3 for 5 minutes (0.1 generations) and then transferred to medium containing 5 BU instead of thymine. Samples were taken at intervals, lysed, and placed in cesium chloride solution. They were then examined in the analytical ultracentrifuge for percent total hybrid DNA (abscissa) and in the preparative centrifuge for the percent radioactivity in the hybrid band (ordinate). The number next to the points indicate the time of sampling after transfer to 5 BU medium.

Fig. 4 are performed on cell populations rather than individual cells and thus yield a minimum rather than maximum estimate of the degree to which sequence is followed in chromosome replication. Any correction for heterogeneity of replication rate in the cell population will emphasize the sequential nature of the replication pattern observed.

Genetic measurements of sequential chromosome replication were made by Nagata (1963a). Using the filtration technique developed by Maruyama and Yanagita (1956), he synchronized the cell division cycles of an F^- (recipient) and of 2 different Hfr (donor) strains of *E. coli* K-12. The 2 Hfr strains (Hfr H and Hfr CS 101) were chosen such that the origin and direction of transfer of genetic material was different (Fig. 2). In all cases, DNA was synthesized throughout 90% of the division cycle. All of these strains carried 2 lysogenic bacteriophages λ and 424 in the prophage state, i.e., as genetic markers integrated into the bacterial chromosome at defined loci. Upon irradiation of the bacterial host (induction) these 2 phages will develop independently into vegetative bacteriophages.

The relative quantity of a phage produced could be shown to depend on the relative number of vegetative phage genomes originally present (also Bertani, 1954). For his experiments, Nagata assumed that the relative yield of a particular temperate phage would also be proportional to the relative number of prophage genomes present at the time of irradiation. Following irradiation of the synchronous cultures at various times in the division cycle, Nagata measured the relative quantities of the 2

TABLE I

CORRELATION BETWEEN DUPLICATION TIME OF TWO PROPHAGES EXPRESSED IN TERMS OF PERCENT INCREASE OF DNA AT THE MOMENT OF DUPLICATION AND THE RELATIVE POSITION OF THE PROPHAGES FROM THE F FACTOR ON EACH HFR LINKAGE MAP[a]

Strain	Prophage duplication	% increase of DNA	% distance from F factor
Hfr H	424	48.3	52.8
	λ	80.5	85.0
Hfr CS 101	424	36.7	38.0
	λ	6.7	5.5

[a] From Nagata (1963b).

bacteriophages produced. He was able to demonstrate that the ratio of the burst sizes varied discontinuously by a factor of 2 in Hfr cultures, increasing and decreasing at times consistent with the replication of the individual prophage loci.

These times corresponded to an increase in DNA equal to the genetic distance between the markers. The time of change in the relative proportion of phages produced indicated that replication proceeds sequentially from that end of the chromosome which is transferred last in bacterial conjugation (i.e., from the end containing the F factor) toward the end transferred first (the chromosome origin) (Table I). The replication of the F^- strain did not show this ordered behavior. That is, the ratio of the 2 prophages remained constant throughout the division cycle. This led Nagata to the conclusion that the origin from which sequential replication proceeds is not fixed in the F^- strain as it is in Hfr strains. The extent to which this origin can shift will be discussed later.

Nagata's observations, while constituting strong supportive evidence for sequential replication of the chromosome, are based on the indirect measurement of vegetative phage production and could represent the order in which information to be transcribed is read out of the bacterial chromosome, a problem related to that discussed by H. O. Halvorson *et al.* (Chapter 6).

Perhaps the most elegant experiment demonstrating sequential replication of the *E. coli* chromosome is that of Cairns (1963). He labeled the chromosome of *E. coli* K-12 Hfr H for 1½ generations with radioactive thymine. Subsequently, he isolated the intact chromosome and overlaid it with photographic film. The resulting autoradiograph revealed a circular structure (Fig. 5) containing a single replication loop. The circumference of the structure was between 1000 and 1400 μ. One side of the replication loop was twice as radioactive as the other, a result to be

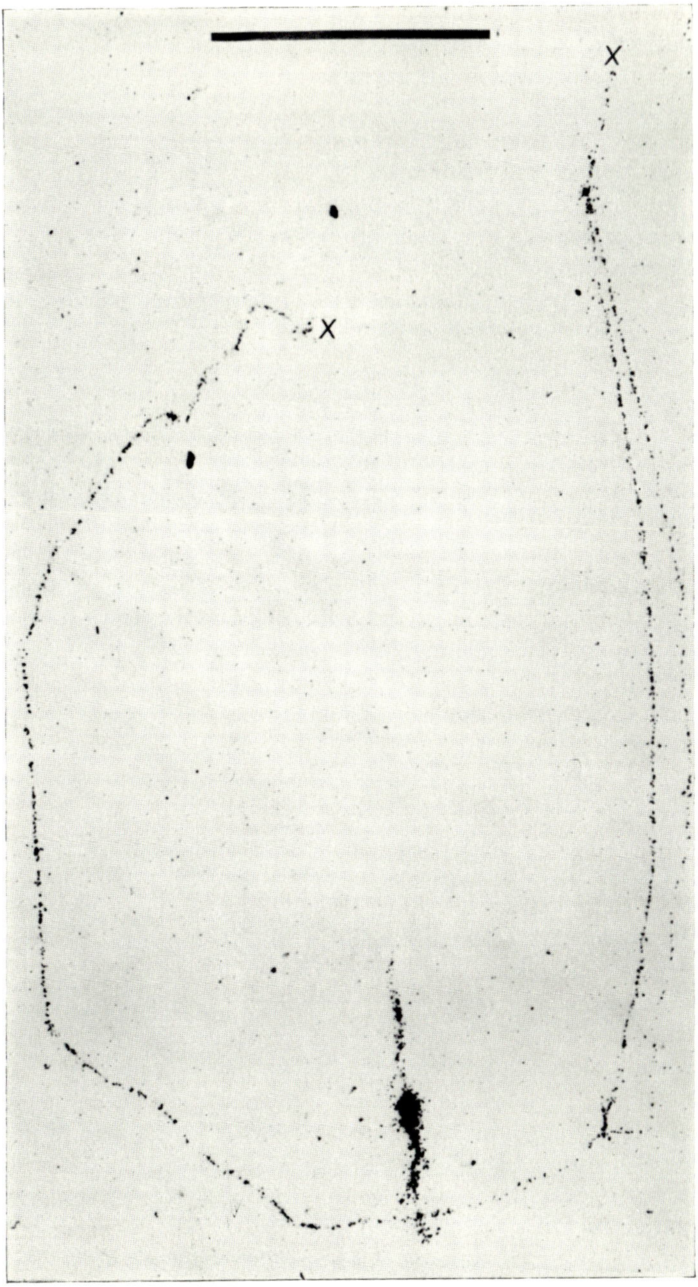

FIG. 5. Autoradiograph of the chromosome of *E. coli*. (From Cairns, 1963.) The bar represents 100 μ. The chromosome in this preparation is broken at the point marked with an X.

expected from the length of time over which label was administered and the semiconservative and sequential nature of DNA replication. He also subjected cells to short periods of thymine labeling (10% of the generation period) and demonstrated that the length of chromosome replicated equaled that expected from only *one* replication point per chromosome. Thus Cairns' experiment demonstrated the circularity of the replicating *E. coli* chromosome, its sequential replication and, finally, the existence of a single locus of replication on the chromosome.

This last point was independently confirmed by Bonhoeffer and Gierer (1963). In their experiments, a culture of *E. coli* was exposed for a very short time to radioactive 5-bromouracil yielding segments of chromosome with both a radioactive and density label. The total sum of all of the radioactive segments per cell was calculated from the net amount of radioactivity incorporated per cell. The actual length of any individual segment was then estimated by comparing the density profile of the radioactive material, after the chromosome DNA had been carefully sheared into segments of measurable length, with the average molecular weight of the material which they obtained. Their results indicated that the total length of radioactive material was contained in a single continuous unit in each chromosome—that is, only one point of replication existed on each chromosome. Their technique is of general interest since it represents the most feasible approach to determining the number of replication points on a single chromosome of a higher organism.

IV. REGULATION OF SEQUENTIAL CHROMOSOME SYNTHESIS

A. Amino Acid Starvation

In 1961, Maaløe and Hanawalt (also Hanawalt *et al.*, 1961) proposed a hypothesis based, in part, upon the earlier work of Pardee and Prestidge (1956) in which it had been shown that amino acid auxotrophs of *E. coli* deprived of essential amino acids could continue to synthesize DNA to a limited extent, the DNA content of the culture increasing by a factor of 1.5. Maaløe and Hanawalt interpreted this finding, which they reconfirmed, to mean that each cell had finished replicating its chromosome in the absence of amino acids but was not able to initiate a new cycle of chromosome replication as long as amino acids were withheld. They supported their hypothesis by establishing that cells starved of amino acids ceased to synthesize DNA on an all or none basis. That is, radioautographs of bacteria starved of amino acids and pulse labeled with thymine-H^3 revealed that as starvation progressed more and more cells became incapable of synthesizing DNA, presumably as they reached the end of their synthetic cycle.

Proof of this hypothesis was eventually provided by the following experiment (K. G. Lark *et al.*, 1963; Pritchard and Lark, 1964): A quadruple auxotroph of *E. coli* requiring thymine and 3 amino acids was starved of amino acids for 80 minutes (almost 2 generations). During this period, DNA synthesis continued to a limited extent and ceased after the DNA content of the culture increased by a factor of between 1.4 and 1.5. Starvation was ended by the addition of amino acids and at the same time DNA synthesis was reinitiated. Radioactive thymine-H^3 was also added to label that portion of the chromosome replicated immediately after the resumption of DNA synthesis. After a short period during which the DNA content of the culture increased by a factor of 7 to 10%, the cells were placed in a nonradioactive medium and grown for several generations. They were then again deprived of amino acids, which were subsequently restored in the presence of a density label, 5-bromouracil. The conversion of the total DNA and of the radioactive DNA into hybrid density material was measured. The results were analyzed in the graph shown in Fig. 6. We see that the bulk of the radio-

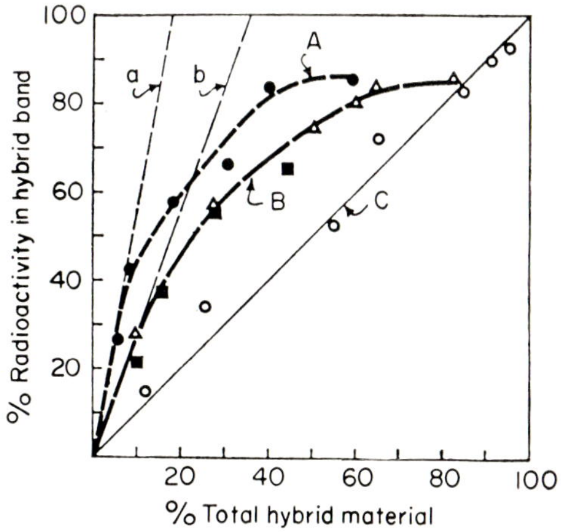

FIG. 6. The replication pattern of DNA labeled with radioactive thymine after amino acid starvation. (From Pritchard and Lark, 1964.) A culture of *E. coli* $15T^-$ was starved of 3 essential amino acids in the presence of nonradioactive thymine for 90 minutes and then transferred to medium containing radioactive thymine and the required amino acids. After 10% of the chromosome had replicated, the cells were transferred to nonradioactive medium and grown for several generations. Culture A was transferred to medium lacking amino acids and, after 90 minutes starvation, into medium containing bromouracil plus amino acids. Culture B was transferred to medium lacking thymine and starved for 30 minutes after which bromouracil was added. Culture C was transferred to bromouracil medium without any starvation. Samples were taken from A, B, and C at intervals, lysed, and analyzed as in Fig. 4. The dashed lines a and b are the results expected for (A) and (B) respectively if all of the *cells* responded with complete homogeneity to the starvation treatments.

active DNA is replicated before the rest of the DNA. (Up to 60% of the radioactivity is in the hybrid form before 25% of the total DNA has replicated.) This must mean, that in almost all of the cells radioactive thymine was inserted after amino acid starvation into a special region of the chromosome, which after several generations of growth could be located again by another starvation treatment. Subsequently, this region was the first to replicate in the presence of bromouracil. This particular region of the chromosome can be revealed by amino acid starvation after 2, 4, or 6 generations of growth. The replication which commences after amino acid starvation has been shown to be sequential (K. G. Lark *et al.*, 1963; Pritchard and Lark, 1964). These data fulfill the predictions of the Maaløe-Hanawalt hypothesis and, indeed, constitute its proof.

It is interesting to note, that the strain used in these experiments, 15T$^-$ is an F$^-$ strain, although an exceedingly poor recipient (K. Rasmussen, personal communication, 1964). Data of the type shown in Fig. 6 are unable to distinguish a difference in the location of the special region of the chromosome between cultures in which 2, 4, or 6 generations of growth have intervened between the 2 amino acid starvations. This would indicate that any possible relocation of this region (postulated by Nagata for K-12 F$^-$ strains) would be rare, occurring with a frequency which could not be detected by our experiments—i.e., in less than about 4% of the cells per generation.

B. The Effect of Thymine Starvation

Barner and Cohen (1956) found that cells starved of thymine would divide synchronously upon restoration of thymine and would also synthesize DNA in a synchronous manner. This finding, together with observations on the induction of lysogenic bacteria, led us (Pritchard and Lark, 1964) to investigate the effect of thymine starvation upon sequential DNA synthesis. In an experiment also shown in Fig. 6, we labeled the chromosome with thymine-H^3 immediately following a period of amino acid starvation. After several generations of growth, the cells were starved of thymine and then DNA synthesis was reinitiated by the addition of bromouracil.

The replication of radioactive DNA was followed. It was observed that thymine starvation led to the preferential replication of the radioactive DNA (Fig. 6). This result could only be interpreted to mean that thymine starvation had initiated a cycle of DNA replication prematurely and that this replication was initiated in that same portion of the chromosome which is replicated immediately after amino acid starvation. It seems reasonable to define this portion of the chromosome as its beginning.

A summary of the results of experiments on thymine starvation (Prit-

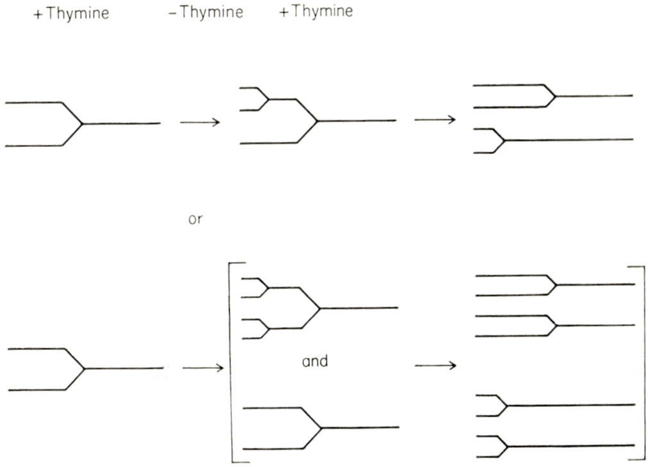

FIG. 7. A schematic representation of the effect of thymine starvation on chromosome replication. Of the 2 possibilities shown (discussed by Pritchard and Lark, 1964) the upper 1 has now been established as correct. (K. G. Lark and Bird, 1965.)

chard and K. G. Lark, 1964) follows: (*a*) Thymine starvation initiates replication prematurely at the chromosome origin. (*b*) Replication also proceeds at the original locus of replication at which it stopped when thymine was removed. (*c*) Premature replication is induced at the beginning of only 50% of the 2 partial chromosomes ("daughters") which are present at the moment thymine is removed. Recent experiments have shown that there is one such premature replication point on every replicating chromosome (K. G. Lark and Bird, 1965).

This replication pattern is shown schematically in Fig. 7, and is similar to the pattern observed by Sueoka and his associates in cultures of rapidly dividing *B. subtilis*, or following spore germination in *B. subtilis*.

The existence of 2 replicating points on one chromosome following thymine starvation does not persist indefinitely. At the end of the replication cycle, the situation is resolved (Fig. 7) to give 2 chromosomes with only one point of replication each.

V. MEASURING CHANGES IN THE SEQUENTIAL PATTERN OF REPLICATION

In the course of the experiments on amino acid and thymine starvation, described above, we developed a method for measuring deviations from the normal pattern of sequential replication. In sequential replication, the conversion of a short piece of radioactive DNA to the hybrid form follows a typical pattern as was described in Fig. 4. The essential characteristics of this pattern (i.e., the synthesis of a large amount of nonradioactive DNA before the radioactive material begins to replicate) is independent of the growth conditions of the culture (whether division is synchronous or the growth rate is slow or fast) provided (*a*) that nothing changes the relative sequence of replication between administration of the radioactive label and the subsequent administration of the density label and (*b*) that all of the cells of the culture continue to behave homogeneously.

If a treatment which changes the sequence of replication is interposed between the incorporation of radioactive label and the subsequent replication in density label, one may expect a change in the pattern of replication. For example, thymine starvation, while not allowing DNA synthesis to proceed, will induce a premature replication cycle in which the restoration of thymine results in the situation shown in Fig. 7. If a portion of the chromosome is labeled with thymine-H^3 immediately prior to thymine starvation and bromouracil is then added to end starvation, the radioactive portion of the chromosome will be converted to a hybrid density form prematurely by the new replicating locus induced by the thymine starvation (Pritchard and Lark, 1964; Billen, 1964).

This technique has been used by Hewitt and Billen (1964) and by Billen *et al.* (1965) to study the effect of ultraviolet and x-irradiation on sequential DNA synthesis.

In their experiments they labeled a portion of the chromosomes of different cells with a pulse of thymine-H^3 and then irradiated the cell population. After irradiation the cells were grown in bromouracil. The results of their experiments are shown in Fig. 8.

It may be seen that both types of irradiation disrupt sequential synthesis; this results in the premature replication of the radioactive portion of the chromosome.

Whether this replication is initiated at random in different chromosomes could eventually be determined by studying the effect of irradiation on chromosomes labeled in one position—i.e., at their origin following amino acid starvation.

The quantitative analysis of such curves could yield some information as to the nature of the change which has occurred in sequential replication.

VI. STUDIES ON THE MECHANISM CONTROLLING INITIATION OF DNA SYNTHESIS: USE OF INHIBITORS

We have considered some of the situations that regulate the initiation of DNA synthesis and which lead to an alteration of the sequential pattern of chromosome replication. The reinitiation of DNA synthesis

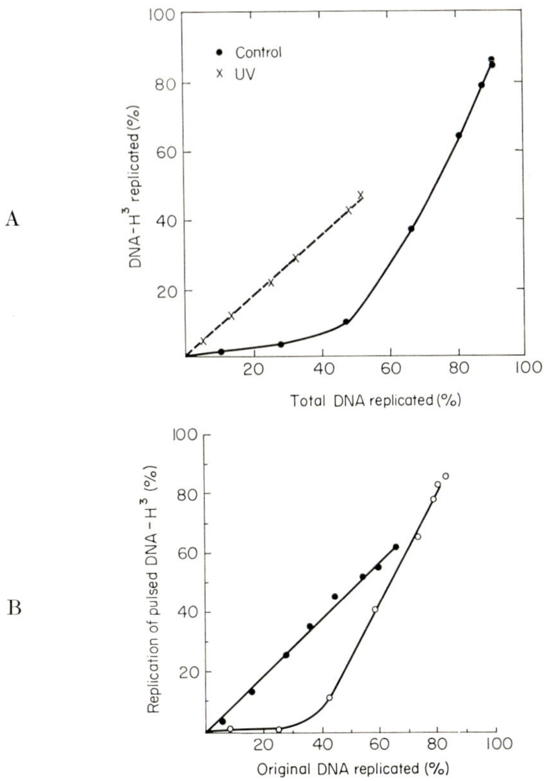

FIG. 8. Replication pattern of DNA in cells irradiated with (A) ultraviolet or (B) x-radiation. (From Billen *et al.,* 1965.) (A) Cells were labeled for about 7% of a generation period with thymine-H^3, washed and irradiated with ultraviolet light. A control (●) was not irradiated. Both control and irradiated cells were grown in bromouracil medium; samples were taken at intervals and analyzed in a manner similar to that used in Fig. 1. Note that the abscissa represents the percent replicated DNA rather than the percent of hybrid material. (B) Essentially similar to (A) except that x-irradiation rather than ultraviolet radiation was used. (○) irradiated culture; (●) control culture.

following amino acid starvation can be differentiated from the premature initiation which results from thymine starvation by the use of appropriate inhibitors.

Thus, very low concentrations of chloramphenicol (12–25 μg/ml) will inhibit premature initiation of DNA synthesis by thymine starvation, if added during the period of thymine deprivation (C. Lark and Lark, 1964). This result, is not unexpected since simultaneous amino acid and thymine starvation does not result in the initiation of an extra locus of replication. Again, 5-fluorouracil when present throughout the period of thymine starvation will prevent the induction of an extra replication point. This effect of 5-fluorouracil can be overcome by addition of uracil which indicates that the inhibitor is blocking some aspect of the thymine starvation mechanism which involves RNA synthesis. Both the chloramphenicol and 5-fluorouracil results implicate RNA mediated protein synthesis in the induction of premature chromosome replication by thymine starvation.

This mechanism could, at first thought, be related to the inhibition of the initiation of DNA synthesis by amino acid starvation. In amino acid starvation some protein necessary for initiation of replication is *not* synthesized, whereas during thymine starvation an excess of this material is synthesized but not used in the absence of DNA synthesis.

This hypothesis appears to be too simple to explain all of the facts. Following amino acid starvation, chromosome replication is reinitiated in the presence of either chloramphenicol (12–25 μg/ml) or 5-fluorouracil (C. Lark and Lark, 1964). Moreover, this reinitiation occurs at the chromosome beginning. The initiation of chromosome replication is therefore more complex than the formation of a single protein initiator. These results (see section on regulation of chromosome replication) have been explained by assuming that a structural protein is required for replication in addition to an initiator protein whose synthesis is inhibited by chloramphenicol or fluorouracil.

Another aspect of chromosome replication is revealed by the use of phenethyl alcohol (Berrah and Konetzka, 1962; Treich and Konetzka, 1964). In appropriate doses, this chemical will inhibit division but not protein or RNA synthesis. DNA synthesis will continue to a limited extent—i.e., an increase of 40–50% in the DNA content of the culture. If the agent is removed, DNA synthesis and cell division will resume. If it is only removed temporarily—i.e., long enough for DNA synthesis to be initiated (about half of a generation period), its readdition will restrict synthesis to an exact doubling of the DNA content of the culture. This process can be repeated to yield twofold increments in the DNA content of the culture (Fig. 9). Since neither DNA, RNA, or protein synthesis are directly inhibited by this agent, it seems logical to attribute its action to interference with the replication cycle by some means other than interference with RNA or protein synthesis (implicating a third factor in the regulation of chromosome replication). A suggestion for the mode of action of phenethyl alcohol will be discussed below.

VII. CELL DIVISION AND THE REGULATION OF CHROMOSOME REPLICATION

During balanced growth, the DNA content of single *E. coli* cells remains constant and does not increase beyond the requirements for repli-

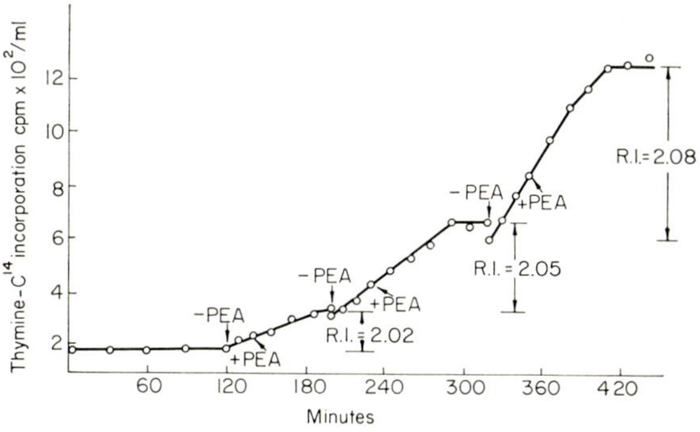

FIG. 9. Control of DNA replication by phenethyl alcohol. (From Treich and Konetzka, 1964.) Maximum stationary phase cells were exposed to phenethyl alcohol (PEA) for 2 hours. Removal of PEA permitted the initiation of a new cycle of replication. Further addition of PEA at intervals restricted the synthesis of DNA to a relative increase (R.I.) of 2. Removal of PEA allowed initiation of a new cycle of replication.

cation. Obviously DNA replication and cell division are related, maintaining a type of homeostasis. Although the nature of this relationship is still unclear, some attempts have been made to study it by use of synchronously dividing cells (Maaløe, 1962).

The first attempts to synchronize cell division in bacteria used some type of metabolic shock, such as temperature shifts, thymine starvation, or glucose starvation (Maaløe, 1962). While the results of such methods may not be typical of the normal cell cycle, certain conclusions can be drawn from them.

Thus, the synchronization of cell division which follows thymine starvation (Barner and Cohen, 1956; K. G. Lark, unpublished data, 1965) suggests a direct relation between chromosome replication and cell division. The data indicate that completion of the chromosome must occur before cell division since no division takes place until 40 minutes after thymine starvation is ended. This is just the time required for the completion of the newly initiated round of chromosome replication induced by thymine starvation.

Under certain conditions of metabolic shock, DNA synthesis can be confined to a short period of the division cycle, a phenomenon also observed in other gram negative organisms. When this occurs, several cycles of DNA synthesis and cell division may result before discontinuous synthesis is lost (Maruyama and Yanagita, 1956; K. G. Lark, 1960).

Newer methods of synchrony have tried to avoid metabolic shock and have relied instead on differences in surface properties of cells and on cell size as a basis of selection (Helmstetter and Cummings, 1963; Maruyama and Yanagita, 1956). In such systems *E. coli* grown in glucose synthetic medium appears to synthesize DNA during at least 90% of the division cycles (Abbo and Pardee, 1960; Nagata, 1963a; Cummings, 1964). This observation agrees with earlier results based on radioautography (Schaechter *et al.*, 1959) and P^{32} decay (McFall and Stent, 1959) in random cultures.

The low concentrations of cells produced by the best (most reproducible) methods of synchrony have made it difficult to investigate the biochemical or physiological events controlling chromosome replication during the cell division cycle. At the moment, data from such systems have revealed little beyond the fact that DNA synthesis occupies most of the division cycle but can also on occasion, occupy only a small fraction of it. It is hoped, that in the near future, more information using synchronous systems will be available.

Additional information has been obtained using nonsynchronous random populations in which changes in cell division rate have been induced by changing the growth rate.

Cultures growing rapidly in broth or in minimal salts glucose media synthesize DNA throughout their entire cell cycle (Schaechter *et al.*, 1959; McFall and Stent, 1959). Cells growing more slowly, for example in succinate, aspartate, or acetate media, have a smaller DNA content (C. Lark, 1966) but a longer division cycle. Obviously, they must either synthesize DNA more slowly or replicate their chromosome throughout only a portion of the cell cycle. Maaløe and Kjeldgaard (personal communication, 1964) have postulated that the latter is the case.

We have studied chromosome replication in cells growing in succinate medium (C. Lark and Lark, 1965) where the generation period is almost double that in glucose. Cells growing in succinate have only 20% less DNA than glucose grown cells. Radioautographs have shown that almost all (more than 90%) of the cells, under these conditions are synthesizing DNA at any instant. On the other hand, it has been possible to show that the number of conserved units being synthesized at any instant is only half the number found in glucose cultures. This was demonstrated by pulse labeling glucose- or succinate-growing cells with thymine-H^3 and letting them multiply in nonradioactive medium. At the end of 4 generations of growth, there were twice as many radioactive cells in the culture labeled while growing in glucose medium as in the culture labeled while growing in succinate medium.

The distribution of radioactive label incorporated during growth in succinate in the absence of amino acids as compared to glucose medium was also examined after subsequent growth of the labeled cells into microcolonies on agar (32–64 cells) using the technique of Forro and Wertheimer (1960). Again the majority of cells labeled in succinate had

2 conserved units, whereas those grown in glucose had 4. These data were interpreted to mean that cells growing in succinate had 2 chromosomes only 1 of which was replicating at any instant. In glucose, both were replicated simultaneously. Experiments with density label demonstrated that chromosome replication in cells from succinate cultures was sequential, from which it follows that one chromosome was synthesized after the other.

Deprivation of amino acids was shown to allow completion of replication of that chromosome which had already embarked on a replication cycle, but replication of other chromosomes could not be initiated. Thymine starvation initiated replication prematurely, but only on that chromosome which was in the process of replication. These data indicate (*a*) the existence of a cellular control over the selection of a given chromosome for replication and (*b*) that the mechanism of premature initiation of chromosome replication operates at the level of the individual chromosome.

VIII. HETEROGENEITY IN THE RESUMPTION OF DNA SYNTHESIS AFTER THE COMPLETION OF A CHROMOSOME REPLICATION CYCLE

Attempts to synchronize the cell division cycle of *E. coli* by first starving of amino acids have failed (Maaløe and Hanawalt, 1961; Hanawalt *et al.*, 1961; K. G. Lark *et al.*, 1963). This result, has been explained by the finding that when amino acids are restored to an amino acid starved culture, DNA synthesis does not resume immediately. Instead, individual cells may regain the capacity for chromosome replication at different times after addition of amino acids (Hanawalt *et al.*, 1961; K. G. Lark *et al.*, 1963). Heterogeneity in the time of reinitiation of chromosome replication for different cells also is encountered after phenethyl alcohol treatment (W. A. Konetzka, personal communication, 1965).

In both cases, the heterogeneity among cells with respect to the time at which replication is reinitiated has not as yet been correlated with any known parameter. In particular there is *no* correlation between when cells reinitiate replication and when they ceased replication during amino acid starvation. The synchronous cell division observed after thymine starvation indicates however that the heterogeneity encountered after either amino acid starvation or phenethyl alcohol treatment may depend on the treatment used to block reinitiation of the replication cycle. The significance of this will be discussed below.

This situation is complicated by the finding that the portion of the chromosome synthesized during amino acid starvation will not replicate normally upon restoration of amino acids, but instead, requires a period of protein synthesis before it will replicate again at a normal rate (K. G. Lark *et al.*, 1963; and K. G. Lark, 1966). This result indicates that some type of protein is normally associated with DNA which assists in its replication—possibly a protein or histone which lends stability to the replicative state.

Because of the problems outlined above, it is not possible at the present time to achieve prolonged chromosomal synchrony in *E. coli* by a regime of amino acid starvation and refeeding.

IX. RATE OF CHROMOSOME REPLICATION

The results of experiments with cells growing at different rates and with cells previously starved of thymine indicate that the rate at which the cells' chromosome content is duplicated reflects the number of replication points per cell. However, if the growth rate is made sufficiently low, the rate at which the replication point moves along the chromosome i.e., the linear distance of chromosome synthesized per growth point is reduced (C. Lark, 1966).

Maaløe and Rasmussen (1963) found that DNA is synthesized very slowly in the presence of low concentrations of thymine. It could be assumed that this is due to the inability of thymine to penetrate the cell in sufficient quantities for DNA synthesis. However, this may not be true. If cells are starved of thymine to induce a new replication point and then put into a low concentration of thymine, they will synthesize twice as much DNA as if they had not been starved previously (K. G. Lark, unpublished results, 1965). Thus even though the rate of DNA synthesis is restricted, the quantity of thymine available within the cell is sufficient to supply at least *two* growth points on a chromosome, each replicating at the same rate as a single one or twice the total rate to which synthesis had been restricted in a similar culture not starved previously of thymine.

Thus, the rate of linear travel of the chromosome replication point may be restricted by the intracellular concentration of thymine or thymine derivatives.

X. ATTEMPTS TO CHARACTERIZE THE PHYSICAL STATE OF REPLICATING DNA

Previously (K. G. Lark, 1963), we detailed arguments which led us to believe that replication was initiated by the conversion of DNA to a prereplicative, primer state. At that time, experiments with *Alcaligenes faecalis* (K. G. Lark, 1963) and later with *E. coli* (Rosenberg and Cavalieri, 1964) revealed that a small proportion of DNA existed in a potentially denaturable state which could be revealed as denatured DNA if cells were lysed at 60°C. We have now carried out an experiment which appears to rule out this particular material as a prereplicative form of DNA (Lark, unpublished data, 1965).

Cells were labeled with thymine-H^3 for a short period after amino acid starvation, thus labeling the beginning portion of the chromosome. After several generations of further growth, the cells were placed in medium lacking required amino acids and samples analyzed at intervals for the amount of "denaturable" material present. The total amount of such material was found to be 12%, 8.7%, 9%, and 5.5% respectively after 0, 15, 30, and 45 minutes of amino acid starvation. On the other hand, the amount of radioactive material which was potentially denaturable at these times was 9.7%, 9.4%, 8.2%, and 5.7%. Thus, the beginning region of the chromosome could be shown to produce the same proportion of "denaturable" material as the bulk of the DNA *despite the fact that this region of the chromosome was not and could not be replicated during the later portion of the starvation period.* (A fact confirmed on another aliquot, starved of amino acids for the second time in the presence of 5-bromouracil.) It seems likely, therefore, that this material results from some condition within the cell not directly associated with the act of chromosome replication.

An approach which may prove more profitable is that undertaken by Hanawalt and Ray (1964) who have succeeded in isolating the replication fork of the *E. coli* chromosome. This has been done by growing bacteria for about 0.75 of a generation in medium containing a non-radioactive density label and transferring them for an extremely short period to medium containing radioactivity but no density label. Material in which the radioactivity was associated with the density label was then isolated. Further characterization of this material indicated that such replicating material is associated with large protein particles which are presumably part of the bacterial membrane (Smith and Hanawalt, 1965). It may eventually prove possible to examine the physical state of the DNA in this replication "fork," although results obtained indicate that it is extremely sensitive to shear degradation.

XI. THE REGULATION OF CHROMOSOME REPLICATION

In 1962, I discussed the possibility of applying regulatory principles, proposed for the control of protein synthesis, to the control of DNA synthesis (K. G. Lark, 1963). It was noted at that time, that the application of such principles would depend upon chromosome replication being linearly sequential. During the following year the experiments of Nagata 1963a), of Sueoka and Yoshikawa (1963), and of Cairns (1963) demonstrated that such sequential synthesis existed, and Jacob *et al.* (1963)

proposed a regulatory mechanism for chromosome replication analagous to mechanisms proposed for the control of protein synthesis. They supported their model with experiments studying the maintenance of episomes in certain temperature sensitive mutants.

Their model (Jacob *et al.*, 1963) assumes that 2 components are involved in regulating chromosome replication. One component is structural, possibly a site of attachment of the chromosome to the cell membrane. The other involves a positive control mechanism, the synthesis of an initiator protein which serves to initiate a round of DNA replication.

They conceive of a DNA-replicating unit, the replicon, which initiates replication as the result of the interaction of the initiator with the structural component, the replicator, to begin a cycle of replication. When initiator is not present, the replicator cannot be activated. They proposed that in the bacterial cell, the chromosome and the episome were independent replicons with distinct initiators and replicators.

They produced mutants which, when grown at a certain temperature, lost their episome and suggested that such mutants could not produce initiator protein for the episome at this temperature. They then observed that introduction of a temperature insensitive episome rescued the temperature sensitive one presumably because the former produced the necessary initiator substance.

Finally, they proposed that the act of conjugation would activate the replicator in the donor Hfr cell to initiate a new round of replication in which the freshly synthesized chromosome is transferred as a consequence of replication into the recipient or F^- cell. The recent radioautographic experiment of L. Caro and Gross (personal communication, 1965) supports this in so far as they demonstrate that only DNA synthesized immediately before mating is transferred from the Hfr into the F^- cell.

The role of thymine deprivation in the formation of new initiator is not clear. The number of cells with premature replication cycles increases linearly with time in the absence of thymine reaching a maximum value in glucose medium after about 40 minutes or one generation period (Pritchard and Lark, 1964). P. C. Hanawalt has suggested (personal communication, 1964) that this is the result of the progression of the replication point in the absence of thymine, i.e., of DNA synthesis, (phantom replication point) which ceases when the end of the chromosome is reached. Thus, chromosomes are accumulated which are prepared to commence replication at the chromosome origin.

Two results contradict this hypothesis: (*a*) Chromosome replication can proceed to the end of the chromosome in the absence of amino acids whereas starvation of both thymine and amino acids does *not* result in induction of premature replication. (*b*) It can be shown (Pritchard and Lark, 1964) that after thymine starvation DNA replication proceeds *also* from the point on the chromosome at which replication ceased.

Another possible explanation is that sequential transcription of RNA proceeds in the absence of DNA synthesis thus leading to a synthesis of initiator protein in increasing numbers of cells. This, however, is also unlikely since incubation with cytosine arabinoside does not lead to a premature replication cycle although chromosome replication is inhibited enough to reduce the rate of DNA synthesis to 5% of normal. Deprivation of thymine in the presence of cytosine arabinoside, however, leads to a premature replication cycle (C. Lark and Lark, 1964).

At the moment, it seems most profitable to assume that thymine starvation results in the derepression of initiator protein synthesis.

In its most general form, the "replicon" model is consistent with the data which we have discussed in previous sections, most of which was obtained after the model had been proposed.

We have used the model in Fig. 10 to explain our results with amino acid and thymine starvation. The 2 mechanisms regulating chromosome replication, which may be distinguished by inhibitors such as chloram-

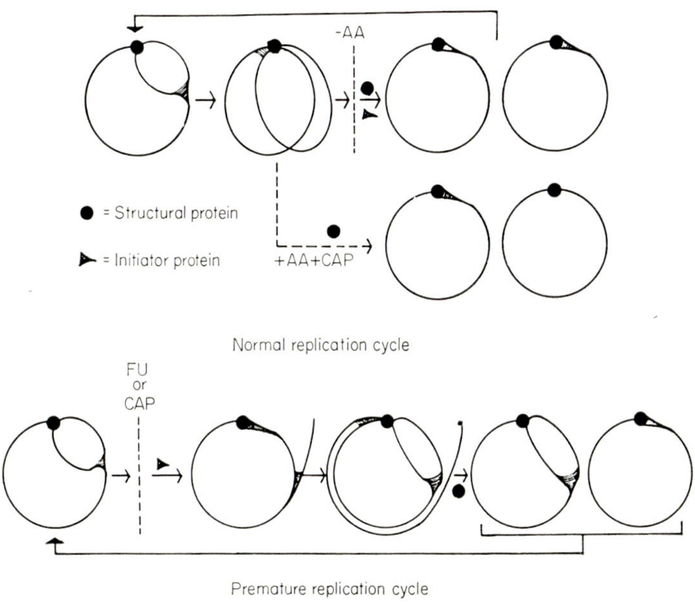

FIG. 10. Schematic representation of the effect of inhibitors on chromosome replication in *E. coli*. (From C. Lark and K. G. Lark, 1964.) −AA = amino acid starvation; CAP = chloramphenicol; FU = fluorouracil. The premature replication cycle is induced by thymine starvation.

phenicol or fluorouracil, are identified as involving a structural and an initiator protein. In the normal replication cycle, the circular chromosome is replicated from a structural site which accommodates the 2 partial replicas. Replication is initiated by an initiator which proceeds along the chromosome, accompanying the point of replication. When replication is completed, a new structural site and initiator are formed. Amino acids are required for the formation of each of these. Once formed, a new cycle of replication commences. In the presence of chloramphenicol a new structural site is formed, but no new initiator. As a result, replication can proceed on only 1 of the 2 daughter chromosomes.

In the premature replication cycle, deprivation of thymine derepresses the formation of an initiator, which displaces one of the partial replicas from the structural site and initiates replication prematurely on the other. This situation is resolved when the original locus of replication reaches the end of the chromosome. At that time, the structural site replicates but the initiator does not. Thus, one chromosome finishes its partially completed replication, while the other begins a new cycle as a separate structure. Fluorouracil or chloramphenicol inhibit the formation of initiator.

The stoichiometric relation of the structural protein to the chromosome beginning is extremely important in the regulatory process. Thus, we account for the inability of thymine starvation to induce more than one extra replication fork by assuming that only the original structural protein is available to accommodate chromosome replication. Displacement of one of the chromosome beginnings from this structural site allows the other to commence replication.

We can extend this model to the studies on chromosome replication in slowly growing cells (cells grown in succinate medium) where replication of chromosomes occurs in alternating sequence. This is seen in Fig. 11 which illustrates a mechanism which will lead to such alternate replication.

Immediately prior to division, the cell contains 2 replicated chromosomes each set of 2 still bound to a single structural protein and a single initiator protein. During division each pair is separated to give one chromosome complete with structural protein and initiator protein and one chromosome with neither. Each daughter cell will thus receive one chromosome prepared for replication and one which is not. Consequently, only the one chromosome will commence replication. While this chromosome is replicating, another structural protein is synthesized which becomes occupied by the second chromosome. When the first chromosome reaches the end of its replication cycle, information for the formation of initiator protein is transcribed. This initiator protein activates the second

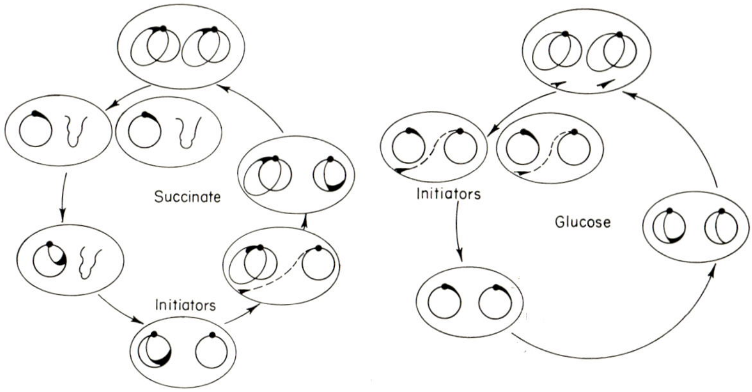

FIG. 11. Hypothetical effect of changing growth rate on the regulation of chromosome synthesis in *E. coli*.

chromosome which thus begins replication. This initiator cannot reinitiate replication of the first chromosome since this requires chromosome separation and the participation of another structural protein which is not available. Certain aspects of this model have been modified as a result of more recent data (see K. G. Lark, 1966).

Initiator protein may or may not be made upon completion of the second chromosome. Lack of synthesis would imply that transcription of the information for synthesis of initiator can only be successful if a non-replicating single chromosome is present in the cell, or the cell contains more structural than initiator proteins. It is also possible that the initiator is extremely unstable unless combined with a chromosome.

In a glucose culture, structural protein would be synthesized much more rapidly allowing more chromosomes to replicate at any time and consequently more initiator protein to be made. A prediction of this model is that upon shifting from succinate to glucose medium, about a generation will be required before the synthesis of extra initiator protein will catch up with the extra structural proteins available. Thus, DNA synthesis should continue at a slow rate for a period after a shift up. This is what is observed (Kjeldgaard *et al.*, 1958; K. G. Lark, 1966).

The separation of the replicated chromosome from the structural protein is represented in the model as a crucial step in the replication cycle, since it allows one of the pair to replicate. If it is coupled in some manner to the division process, it would allow the rate of chromosome replication to increase with division rate.

As we have already discussed, the rate of chromosome replication is regulated by the number of replication points available within a cell. If, in rapidly growing cells (20-minute generation period) 3 replication points exist for each chromosome as found in *B. subtilis* by Sueoka and Yoshikawa (1963), the rate of replication would again be commensurate with the rate of cell division. This would imply, that in such cells sufficient structural sites exist to accommodate the extra chromosome origins. At the moment, it would appear that such structures would have to be different from those at slower growth rates since they could accommodate 4 beginnings and 1 end rather than 2 beginnings and 1 end.

At the moment, it is difficult to envision the nature of such a site. It is possible that it represents a special region of the cell wall-cell membrane complex which is formed during cell division. If synthesized independently of DNA replication, it would serve as a regulatory mechanism which restricts chromosome replication to conform to the cell division cycle.

W. A. Konetzka (personal communication, 1965) has suggested that phenethyl alcohol inhibits the initiation of DNA replication by preventing the attachment of the chromosome to the structural site or cell membrane. He suggests that the heterogeneity observed in the time of reinitiation of replication following removal of this inhibitor is a reflection of the probability that any individual chromosome will reattach to the membrane. Certainly, his hypothesis is in agreement with the models presented in Figs. 10 and 11 and could be used to explain the heterogeneity observed when chromosome replication is reinitiated after amino acid starvation.

At present, then, it seems most profitable to assume that the regulation of the initiation of chromosome replication is adjusted to the cell division cycle by any one, or a combination of, 3 events: chromosome separation, the synthesis of structural site, or the attachment of the chromosome to the structural site.

It is not clear how the initiator would act, but it seems safe to presume that it facilitates the conversion of the chromosome origin to a primer state either directly or by aiding in its attachment to the structural site. Its regulatory role would be to relate the rate of transcription and/or replication to initiation insuring that a new replication cycle only was initiated if a previous cycle was completed. This could easily occur if the sequential transcription discussed by H. O. Halvorson, (Chapter 6) can be related to chromosome replication.

Acknowledgments

I am grateful to my wife, Cynthia Lark, and to Dr. James Friesen for reading and criticizing the manuscript. Some of the work reported here was supported by grants from the National Institutes of Health (AI 05711) and from the National Science Foundation (GB 993). The author is a recipient of a Career Development Award from the National Institutes of Health, Bethesda, Maryland.

REFERENCES

Abbo, F. E., and Pardee, A. B. (1960). *Biochim. Biophys. Acta* **39,** 478.

Barner, H., and Cohen, S. S. (1956). *J. Bacteriol.* **72,** 115.

Berrah, G., and Konetzka, W. A. (1962). *J. Bacteriol.* **83,** 738.

Bertani, G. (1954). *J. Bacteriol.* **67,** 696.

Billen, D. (1964). *Exptl. Cell Res.* **34,** 396.

Billen, D., Hewitt, R., and Jorgensen, C. (1965). *Biochim. Biophys. Acta* **103,** 440.

Bonhoeffer, F., and Gierer, A. (1963). *J. Mol. Biol.* **7,** 534.

Cairns, J. (1963). *J. Mol. Biol.* **6,** 208.

Campbell, A. (1962). *Advan. Genet.* **11,** 101.

Caro, L., and Forro, F. (1961). *J. Biophys. Biochem. Cytol.* **9,** 555.

Cummings, D. J. (1964). *Biochim. Biophys. Acta* **85,** 341.

Forro, F., and Wertheimer, S. A. (1960). *Biochim. Biophys. Acta* **40,** 9.

Hanawalt, P. C., and Ray, D. S. (1964). *Proc. Natl. Acad. Sci. U.S.* **52,** 125.

Hanawalt, P., Maaløe, O., Cummings, D. J., and Schaechter, M. (1961). *J. Mol. Biol.* **3,** 156.

Hayes, W. (1952). *Nature* **169,** 118.

Hayes, W. (1964). "Genetics of Bacteria and Their Viruses." Wiley, New York.

Helmstetter, C. E., and Cummings, D. J. (1963). *Proc. Natl. Acad. Sci. U.S.* **50,** 767.

Hewitt, R., and Billen, D. (1965). *J. Mol. Biol.* **13,** 40.

Jacob, F., and Wollman, E. L. (1958). *Ann. Inst. Pasteur* **95,** 497.

Jacob, F., and Wollman, E. L. (1961). "Sexuality and the Genetics of Bacteria." Academic Press, New York.

Jacob, F., Brenner, S., and Cuzin, F. (1963). *Cold Spring Harbor Symp. Quant. Biol.* **28,** 329.

Kellenberger, E. (1960). *Symp. Soc. Gen. Microbiol.* **10,** 39.

Kellenberger, E., and Ryter, A. (1958). *J. Biophys. Biochem. Cytol.* **4,** 671.

Kjeldgaard, N. O., Maaløe, O., and Schaechter, M. (1958). *J. Gen. Microbiol.* **19,** 607.

Lark, C. (1966). *Biochim. Biophys. Acta* (in press).

Lark, C., and Lark, K. G. (1964). *J. Mol. Biol.* **10,** 120.

Lark, C., and Lark, K. G. (1965). *J. Mol. Biol.* **13,** 105.

Lark, K. G. (1960). *Biochim. Biophys. Acta* **45,** 121.

Lark, K. G. (1963). *In* "Molecular Genetics" (J. H. Taylor, ed.), Pt. I, p. 163. Academic Press, New York.

Lark, K. G. (1966). *Bacteriol Rev.* (in press).

Lark, K. G., and Bird, R. (1965). *J. Mol. Biol.* **13,** 607.

Lark, K. G., Repko, T., and Hoffman, E. (1963). *Biochim. Biophys. Acta* **76,** 9.

Lederberg, J. (1947). *Genetics* **32,** 505.

Maaløe, O. (1962). *In* "The Bacteria" (I. C. Gunsalus and R. Y. Stanier, eds.), Vol. 4, p. 1–32. Academic Press, New York.

Maaløe, O., and Hanawalt, P. C. (1961). *J. Mol. Biol.* **3,** 144.

Maaløe, O., and Rasmussen, K. (1963). C.N.R.S., *Paris Symp., Marseille, 1963.*

McFall, E., and Stent, G. (1959). *Biochim. Biophys. Acta* **34,** 580.

Marmur, J., Rowne, R., Falkow, S., Baron, L. S., Schildkraut, C., and Doty, P. (1961). *Proc. Natl. Acad. Sci. U.S.* **47,** 972.

Maruyama, Y., and Yanagita, T. (1956). *J. Bacteriol.* **72,** 821.

Meselson, M., and Stahl, F. (1958). *Proc. Natl. Acad. Sci. U.S.* **44,** 672.

Nagata, T. (1963a). *Proc. Natl. Acad. Sci. U.S.* **49,** 551.

Nagata, T. (1963b). *Cold Spring Harbor Symp. Quant. Biol.* **28,** 55.

Pardee, A., and Prestidge, L. (1956). *J. Bacteriol.* **71,** 677.
Pritchard, R., and Lark, K. G. (1964). *J. Mol. Biol.* **9,** 288.
Rosenberg, B. H., and Cavalieri, L. F. (1964). *Proc. Natl. Acad. Sci. U.S.* **50,** 826.
Schaechter, M., Maaløe, O., and Kjeldgaard, N. O. (1958). *J. Gen. Microbiol.* **19,** 592.
Schaechter, M., Bentzon, M. W., and Maaløe, O. (1959). *Nature* **183,** 1207.
Silver, S. (1963). *J. Mol. Biol.* **6,** 349.
Skaar, D., and Garen, A. (1955). *Genetics* **40,** 596.
Smith, D. W., and Hanawalt, P. C. (1965). *Abstr. 9th Meeting Biophys. Soc., San Francisco* p. 162.
Sueoka, N., and Yoshikawa, H. (1963). *Cold Spring Harbor Symp. Quant. Biol.* **28,** 47.
Treich, R. W., and Konetzka, W. A. (1964). *J. Bacteriol.* **88,** 1580.
Weidel, W., Frank, H., and Martin, H. H. (1960). *J. Gen. Microbiol.* **22,** 158.
Witkin, E. (1951). *Cold Spring Harbor Symp. Quant. Biol.* **18,** 357.

CHAPTER 5

Nuclear Events in Synchronously Dividing Yeast Cultures

D. H. Williamson

I. INTRODUCTION

For a variety of reasons, the yeast *Saccharomyces cerevisiae* has attracted the attention of biologists in a number of different fields, with the consequence that we now know a great deal about this interesting organism. However, we shall not achieve a unified picture of the yeast cell or understand its relationships with other organisms until the structure and mode of division of its nucleus has been elucidated. Examination of the literature [for a comprehensive review, see McClary (1964)] shows that areas of agreement on this topic are small. This is not altogether surprising, for the yeast cell suffers from twin disadvantages: Its size and refractility make it an unsatisfactory subject for microscopic examination in the living state; and it is a difficult object to fix and stain in a reproducible fashion. The situation is aggravated by the low DNA content of the nucleus and the intense basophilia of the cytoplasm.

These technical difficulties have not only produced uncertainty about details of the structure and mode of division of the nucleus; they have also led to confusion even as to the true identity of this organelle. However, the growing use of the electron microscope in this field promises an early solution to some of these problems. Few observers now dispute that the basophilic, Feulgen-positive, extravacuolar body which, in one guise or another, has played a prominent part in nearly all interpretations of the yeast's nuclear structure, is in fact the entire nucleus. In an electron microscopic study of osmium-fixed material, Agar and Douglas (1957) were the first to show that this organelle is bounded by a double membrane which carries pores similar to those seen in the nuclei of higher cells. These observations have since been amply confirmed, most recently by Moor and Mühlethaler (1963) using the elegant technique of freeze-etching. Surprisingly however, the procedures used by both the groups mentioned above failed to reveal any internal structures within the nucleus, the division process being visualized as a simple constriction and "pinching off" resulting in the production of two vesicular and apparently structureless nuclei. This is, of course, no more than the picture favored by some authors after examining stained cells with the light microscope. However, genetic evidence points clearly to the existence in *S. cerevisiae* of at least 14 chromosomes in the haploid state (Mortimer and Hawthorne, 1966), and in the light of modern knowledge, it is hard to believe that these would not take the form of DNA-containing structures capable of detection by electron microscopy.

The solution to this problem seems to lie in the way the material is prepared for examination. Using permanganate fixation, Yotsuyanagi (1960) reported the presence in yeast nuclei of electron-transparent areas containing fibrillar material. After successive treatments with ribonuclease and uranyl acetate, these areas became denser than the surrounding nucleoplasm, and comparison with onion cells treated in the same way led Yotsuyanagi to interpret the structures thus "stained" as chromosomes. These bodies have since been seen by a number of authors (Hashimoto *et al.*, 1960; Koehler *et al.*, 1961; Vitols *et al.*, 1961; Hirano and Lindegren, 1961; Hirano, 1962; Yuasa, 1962; Polakis *et al.*, 1964), but as yet there has been no further evidence about their composition or role in the cell. They obviously deserve further investigation, for they are the only intranuclear structures reproducibly detectable in yeast under conditions which lead to seemingly good preservation of other cellular components. More importantly however, Yotsuyanagi's discovery should, for the first time, permit us to reconstruct the mitotic process of this organism in considerable detail.

Such a reconstruction, from electron micrographs of fixed material, presents a number of problems. There is little in the literature to suggest that the division of the yeast nucleus is likely to follow closely the familiar pattern of mitosis seen in higher cells. Consequently there is little to guide one in ruling out possible artifacts. Moreover, since nuclear division is likely to be a rapid process, sampling a randomly dividing culture will only yield a small proportion of cells actually caught in the act. These technical problems become particularly troublesome when using the electron microscope, for this instrument limits one to examining randomly oriented thin sections. This severely reduces the chance of detecting division figures, for only a proportion of the sections through a dividing nucleus will yield recognizable patterns. A more subtle problem is that most sections through a yeast cell are circular or ellipsoidal in shape. Therefore, it is rarely possible to judge the cell cycle age of the cell from which a particular section was obtained, and it is clearly unwise to give credence to an unusual nuclear configuration unless it only occurs in cells of a particular age group.

Fortunately most of these difficulties can be overcome, or at least alleviated, by using properly synchronized cultures. Ideally such populations permit the preparation of samples in which a majority of the cells are engaged in nuclear division at the same time. Moreover, the age of every section in a sample is known with some degree of certainty. Thus the use of such cultures not only improves one's chances of observing the division process, but at the same time permits one to reject, as irrelevant, nuclear configurations that are not restricted to cells of a single age group.

Fundamentally, however, the value of synchrony in approaching this problem has a broader basis than is implied by these practical considerations. This lies in the fact that it permits one to relate any observed structural changes simply, and directly, to the accompanying physiological events in the cell cycle. Nuclear division is only one aspect of the coordinated sequence of changes which comprise the growth and reproduction of the cell, and it cannot be considered in isolation. Part of the purpose of this article is to show that a better understanding of the nucleus of the yeast cell can be obtained by bringing together information about as many of its activities during the cell cycle as possible. In acquiring the relevant information, synchrony plays a unique, and perhaps indispensable role.

II. NUCLEIC ACID SYNTHESIS IN SYNCHRONOUSLY DIVIDING CULTURES

The synchronized cultures used for all the experiments referred to in this article were prepared by the method of Williamson and Scopes

(1962). This gives rise to a population of uniformly sized resting cells which, when inoculated into a suitable nutrient medium, undergo several cycles of growth and division with a marked degree of synchrony. Immediately after inoculation there is a period of about an hour during which little net synthesis of macromolecules takes place (Williamson, 1964), though there is some incorporation of adenine into ribonucleic acid (RNA) (Williamson, 1965). This interval is referred to as a lag phase, and is presumably a period of preparation for active growth. The lag phase ends, and the first generation period starts when the first buds emerge, an event which also marks the start of extensive synthesis of nucleic acids and proteins. At 25°C, the generation period in a favorable medium occupies about two hours and ends when the mature bud, having received a nucleus and grown to about the same size as the parental part of the cell, separates from the latter and assumes an independent existence. In the experimental conditions used in these studies, the act of division is accompanied by the almost simultaneous appearance of a new bud on both the parent and daughter cells, there being no lag between the end of one generation and the start of the next. The overall behavior of the system, which retains a useful degree of synchrony for at least 3 generations, is illustrated in Fig. 1.

A. Synthesis of Deoxyribonucleic Acid (DNA)

The first important step in the nuclear cycle of any organism is the replication of its DNA. It is now well established that in *Saccharomyces cerevisiae* net synthesis of DNA is restricted to the first quarter of the cell cycle, starting immediately after the initial appearance of the bud (Ogur *et al.*, 1953; Williamson and Scopes, 1960; Williamson, 1965). In the individual cell growing at 25°C therefore, DNA replication lasts about 30 minutes. In synchronous cultures, which of course are never in perfect synchrony, it lasts a little longer, the first round of DNA synthesis starting around 60 minutes after inoculation and ending at 110 minutes (Fig. 2). Strictly speaking, no formal proof has been published that this synthesis, which entails a doubling in the DNA content of the cell, actually takes place in the nucleus. In principle this should be demonstrable autoradiographically but, unfortunately, the geometry of the ovoid cell of *S. cerevisiae* in light microscope autoradiographs does not permit one to distinguish between radioactivity located in the nucleus and the cytoplasm. However, on applying an autoradiographic procedure for the detection of DNA synthesis (Williamson, 1965) to the cylindrical cells of a strain of *S. carlsbergensis*, photographic grains were clearly concentrated over the presumed nuclear area (Fig. 3), and there seems little doubt that in other species also, synthesis of DNA can be correctly described as a nuclear activity.

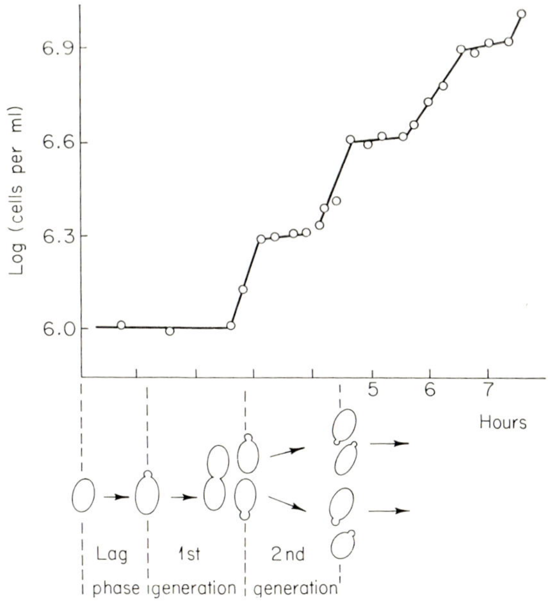

FIG. 1. Diagrammatic representation of the growth and division of an initially resting cell on inoculation into a growth medium, together with the growth curve of a synchronized population of such cells.

The replication of DNA has the significance of being the only biochemically detectable sign of chromosome reproduction, and in the normal growth cycle of any cell, this event must be completed before the mitotic segregation of the chromosomes can take place. This is not to suggest that DNA replication triggers mitosis, for the interval between the two events may be considerable (Nygaard *et al.*, 1960); it is simply that DNA synthesis is normally an essential prerequisite of mitosis. It follows that knowledge of the timing of the phase of DNA synthesis in the cell cycle permits one to set a limit to the portion of the cell cycle in which the mitotic process should occur. In the case of the yeast cell, this reasoning is evidently of restricted value, since DNA synthesis occurs at the start of the cell cycle. Nevertheless, it is at least permissible on this basis to conclude that mitosis should occur at some time during the last three quarters of the cell cycle, and cytological evidence described below will enable us to refine this prediction.

B. Synthesis of Ribonucleic Acid (RNA)

Synthesis of RNA seems to be an important function of the nucleus in all cells, and there is a good deal of indirect evidence that some of the RNA made there, the so-called "rapidly labeled" fraction, is made on a

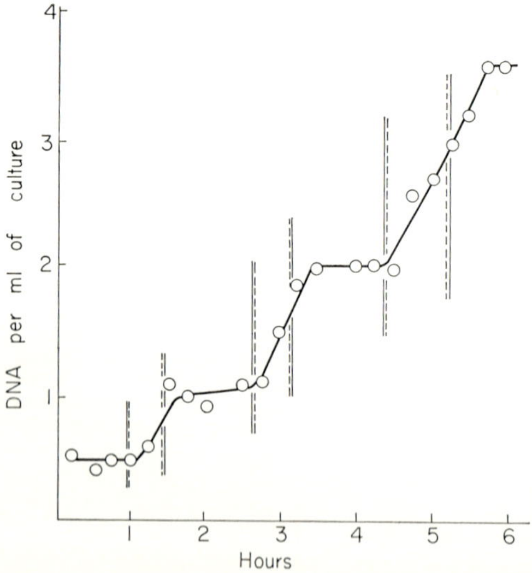

FIG. 2. The course of DNA synthesis in a synchronously dividing culture of *S. cerevisiae*. DNA is recorded in arbitrary units. The pairs of vertical lines indicate, respectively (from left to right), the period occupied by the emergence of the first generation of buds, and the first and second periods of cell division. (Reproduced from Williamson, 1964.)

template of DNA (Davidson and Cohn, 1963). The formation of rapidly labeled RNA in the nucleus may be detected by briefly exposing cells to a radioactive precursor and then examining them autoradiographically. In the case of *S. cerevisiae*, the resolution needed to distinguish between nuclear and cytoplasmic label can be attained by employing the technique of electron microscope autoradiography. Figure 4 shows an autoradiograph of this type, the cell concerned having been obtained from a synchronously dividing culture exposed briefly to adenine-H^3 during a period when no DNA was synthesized. Although the resolution of the detail in the section itself leaves a lot to be desired, the concentration of photographic grains over the nuclear area is clearly apparent, and constitutes evidence for nuclear synthesis of RNA in this species.

The significance of nuclear RNA synthesis in the present context lies in its relationship to the mitotic separation of the chromosomes. There is a good deal of evidence that in the cells of higher plants and animals, synthesis of RNA on the chromosomes is either greatly depressed or eliminated during mitosis (Feinendegen *et al.*, 1960; Taylor, 1960; Baserga, 1962; Prescott and Bender, 1962). The reason for this is not clear, though it has been suggested that it may be due to the inability of the DNA in the highly condensed mitotic chromosome to act as a template for RNA synthesis. In any event, if the nucleus of the yeast cell

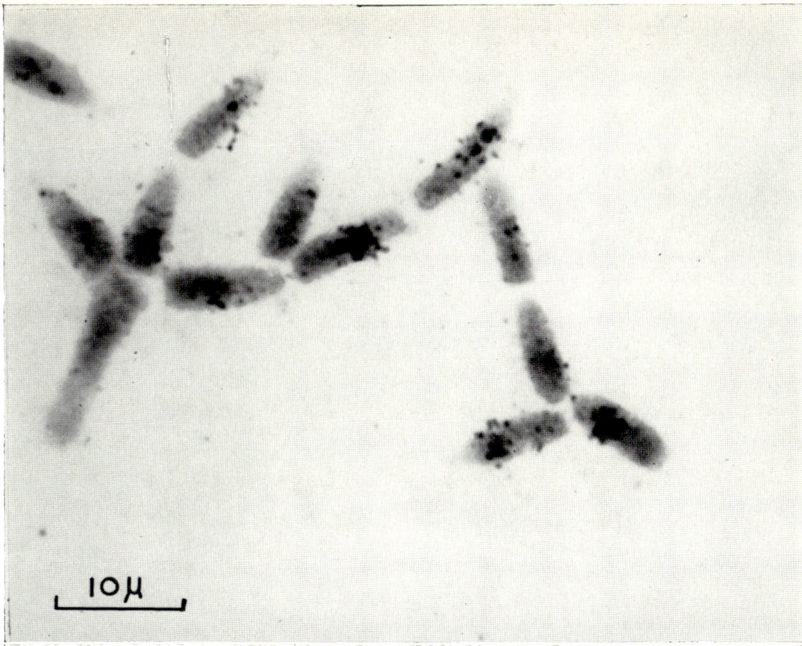

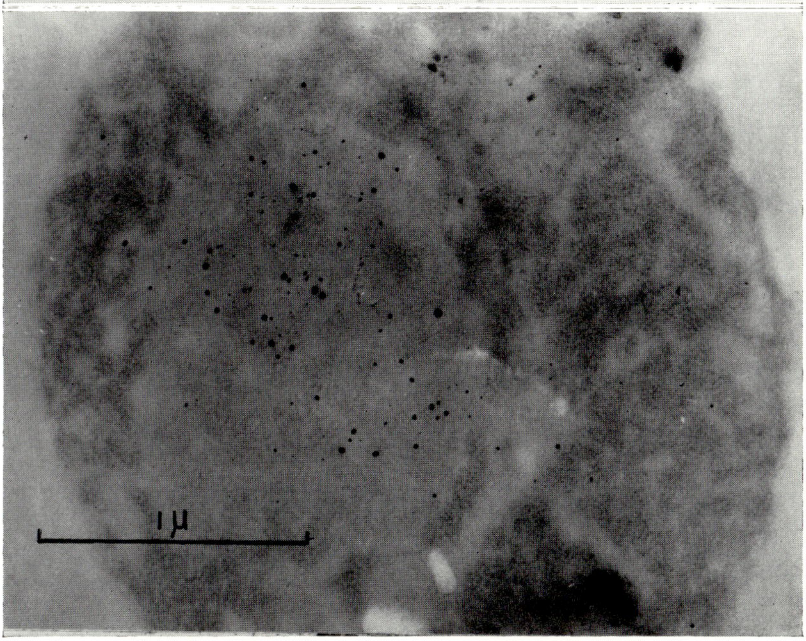

FIG. 3. Autoradiograph of cells of *S. carlsbergensis* (National Collection of Yeast Cultures, strain 74). The cells were from a randomly dividing culture labeled with adenine-C^{14} for 15 minutes and subsequently treated so as to remove all radioactivity other than that in DNA (Williamson, 1965). Grains are clearly concentrated in one area, believed to be the location of the nucleus.

FIG. 4. Electron microscope autoradiograph (made in collaboration with L.F. LaCour) of a section of a cell from a synchronously dividing culture of *S. cerevisiae* exposed to adenine-H^3 for 15 minutes during a period when no DNA was synthesized. Photographic grains are clearly concentrated over the nucleus.

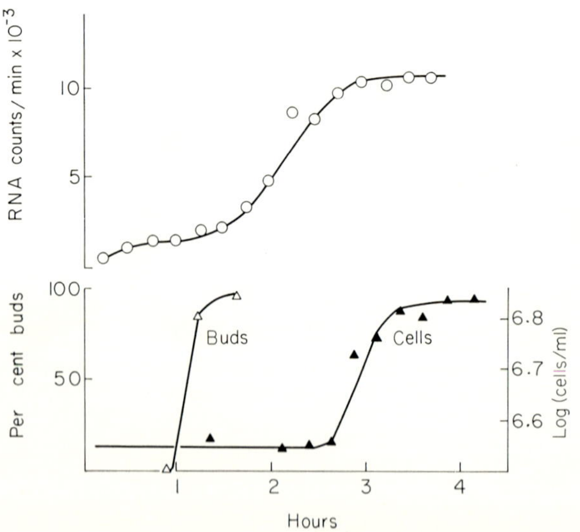

FIG. 5. Changes in the incorporation of adenine-C^{14} into RNA during the growth of a synchronously dividing culture of *S. cerevisiae*. Samples were withdrawn at intervals from the culture, incubated for 15 minutes in the presence of the precursor, and the amounts of radioactivity in cell RNA then determined (Williamson, 1965). The course of cell division and the emergence of the first generation of buds are also shown.

behaves analogously, the rate of incorporation of exogenously supplied RNA precursors should drop during mitosis, and this phenomenon might be used as a biochemical indicator of the passage of a yeast cell through this phase in the cell cycle.

This possibility was examined by sampling a synchronously dividing culture at intervals, incubating the samples for a few minutes in adenine-C^{14} and then determining the amounts of radioactivity that had entered the RNA fraction in each sample (Williamson, 1965). The results (Fig. 5) were at first sight perplexing, for at no stage of the cell cycle did the incorporation of radioactivity either cease or even slow down. It is of course true that a brief period of reduced synthesis in the individual cells could have been missed simply because of the asynchrony of the system. However, it is estimated that to have escaped detection, any such "quiet" period must have occupied less than 3% of the individual cell cycle, and it therefore seems reasonable to suppose that nuclear RNA synthesis in this organism continues throughout the growth of the cell.

Clearly this experiment did not lead to the hoped for location of mitosis in the yeast cell cycle, but the results are of interest in that they revealed a distinct difference in behavior between the yeast nucleus and that of higher cells. It is perhaps unwise to read too much into this difference, for the cause of reduced RNA synthesis during mitosis in higher cells is not certain, and there is in any case no firm evidence that nuclear RNA in yeast is made on a template of DNA. Nevertheless, this experiment at least raises the possibility that, unlike higher organisms, the chromosomes of the yeast cell do not undergo an extensive condensation or coiling up during nuclear division. As it happens this suggestion finds support in the electron microscope observations described below.

III. CYTOLOGICAL OBSERVATIONS ON SYNCHRONOUSLY DIVIDING CULTURES

A. Optical Microscopy

Although the details of nuclear division in *Saccharomyces cerevisiae* are obscure, there is a measure of agreement among optical microscopists as to the main feature of the process. Numerous authors have described how the nucleus migrates into the neck of the bud, where it assumes what might be called an "hourglass" configuration. It then splits into two, one of the resultant nuclei passing into the bud, the other retracting into the parental part of the cell. Some descriptions carry more detail (e.g., Ganesan, 1959; Ramirez and Miller, 1962), but the overall outline is generally agreed, and it is worth pointing out that this aspect of the nuclear division process was described even by earlier authors who identified the structure in question as something other than the whole nucleus (Lindegren *et al.*, 1956).

Swaminathan and Ganesan (1958) suggested that some of the confusion surrounding the topic of nuclear division in yeast has arisen from the assumption of certain authors that the migration and fission of the nucleus represent the entire mitotic process. They rightly pointed out that this need not be so and that the actual separation of the chromosomes into two groups within the nuclear membrane might take place some time previously. The division of the nucleus in the neck of the bud would then merely represent the final distribution of the two groups of chromosomes to their respective ends of the growing cell. This possibility receives indirect support from the observations of Robinow (1957a) on *Mucor* species, and an intranuclear separation of chromatin well in advance of nuclear fission may be a common mode of nuclear replication among fungi.

Whether or not this is so, it is clear that the separation of the chromosomes in the yeast cell must take place before the nucleus finally splits into two. Consequently, knowledge of the timing of nuclear migration and fission in relation to the growth of the cell should at least permit us

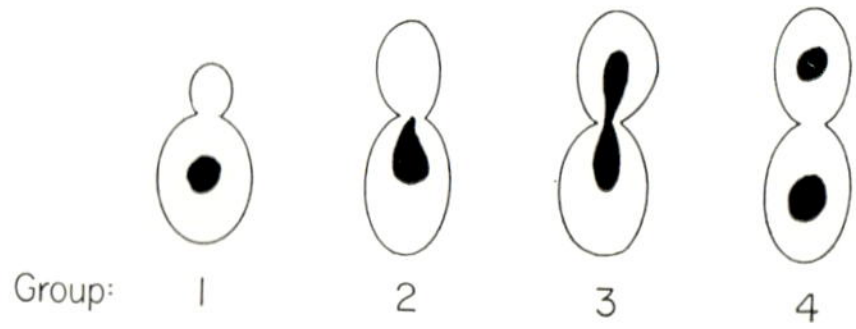

FIG. 6. Diagrammatic illustration of the four classes of cell which can readily be distinguished in Giemsa-stained preparations, arranged in the sequence in which they occur in the cell cycle.

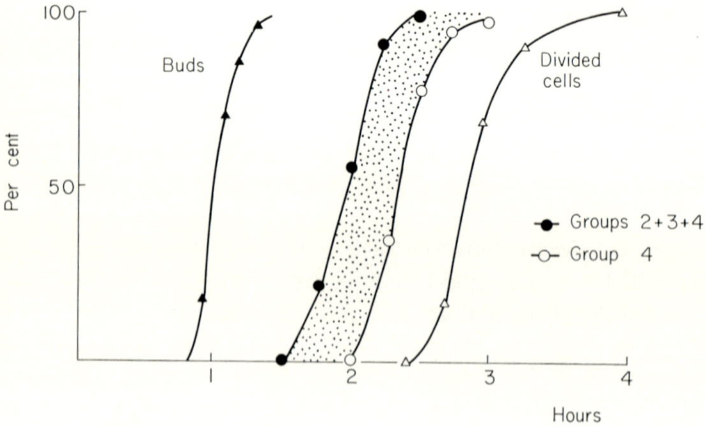

FIG. 7. The timing of nuclear migration and fission in a synchronously dividing culture of *S. cerevisiae*, determined using Giemsa-stained preparations of formalin-fixed material. The passage of the culture through the phase of nuclear migration and fission is represented by the stippled area, arrived at by determining the frequencies of the different classes of cell indicated in Fig. 6.

to set a late limit on the portion of the cell cycle in which chromosome separation should occur.

The migration of the nucleus is readily timed by examining suitably stained samples from synchronously dividing cultures (Williamson, unpublished observations, 1965). In conventional Giemsa-stained preparations, the cells in such samples can readily be classed into one of the four groups, based on the attitude of the nucleus in the cell, shown in Fig. 6. By determining the percentage contribution of these classes to each of a series of samples from a synchronously dividing culture, the migratory stage can be timed with some accuracy (Fig. 7). Although there is some variability between different cultures, it is fair to say that in general, the nucleus starts to migrate into the neck of the bud about half way through the cell cycle, and the entire process of migration and fission occupies between 10 and 20% of the cell cycle. Thus, even if the

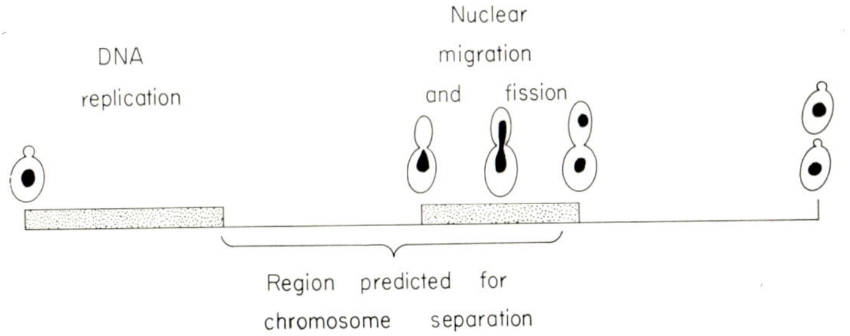

FIG. 8. Diagrammatic illustration of the cell cycle in *S. cerevisiae,* showing the region predicted for the occurrence of chromosome separation. For further explanation, see text.

separation of the chromosomes is delayed until the last possible moment, it must be completed by the time the cell is about two thirds of the way through the cell cycle. Coupling this information with our previous considerations about the time relationship of DNA synthesis and mitosis, we are now in a position to predict with reasonable confidence that the mitotic separation of the chromosomes must take place within the sector of the cell cycle indicated in Fig. 8.

B. ELECTRON MICROSCOPY

The above considerations clearly indicated that the electron microscopic search for sections showing separation of chromatin into two groups could profitably be confined to a limited region of the cell cycle in synchronous cultures. However, much interest attached to the behavior of the chromatin during other phases of the cell's life history, and the description that follows covers the main events in samples taken throughout the whole of the lag phase and first cell cycle. The samples were harvested on a membrane filter, washed with water, fixed briefly in aqueous potassium permanganate (1.5%, w/v) and then for a longer period in aqueous osmium tetroxide (2%, w/v). After dehydration in alcohol and embedding in cross-linked methacrylate (Kushida, 1961), sections were cut with a diamond knife and stained in uranyl acetate (Stoeckenius, 1961) followed by lead citrate (Reynolds, 1963).

A section through a resting cell prior to inoculation is shown in Fig. 9. It is characteristic of resting cells that the chromatin is aggregated into a single mass occupying a small part of the nuclear volume. Although serial sections were not made, examination of a large number of sections suggested that the chromatin of the resting cell was disposed in the form of a flat disc or, in some strains, a toroid (*cf.* Yotsuyanagi, 1960).

On inoculation of the cell into a growth medium, the picture soon changed. Within 5 minutes, the aggregated chromatin broke up (Fig. 10).

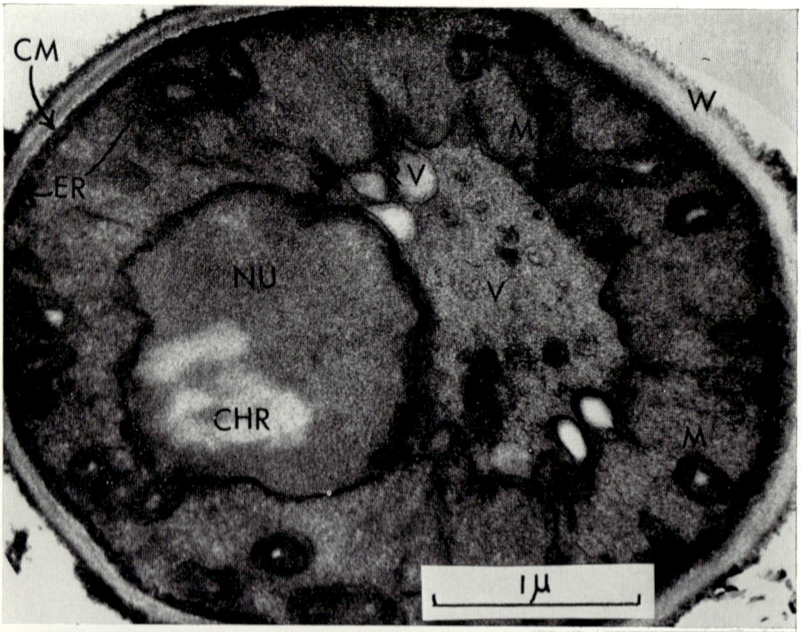

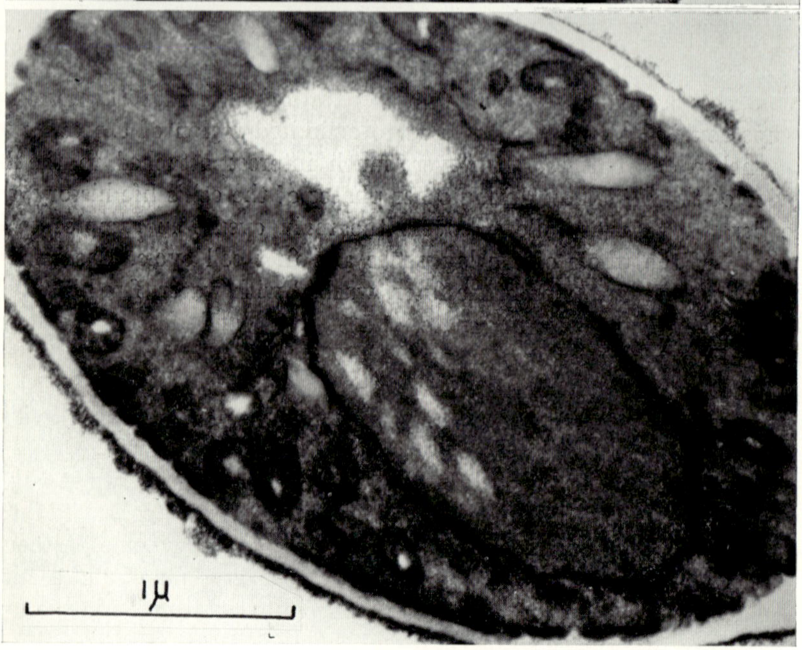

FIG. 9. Section through a resting cell from a synchronized population, prior to inoculation into growth medium. The electron transparent chromatin (CHR) is aggregated into a mass occupying a small part of the nucleus (NU). Other details

The separated pieces, presumably individual chromosomes (Yotsuyanagi, 1960), did not disperse throughout the nucleus, however, but stayed, as far as could be judged, on one *side* of it. There they remained throughout the lag phase, no significant change being detected until after the first generation of buds had appeared (around 60 minutes after inoculation). Then, toward the end of the phase of DNA synthesis, an increasing proportion of longitudinal sections was noted in which the chromosomes were carried at one *end* of the nucleus, apparently the end nearest the developing bud. At the same time, the physical appearance of the chromosomes altered in a subtle way. Although their texture and electron density remained unchanged, they seemed to become broader and slightly better defined (Fig. 11). It is possible that this change was a consequence of DNA synthesis, and might have been indicative of a final stage of chromosome replication.

Immediately following the period of DNA synthesis, in the samples taken between 90 and 150 minutes a small proportion of sections appeared in which the chromosomes were arranged in a new and distinctive configuration. No longer were they restricted to one end of the nucleus. Instead, as in Fig. 12, they occupied the whole nucleus, but were apparently segregated into two roughly equal masses, separated by a zone of seemingly homogeneous nucleoplasm. The nuclear membrane was intact, but had developed a pronounced equatorial furrow, so that the nucleus adopted a dumbbell shape. Although there was no evidence of a spindle mechanism, the visual aspect of these sections strongly suggested that they were in fact the looked for sections through nuclei undergoing chromosomal separation. It might be argued that their extremely low frequency of occurrence (not more than 8% of nucleated sections) was evidence that they were merely random configurations, or even gross artefacts. However, these criticisms can readily be refuted. In the first place, the real frequency of these nuclei in the original population might have been severalfold higher than their apparent frequency in the sectioned material, since only a proportion of sections through the nucleus in this configuration would be recognizable as such. A stronger argument, and one which illustrates the peculiar merit of synchrony in this type of

seen include mitochondria (M), cell wall (W), vacuole (V), and cytoplasmic membrane (CM). Sections through some of a system of vesicles (RV) radiating from the vacuole are also apparent. The function of these vesicles, which are particularly prominent in the resting cell, is unknown. Lying under the cytoplasmic membrane is another heavily stained membrane (ER) which is believed to originate in association with the nuclear membrane, and possibly represents a simple type of endoplasmic reticulum.

FIG. 10. A cell from a synchronized culture after 5 minutes incubation in growth medium. The chromatin has broken up into separate pieces, which remain in a loose group in one part of the nucleus.

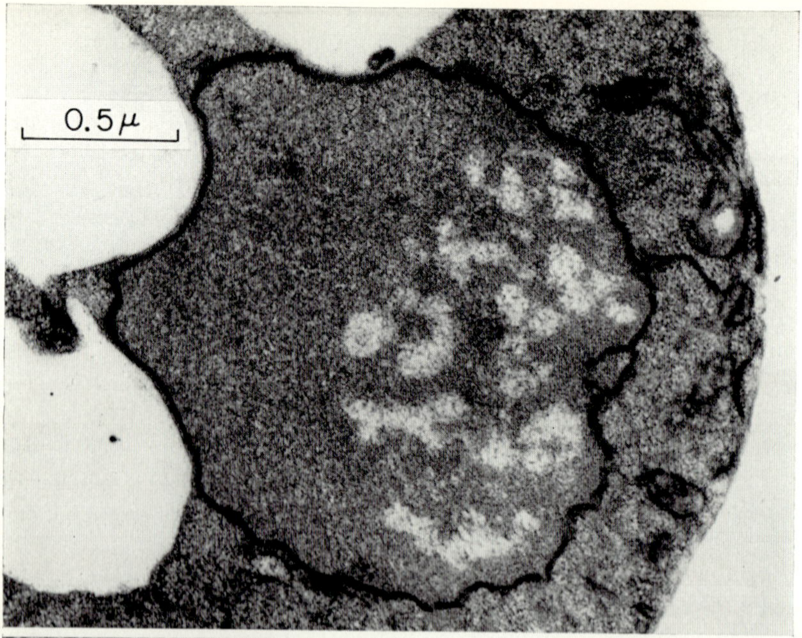

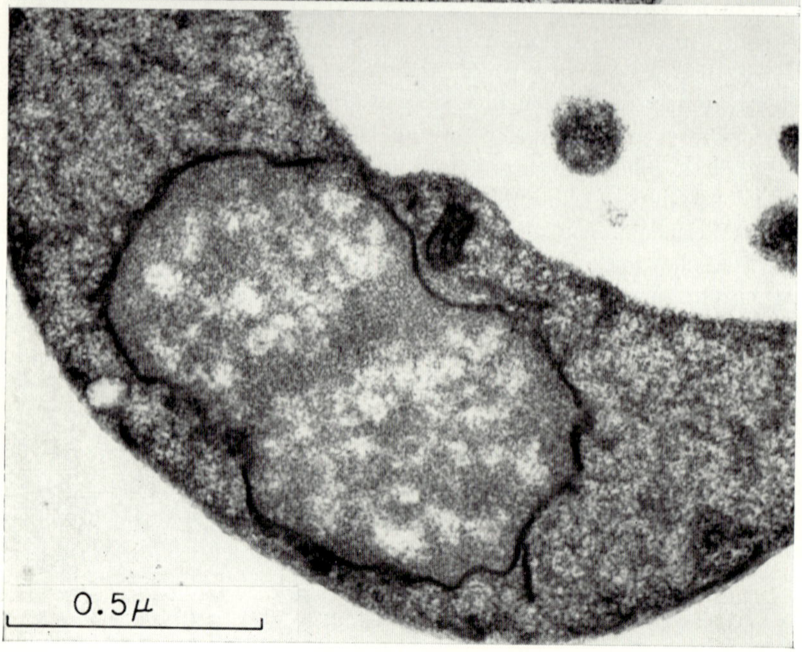

FIG. 11. A cell from a synchronized culture during the later stages of the period of DNA synthesis. The chromosomes are disposed toward one end of the nucleus, and show the characteristic change in appearance referred to in the text.

FIG. 12. An example of a section showing the nucleus undergoing the process of separation of the chromosomes. These are clearly disposed in two discrete masses separated by a zone of seemingly undifferentiated nucleoplasm. The nuclear membrane shows a distinct equatorial furrow.

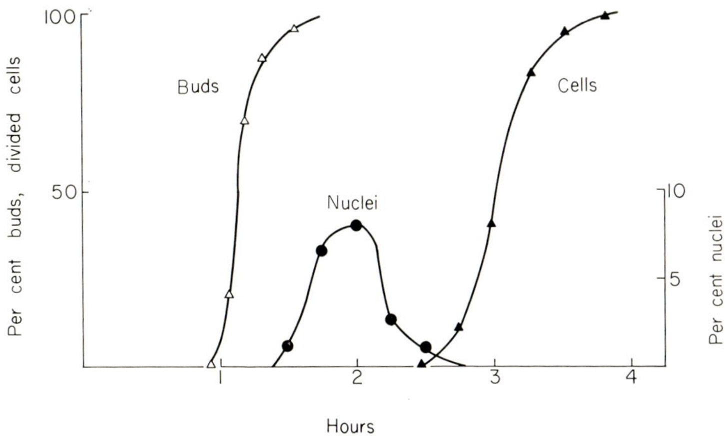

FIG. 13. Changes, during the growth of a synchronously dividing culture, in the frequency of sections showing nuclei in the phase of chromosome separation.

investigation, derives from the way in which the frequency of the sections changed with time. A chance configuration, or for that matter a gross artefact, would have been found in cells of all ages. However, as the data in Fig. 13 show, these sections appeared only in a particular region of the cell cycle. This evidence, which would have been exceedingly difficult to obtain without employing synchronized populations, strongly supports the view that this change in the structure of the nucleus played a meaningful role in the life of the cells. Added to these lines of argument is the knowledge that the change in question occurred at just the region of the cell cycle which, on logical grounds, was predicted for the occurrence of mitotic separation of the chromosomes (Fig. 8). Consequently there seems little doubt that this was in fact its significance.

Immediately following the stage of chromosomal segregation, the nucleus migrates into the neck of the bud. An example of a nucleus in this stage is shown in Fig. 14. At the time of writing, no structures have been detected which might be responsible for the movement of the nucleus. There is also some uncertainty about the orientation of the nucleus as it passes through the neck, and about the movements of the chromosomes within the nuclear membrane. It seems possible that the equatorial zone seen in the separation stage elongates and moves through the narrow neck parallel to the long axis of the cell. This might account for the twin strands of chromatinic material sometimes seen in light microscope preparations of the dividing nucleus (Ganesan, 1959; Ramirez and Miller, 1962). If each mass of chromosomes then moved longitudinally in opposite directions, it would also be possible to account for the frequent appearance of sections like that in Fig. 14, in which the two masses of chromosomes, one in the parental end of the nucleus, the other in the bud, are found on opposite *sides* of the nucleus. This aspect of the division process requires further study. By comparison, the final stage, the fission of the nucleus, brought about by the nuclear membrane invaginating, usually just within the bud, seems fairly straightforward (Fig. 15). Each of the resulting daughter nuclei carries a complement of chromosomal material, and in this we see the conclusion of a cycle of changes involving the apparent replication, separation and partitioning between parent and daughter cells of the bodies identified by Yotsuyanagi (1960) as chromosomes.

IV. GENERAL CONCLUSIONS

Strictly speaking, there remains an element of doubt about the nature of the electron transparent areas observed in these studies. Yotsuyanagi's (1960) identification of them as chromosomes, based as it is on a staining procedure whose mechanism is uncertain, is not absolutely conclusive, and there is still a need for an independent demonstration that these structures alone in the nucleus contain DNA. However the present observations at least show that this electron transparent material is more than merely a physically passive component of the nuclear structure. On the contrary, its distribution within the nucleus was seen to respond to the introduction of the resting cell into growth medium, and once the cell cycle started, it went through an ordered series of changes interpretable as replication, separation, and partition between parent and daughter cells. These changes took place in the sequence required of a mitotic process and, as far as can be judged, at the times in the cell cycle at which the mitotic changes should have occurred. In these circumstances, it is hard to believe that the electron transparent structures are anything other than chromosomes or that the observed changes during the cell cycle represent anything other than the mitotic process of this organism. Consequently it seems permissible to outline, tentatively, the *Saccharomyces* cell cycle shown in Fig. 16.

The timing of these structural changes in the cell cycle is of interest, but a more important general feature to emerge from these observations is that the nucleus of the yeast cell, a eukaryotic organism (Stanier and van Niel, 1962), seems to be organized along somewhat different lines from the nuclei of higher cells taken as the usual model of the eukaryotic state. In reaching this opinion, note is taken of several observations.

First, in contrast with higher cell nuclei, but in common with at least some other fungi (Robinow, 1957a; Conti and Naylor, 1959; Thyagarajan *et al.*, 1962), the nuclear membrane was found to remain intact throughout the division process. Second, no structure resembling a nucleolus was

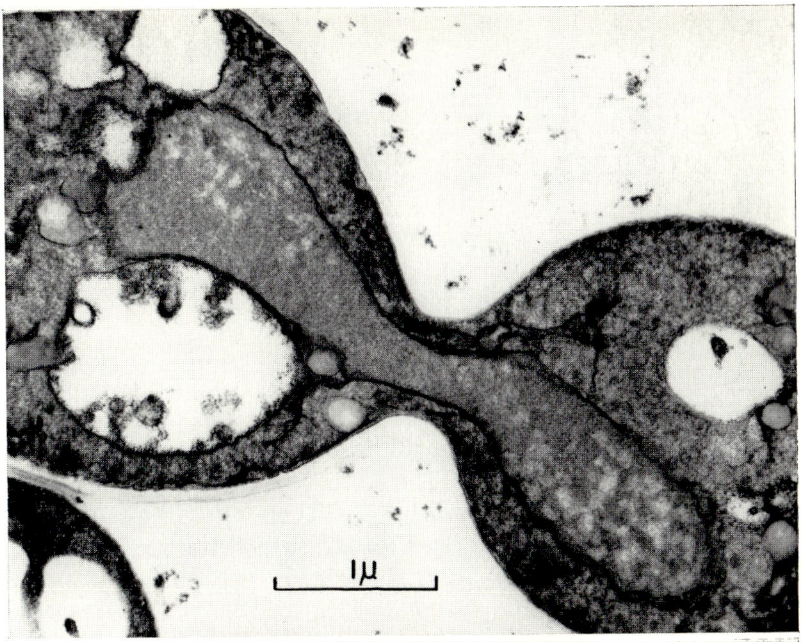

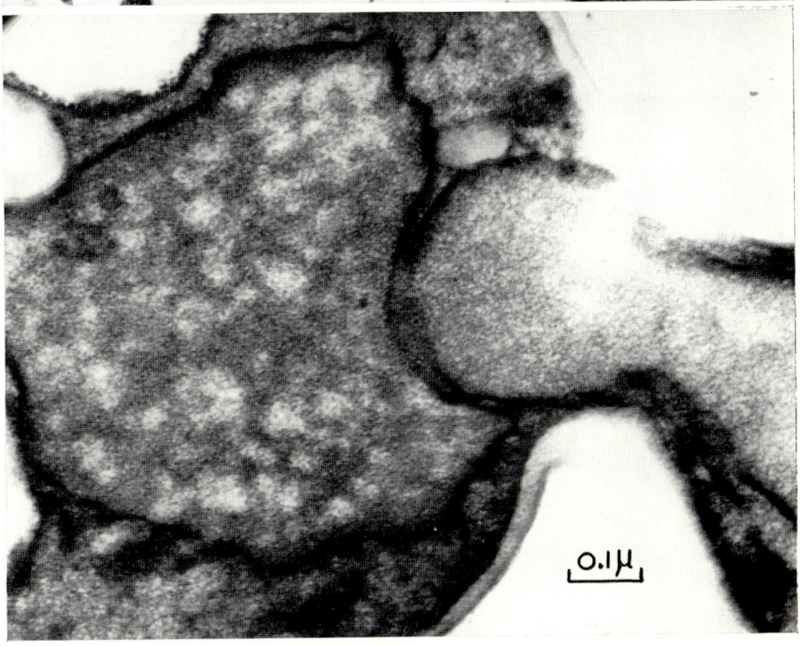

Fig. 14. A section through a cell showing a migrating nucleus. Sections such as this, in which the two groups of chromosomes occur on different *sides* of the nucleus, as well as at different *ends*, are common.

Fig. 15. The invagination of the nuclear membrane which results in the binary fission of the nucleus.

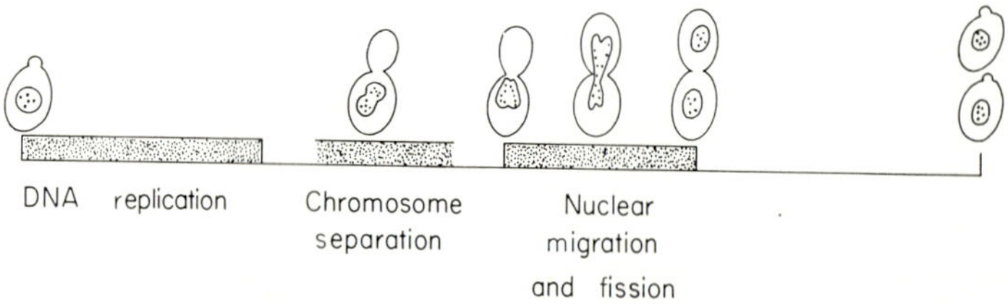

FIG. 16. Diagrammatic representation of the cell cycle in *S. cerevisiae*.

observed. This is thought not to have been a limitation of the method of preparation, since this same method has revealed the nucleolus in electron micrographs of *Lipomyces lipofer* (Williamson, unpublished observations, 1965), an organism known to possess such a body (Ganesan and Roberts, 1959; Robinow, 1961). There is in any case no substantial evidence in the literature for a nucleolus in *S. cerevisiae*, even though structures resembling nucleoli have been clearly observed in some other fungi (McClintock, 1945; Robinow, 1957a,b, 1963; Edwards *et al.*, 1959; Shatkin and Tatum, 1959; Thyagarajan and Naylor, 1962). The absence of a nucleolus in *S. cerevisiae* does not in itself distinguish this organism from the cells of higher animals and plants, for there are several higher cells which lack this structure. Nevertheless, from the available evidence it does seem that these are all specialized cells such as lymphocytes and, in general, only certain of the *protista*, of which *Saccharomyces* is one example, are capab!e of active growth in the absence of a nucleolus.

In the present observations, the nucleus of this organism was also found to differ from that of higher cells in the absence of a spindle or similar mechanism for separating the chromosomes. In this connection, the failure of colchicine to disrupt nuclear division in yeast (Richards, 1938; Beams *et al.*, 1940) may be significant. However, there presumably exists some mechanism for accomplishing the separation in an orderly way, and it is probably only a matter of time before suitable fixation and staining procedures for its demonstration will be devised.*

Finally, the most striking feature of the *Saccharomyces* nucleus lies in the nature of its chromosomes. It is too early to reach firm conclusions about the structure of these bodies; the impression given by the electron micrographs that they are vacuolar structures containing a relatively small amount of material may to a large extent be a consequence of the fixation procedure employed. Nevertheless, their textural resemblance to the interphase chromatin of plant cells (Yotsuyanagi, 1960) is striking, and suggests they may have a rather loose structure. In this connection

* See Addendum.

information about their protein content, if any, would be of interest. Such information should also be relevant to the more general problem of the behavior of the chromosomes during the mitotic process. In any event, it seems clear from the observations described above that they underwent no process of condensation and coiling comparable with that seen in the mitotic chromosomes of higher cells. It is true that they changed slightly in appearance toward the end of the phase of DNA synthesis. This change however, which was interpreted as a consequence of chromosome replication, was subtle in character and involved no noticeable alteration in the texture of the chromosomes. There was certainly no indication at this or any other stage in the cell cycle of a change in chromosome structure of the same magnitude as the coiling process in higher cells. The lack of such a process in the yeast cell was also hinted at by the absence from the cell cycle of a "quiet period" with respect to nuclear RNA synthesis. In this respect the yeast cell bears a superficial resemblance to the gram-negative bacterium *E. coli* (Abbo and Pardee, 1960). Whether this similarity has a common basis in the structure of their genophores (Ris, 1961) is an open question. Nevertheless, the amount of DNA in a chromosome of *S. cerevisiae* [about 0.0014 picogm, calculated from the data in Ogur *et al.* (1952) and Mortimer and Hawthorne (1966)] is probably not more than about one half of the amount in the *E. coli* nucleus (Cairns, 1963), and is many times less than the DNA content of, for instance, an average mammalian chromosome (Vendrely, 1955). Consequently the amount of condensation needed to make the yeast chromosome into a manageable packet for the purposes of mitosis may be very slight. As Mazia (1965) has stated, "fungi probably *could* handle their DNA by the bacterial method." Perhaps they do. At any rate the possibility that the chromosomes of this organism [and perhaps other fungi (McAlear and Edwards, 1959; Thyagarajan *et al.*, 1962)] are structurally less complex than those of higher cells deserves further investigation.

ADDENDUM

At about the time the manuscript for this article was being completed, C. F. Robinow exhibited, at the 1965 London meeting of the Society for General Microbiology, some masterly light and electron micrographs showing the presence in *S. cerevisiae* of a structure having certain of the morphological features of a spindle. Although the precise role of this body in nuclear division has yet to be ascertained, there is no doubt it has a significant one. Thus already some of the statements in this article require revision. In particular the details of nuclear division are no longer as obscure as they were when Section III was written, and the apparent absence of a spindle referred to in Section IV is clearly, as anticipated, a reflection of the inadequacy of the author's preparative procedures.

REFERENCES

Abbo, F. E., and Pardee, A. B. (1960). *Biochim. Biophys. Acta* **39,** 478.

Agar, H. D., and Douglas, H. C. (1957). *J. Bacteriol.* **73,** 365.

Beams, H. W., Zell, L. W., and Sulkin, N. M. (1940). *Cytologia (Tokyo)* **11,** 30.

Baserga, R. (1962). *J. Cell Biol.* **12,** 633.

Cairns, J. (1963). *J. Mol. Biol.* **6,** 208.

Conti, S. F., and Naylor, H. B. (1959). *J. Bacteriol.* **78,** 868.

Davidson, J. N., and Cohn, W. E., eds. (1963). "Progress in Nucleic Acid Research and Molecular Biology" Vols. 1 and 2. Academic Press, New York.

Edwards, M. R., Hazen, E. L., and Edwards, G. A. (1959). *J. Gen. Microbiol.* **20,** 496.

Feinendegen, L. E., Bond, V. P., Shreeve, W. W., and Painter, R. B. (1960). Exptl. *Cell Res.* **19,** 443.

Ganesan, A. T. (1959). *Compt. Rend. Trav. Lab. Carlsberg* **31,** 149.

Ganesan, A. T., and Roberts, C. (1959). *Compt. Rend. Trav. Lab. Carlsberg* **31,** 175.

Hashimoto, T., Gerhardt, P., Conti, S. F., and Naylor, H. B. (1960). *J. Biophys. Biochem. Cytol.* **7,** 305.

Hirano, T. (1962). *J. Ultrastruct. Res.* **7,** 201.

Hirano, T., and Lindegren, C. C. (1961). *J. Ultrastruct. Res.* **5,** 321.

Koehler, J. K., Birnbaum, W., and Hayes, T. L. (1961). *Cytologia (Tokyo)* **26,** 301.

Kushida, H. (1961). *J. Electromicroscopy (Tokyo)* **10,** 194.

Lindegren, C. C., Williams, M. A., and McClary, D. O. (1956). *J. Microbiol. Serol.* **22,** 1.

McAlear, J. H., and Edwards, G. A. (1959). *Exptl. Cell Res.* **16,** 689.

McClary, D. O. (1964). *Botan. Rev.* **30,** 167.

McClintock, B. (1945). *Am. J. Botany* **32,** 671.

Mazia, D. (1965). *In* "Function and Structure in Micro-organisms" (M. R. Pollock and M. H. Richmond, eds.), pp. 379–394. Cambridge Univ. Press, London and New York.

Moor, H., and Mühlethaler, K. (1963). *J. Cell Biol.* **17,** 609.

Mortimer, R. K., and Hawthorne, D. C. (1966). *Genetics* **53,** 165.

Nygaard, O., Guttes, S., and Rusch, H. P. (1960). *Biochim. Biophys. Acta* **38,** 298.

Ogur, M., Minckler, S., Lindegren, G., and Lindegren, C. C. (1952). *Arch. Biochem. Biophys.* **40,** 175.

Ogur, M., Minckler, S., and McClary, D. O. (1953). *J. Bacteriol.* **66,** 642.

Polakis, E. S., Bartley, W., and Meek, G. A. (1964). *Biochem. J.* **90,** 369.

Prescott, D. M., and Bender, M. A. (1962). *Exptl. Cell Res.* **26,** 260.

Ramirez, C., and Miller, J. J. (1962). *Can. J. Microbiol.* **8,** 603.

Reynolds, E. S. (1963). *J. Cell Biol.* **17,** 208.

Richards, D. W. (1938). *J. Bacteriol.* **36,** 187.

Ris, H. (1961). *Can. J. Genet. Cytol.* **3,** 95.

Robinow, C. F. (1957a). *Can. J. Microbiol.* **3,** 771.

Robinow, C. F. (1957b). *Can. J. Microbiol.* **3,** 791.

Robinow, C. F. (1961). *J. Cell Biol.* **9,** 879.

Robinow, C. F. (1963). *J. Cell Biol.* **17,** 123.

Shatkin, A. J., and Tatum, E. L. (1959). *J. Biophys. Biochem. Cytol.* **6,** 423.
Stanier, R. Y., and van Niel, C. B. (1962). *Arch. Mikrobiol.* **42,** 17.
Stoeckenius, W. (1961). *J. Biophys. Biochem. Cytol.* **11,** 297.
Swaminathan, M. S., and Ganesan, A. T. (1958). *Nature* **182,** 610.
Taylor, J. H. (1960). *Ann. N.Y. Acad. Sci.* **90,** 409.
Thyagarajan, T. R., and Naylor, H. B. (1962). *J. Bacteriol.* **83,** 127.
Thyagarajan, T. R., Conti, S. F., and Naylor, H. B. (1962). *J. Bacteriol.* **83,** 381.
Vendrely, R. (1955). *In* "The Nucleic Acids" (E. Chargaff and J. N. Davidson, eds.), Vol., II, pp. 155–180. Academic Press, New York.
Vitols, E., North, R. J., and Linnane, A. W. (1961). *J. Biophys. Biochem. Cytol.* **9,** 689.
Williamson, D. H. (1964). *In* "Synchrony in Cell Division and Growth" (E. Zeuthen, ed.), pp. 351–379. Wiley, New York.
Williamson, D. H. (1965). *J. Cell Biol.* **25,** 517.
Williamson, D. H., and Scopes, A. W. (1960). *Exptl. Cell Res.* **20,** 338.
Williamson, D. H., and Scopes, A. W. (1962). *Nature* **193,** 256.
Yotsuyanagi, Y. (1960). *Compt. Rend.* **250,** 1522.
Yuasa, A. (1962). *Japan. J. Bacteriol.* **17,** 286.

CHAPTER 6

Periodic Enzyme Synthesis in Synchronous Cultures of Yeast

H. O. Halvorson, R. M. Bock, P. Tauro, R. Epstein, and M. La Berge

I. THE USE OF SYNCHRONOUS CULTURES FOR STUDYING CELLULAR BIOSYNTHESIS

The manner in which the biosynthesis of cell components is regulated has been the subject of renewed interest in recent years. Our information has been largely derived from studies with exponentially growing cultures. Such cultures contain a mixture of cells at various stages in the division cycle, and their analysis yields average values of the individual members of the population. We therefore have little information of the way in which these processes are integrated into the overall process of cellular growth and division in individual cells. With the exception of a few enzymes, techniques are not available for examination of the biosynthesis of macromolecules in single cells. The use of synchronously dividing cultures of microorganisms provides a means by which the events occurring in single cells can be amplified and therefore subjected to analysis by the usual biochemical tools.

The use of synchronous cultures for studying biosynthetic properties has been questioned and deserves further comment. Environmental factors are well known to influence the composition of microbial cells as well as the regulation of macromolecular synthesis. Varying patterns of growth are frequently observed when cells are transferred to a new medium. The environmental influence on basic regulatory mechanisms can be minimized by examination of synchronous systems which display balanced growth. As defined by Campbell (1957), balanced growth is the doubling of every biochemical unit of the cells within the time period of a single division without change in the rate of growth. The biochemical processes should also repeat in succeeding generations precisely at the same period of time. Any population of cells which exhibits balanced growth followed by synchronous division for a number of cell cycles is a valid representative of the life cycle of a single cell during each of the cycles. The precision to which the timing of events in a cell cycle can be measured is limited by the degree of asynchrony which exists in the population at the time of measurement.

Two general approaches have been employed for obtaining synchronously dividing populations of yeast. These are (*a*) selection of a synchronous fraction from an exponentially growing culture by physical methods and (*b*) synchronization by forcing the cells to attain a common physiological stage which they cannot pass because of limitations in their environmental or nutritional conditions. Tauro (1965) has compared the biosynthetic properties of synchronous and synchronized cultures of yeast. In cultures displaying balanced growth, morphological changes in the division cycle and timing or extent of synthesis of several enzymes in yeast were identical in synchronous and synchronized cultures. Both of these systems provide valid models for the processes controlling cellular division. For example, as discussed by Williamson in this symposium, nuclear division is a periodic phenomenon in yeast synchronized by these methods.

One curious feature of synchronous cultures is that the time of synthesis of a particular macromolecule is frequently more precisely timed than that of overall cell division. The higher precision for individual macromolecule synthesis over cell division is perhaps not unexpected, since cell division is more complicated than the expression of any single gene.

In synchronous cultures of yeast, the increase in mass, protein, and RNA continue approximately linearly throughout the division cycle, with the rate of synthesis doubling with each generation (Mitchison, 1957; Scopes and Williamson, 1964; Gorman *et al.*, 1964; Mitchison and Cummins, 1964; Tauro, 1965). DNA content, on the other hand, increases in the first portion of the division cycle and then remains constant until the next division period (Williamson and Scopes, 1960; Gorman *et al.*, 1964; Williamson, Chapter 5). The appearance of buds, nuclear plates or DNA synthesis can all be used as markers for characterizing the individual cycles and estimating the degree of synchrony (Hayashibe and Nose, 1958; Scherbaum, 1964; Engelberg, 1964).

II. ORDERED TRANSCRIPTION DURING THE DIVISION CYCLE

There is increasing evidence that in synchronous cultures of bacteria and yeast different classes of RNA are synthesized at various intervals in the division cycle. Maruyama and Lark (1959) observed that although RNA increased continuously in synchronous cultures of *Aerobactor faecalis*, sedimentable RNA showed discontinuous periods of synthesis. In a hybrid yeast the rate of incorporation of uracil-C^{14} into RNA shows an abrupt rise approximately ⅔ through the generation cycle and then declines at the end of the generation cycle. Differences in the distribution of classes of mRNA during the division cycle were detected from double labeling experiments. Cells were briefly labeled with uracil-C^{14} early in the generation and briefly with uracil-H^3 in the middle of the generation cycle. Comparison of the two radioactivity profiles of RNA chromatographed on a MAK column (Halvorson *et al.*, 1964) or fractionated by means of a sucrose density sedimentation (Halvorson, unpublished results, 1965) showed a number of discrete differences, demonstrating that some order exists in the clssses of RNA synthesized during the cell cycle.

Further evidence for ordered RNA synthesis in synchronous cultures has been provided from examination of the base ratios of pulse-labeled RNA. Rudner *et al.* (1964) observed that during the cell cycle of *E. coli* K-12 (Hfr H), both the rate of RNA synthesis and the base composition of RNA show a rhythmic alteration which repeats in successive cell division. Similar shifts in base ratios of pulse-labeled RNA have been observed during the division of a hybrid yeast (Halvorson, Epstein, and La Berge, unpublished results, 1964).

Cutler and Evans (1965) have attempted to determine whether the order of transcription in *E. coli* is sequential. Synchronous cultures were pulse labeled with 5-BUdR at different intervals during a division cycle. Heavy DNA fragments were isolated from each of these cultures by centrifugation in a CsCl density gradient. In a parallel series of experiments, RNA was pulse labeled with P^{32} during these same intervals. Pairwise testing for hybridization between the heavy DNA fragments and pulse-labeled RNA led the authors to conclude that the entire genome is transcribed throughout the replication cycle but that certain regions are transcribed more actively at characteristic times in the cell cycle. These findings provide further evidence that transcription of the genome is nonrandom.

III. THE PERIODICITY OF ENZYME SYNTHESIS

The use of the synchronous cultures can in principle provide an experimental system whereby the accessibility of the genome to transcription can be experimentally tested. One might anticipate two possible situations: (*a*) The gene has a periodic accessibility to transcription; (*b*) the gene is continually accessible to transcription, the degree of transcription depending upon induction and repression.

As illustrated in Fig. 1, the kinetics of enzyme formation permit a decision between these two possibilities. In the first (*a*) the rate of enzyme formation abruptly rises and falls at a given point in the generation cycle, whereas in the second (*b*), the differential rate of enzyme formation is constant and might be expected to show a rise at the time in which the gene doubles, therefore leading to a doubling of gene function.

The timing of enzyme synthesis in both bacterial (Masters *et al.*, 1964; Hanawalt and Wax, 1964; Kuempel *et al.*, 1965; Steinberg, 1965; Don-

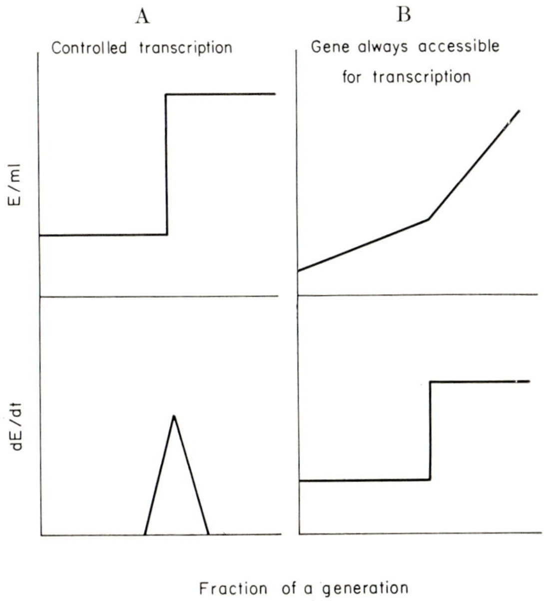

FIG. 1. Predicted kinetics of enzyme synthesis in synchronous cultures. Assuming (A) limited accessibility of the gene for transcription; (B) the gene is continually accessible for transcription.

TABLE I

ENZYMES WHICH ARE PERIODICALLY SYNTHESIZED IN SYNCHRONOUS CULTURES OF YEAST

Enzyme	Reference
Dipeptidase	Sylvén *et al.* (1959)
Catheptic activity	Sylvén *et al.* (1959)
Respiration	Scopes and Williamson (1964)
Fermentation	Ganti (1964)
β-Glucosidase	Gorman *et al.* (1964)
Invertase	Gorman *et al.* (1964)
α-Glucosidase	Gorman *et al.* (1964)
Alkaline phosphatase	Gorman *et al.* (1964)
β-Galactosidase	Halvorson *et al.* (1964)
Aspartic transcarbamylase	J. M. Mitchison (unpublished data, 1966)
Ornithine transcarbamylase	J. M. Mitchison (unpublished data, 1966)

achie, 1965) and yeast synchronous cultures (see Table I) consistently supports the first hypothesis. For example, Fig. 2 illustrates the increases in α-glucosidase activity per milliliter in a synchronously dividing culture of yeast. As can be seen, enzyme increases periodically, one rise per cell generation; and the time at which the rise occurs is repeated over successive cell generations.

The data in Fig. 2 show that α-glucosidase is made only over a very short period of the generation cycle. During most of the generation cycle, this enzyme is not synthesized. These findings are consistent with a model involving limited access of the genome for transcription. In experiments with the highest degree of synchrony, enzyme synthesis is initiated and finished in less than 10% of the generation time. In view of the asynchrony known to be present in such a culture, the actual time used for the formation of these enzymes in a single cell is considerably less and may span only a few percent of the life cycle of the cell.

Periodic formation of enzymes in yeast has been observed in a variety of systems. In each case, the enzyme appears as a burst, and the level then remains constant over the remainder of the generation. Examples are summarized in Table I.

Although there is general agreement that one step of basal (autogenous) enzyme synthesis occurs per generation (Kuempel *et al.*, 1965), there are conflicting reports in bacteria on the behavior of induced enzyme synthesis. Masters *et al.* (1964) reported a periodic synthesis of sucrase, histidase, and aspartate transcarbamylase (ATCase) in synchronously growing cultures of *Bacillus subtilis*. However, in synchronously growing cultures of *E. coli* K-12 Hfr, the synthesis of ATCase was periodic while the synthesis of β-galactosidase was continuous. Enzyme

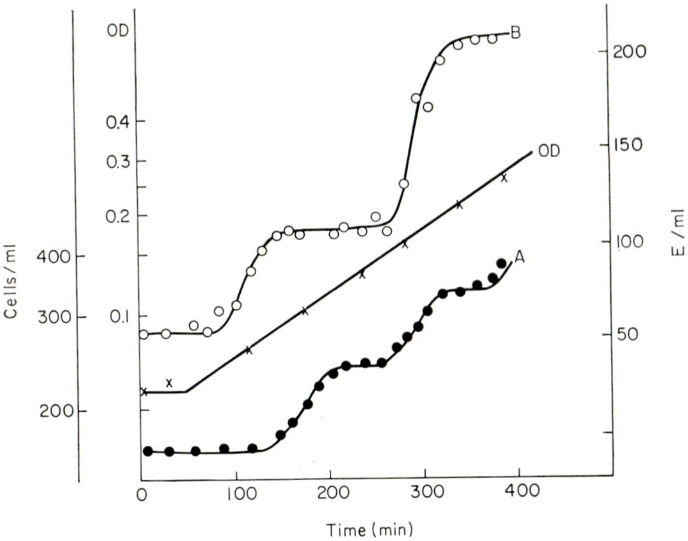

FIG. 2. Order of enzyme synthesis in a synchronous population of the diploid *S. cerevisiae*, Y-55 (M_1m_1). Cells were synchronized by the method of Mitchison and Vincent (1965). The synthetic growth medium contained 1% maltose as the carbon source (data from Tauro, 1965). A. Growth curve; log cell number/ml ($\times 2 \times 10^{-4}$); B. α-glucosidase (E/ml).

synthesis could be induced at any time during the cell cycle; the time of enzyme synthesis under repressed conditions was identical in succeeding generations whereas the synthesis of induced enzymes was not periodic. These authors concluded that the growing bacterium is capable of synthesizing enzyme at any time during the growth cycle and that this capacity doubles with every doubling of the genome. They assumed that as a result of the function of the enzyme, the level of repressor (end product or corepressor) would rise, thereby inhibiting enzyme synthesis. The periodicity in enzyme synthesis in this hypothesis is a function of oscillations in the regulatory components rather than an orderly progression of the transcriptional mechanism.

On the other hand, the time of induced enzyme synthesis in yeast is the same as that for basal enzyme synthesis. As shown in Fig. 3, a fiftyfold increase in the differential rate of α-glucosidase synthesis does not change in the period in the generation in which enzyme synthesis occurs. J. Gorman (unpublished results, 1965) observed similar results in the induction of β-glucosidase in *Streptococcus lactis*. Restricted inducibility is not limited to synchronous cultures of yeast. W. Steinberg (unpublished results, 1965) found that during outgrowth of spores of *Bacillus cereus* T, not only was enzyme synthesis ordered in the same manner as

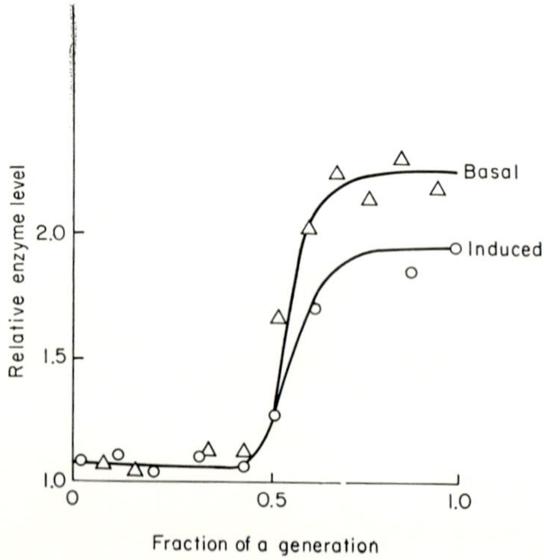

FIG. 3. Comparison of the timing of basal and induced α-glucosidase synthesis in the heterozygous diploid *S. cerevisiae,* Y-55 (M_1m_1). Data from Tauro (1965). The relative rates of the basal and induced enzymes for the second generation of a synchronous culture are normalized and plotted as a function of a generation time.

during synchronous growth, but also the basal and induced α-glucosidase were synthesized in the same brief interval of time. Since basic differences in the mechanism for transcription between *B. subtilis* and *B. cereus* T seem unlikely, further comparisons on the time of induced and basal enzyme synthesis during outgrowth and during synchronous growth are required to interpret the periodicity of enzyme synthesis.

IV. THE RELATIONSHIP BETWEEN GENE POSITION AND THE TIME OF ENZYME SYNTHESIS

One interesting possibility for explaining the timing of basal and induced enzyme synthesis in yeast is that the transcription of the entire genome is sequential. In this model, the order of enzyme synthesis is determined by the order of the structural genes on the chromosome. If it is assumed that the chromosomes of yeast are transcribed in parallel, several predictions of sequential transcription follow:

(*a*) In polyploid cultures, multiple allelic copies of the same structural gene should be transcribed at the same time.

(*b*) The expression of closely linked structural genes should occur at close intervals of the division cycle in synchronous cultures.

(*c*) The expression of structural genes which are scattered throughout a chromosome should occur at different periods of the cell cycle.

The synthesis of α-glucosidase in yeast presents an ideal system to test the above predictions. In *Saccharomyces,* 6 nonallelic polymeric structural genes governing the synthesis of α-glucosidase have been described (Winge and Roberts, 1948, 1950; Gilliland, 1954). No two alleles are found on the same chromosome but several have been centromere linked (Lindgren *et al.*, 1962; von Borstel, 1953). The same enzyme is produced in response to each of these genes (Halvorson *et al.*, 1963); they function independently (Rudert and Halvorson, 1963) and are all subject to inducibility by a common regulatory system. Synchronously growing cultures of *S. cerevisiae* carrying different M genes offer a suitable system to test the predictions of the sequential transcription model.

In a synchronously growing diploid homozygous strain of *S. cerevisiae*, Y-5 (M_1M_1), Halvorson *et al.* (1964) reported that only one period of α-glucosidase synthesis occurred in each generation.

As described earlier (Fig. 2), in a diploid heterozygous strain of *S. cerevisiae*, Y-55 (M_1m_1), carrying one gene M_1 for α-glucosidase, the synthesis of the enzyme was periodic, and only one period of synthesis was observed in each generation. In Fig. 4, the timing during the cell cycle of expression of α-glucosidase in strain Y-5 is compared to strain Y-55. It is clear that the time of synthesis of α-glucosidase in both the strains is identical, consistent with the assumption that the two homologous chromosomes are expressed simultaneously.

Evidence supporting the second hypothesis was provided from synchronous cultures of *S. lactis*. In this organism, the structural gene for

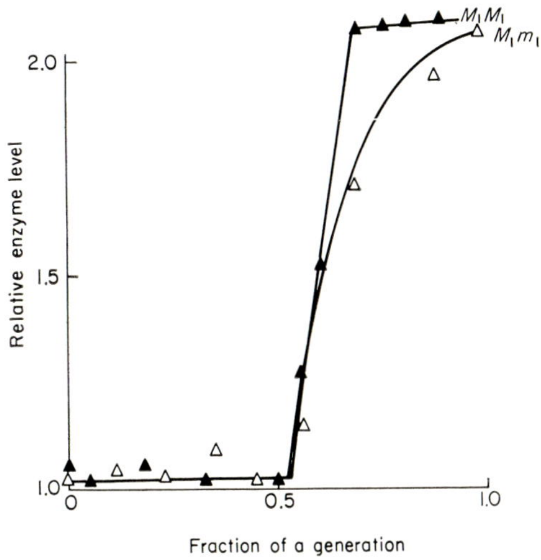

FIG. 4. The timing of α-glucosidase synthesis in diploid heterozygous and homozygous strains of *S. cerevisiae*. The data for *S. cerevisiae* Y-5, diploid homozygous for M_1 is from Halvorson *et al.* (1964). See Fig. 2 for data on *S. cerevisiae* Y-55, diploid heterozygous for M_1. The synthetic growth medium contained 1% maltose as the carbon source.

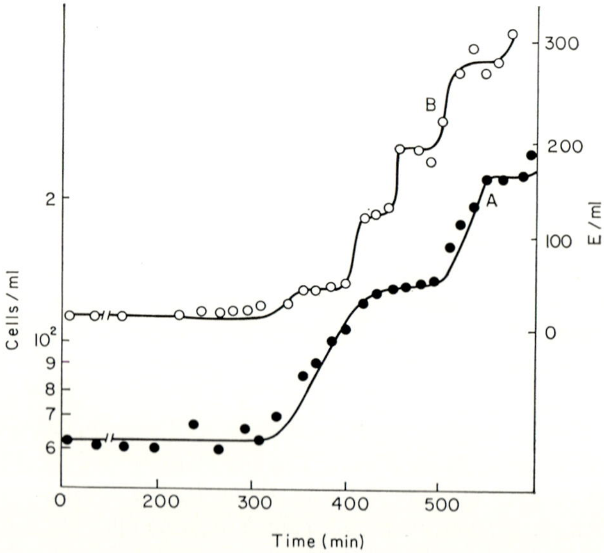

FIG. 5. Timing of α-glucosidase synthesis in synchronously growing diploid *S. cerevisiae*, Y-62 (M_1m_1, M_3m_3). The cells were synchronized by a modification of the method of Sando (1963). The growth media contained 1% maltose as carbon source. A. Log cell number/ml ($\times 4 \times 10^{-4}$); B. α-glucosidase (E/ml).

β-glucosidase is closely linked (1% recombination) to the gene for β-galactosidase (A. Herman, unpublished results, 1964). M. Tingle (see Halvorson *et al.*, 1964) observed that in synchronously growing cultures the initiation of synthesis of β-galactosidase and β-glucosidase differ by only about 3% of the generation time.

The third prediction was tested by examining the synthesis of α-glucosidase in synchronous cultures of four different heterozygous diploid strains of *S. cerevisiae* carrying varying numbers of the *M* genes (Tauro, 1965). As discussed above, in the diploid heterozygous strain Y-55 (M_1m_1), only one period of enzyme synthesis occurs during the middle of the cell generation. In a strain carrying two structural genes, M_1 and M_3 (Y-62), two distinct periods of enzyme synthesis are observed in each generation (Fig. 5). The first period of synthesis occurred at about the beginning of the generation while the second occurred at about the middle of the generation, consistent with the time of expression of M_1. A second typical experiment involving two other *M* genes is shown in Fig. 6. In the diploid heterozygous strain of *S. cerevisiae* Y-62 (M_2m_2, M_4m_4), two distinct periods of enzyme synthesis are noticed in each generation. The first enzyme is synthesized at about 0.25 of the fraction of the generation while the second enzyme is synthesized late in the generation. On the other hand, in synchronous cultures carrying 3 different

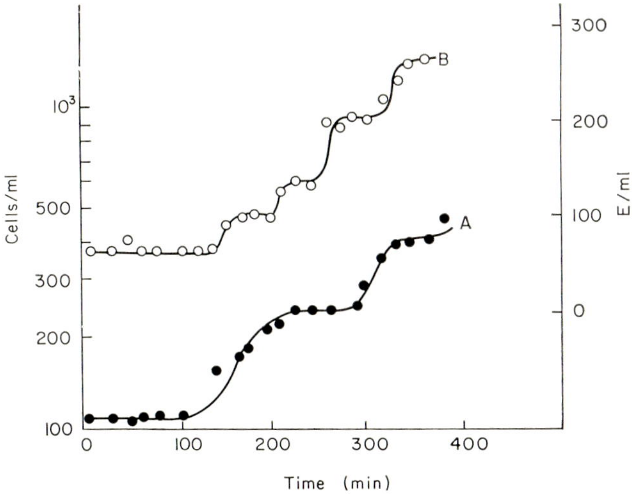

FIG. 6. Timing of α-glucosidase synthesis in synchronously growing diploid *S. cerevisiae*, Y-64 (M_2m_2, M_4m_4). The method of Mitchison and Vincent (1965) was employed to obtain synchronously dividing population of yeast. The synthetic growth medium contained 1% maltose as carbon source. A. Log cell number/ml ($\times 2 \times 10^{-4}$); B. α-glucosidase (E/ml).

M genes, 3 periods of enzyme synthesis are consistently observed in each generation (Tauro, 1965). In each case, the steps correspond to the time at which enzyme synthesis occurs in response to single *M* genes.

In Fig. 7, the timing of the expression of the 4 different *M* genes, as determined from the second generation of synchronous growth, is described. A definite relationship exists between the number of structural genes present and the number of periods of enzyme synthesis. An increase in the number of genes results in a corresponding increase in the number of periods during which the enzyme is synthesized. These results suggest that the genes are expressed independently of each other at different times in the cell cycle.

The probable order of expression of these *M* genes as determined from Fig. 7 is described in Fig. 8, and the timing of their expression is compared with the genetic map of Saccharomyces (von Borstel, 1963). It appears that the four genes examined for the synthesis of α-glucosidase are transcribed in a fixed order during the cell generation with the expression of gene M_3 preceding the expression of the other three genes. The probable order of enzyme synthesis as determined from these experiments is M_3, M_4, M_1, and M_2.

The above four structural genes for α-glucosidase are scattered over four linkage groups in *Saccharomyces*. The position of the gene on the chromosome may determine the timing of transcription of the gene.

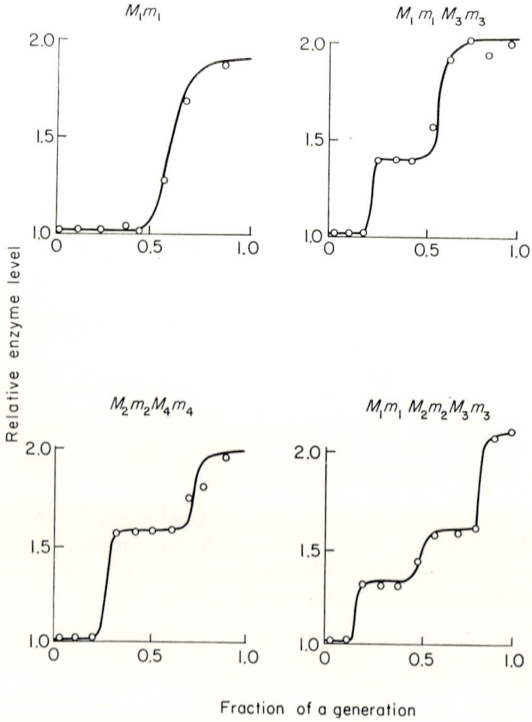

FIG. 7. The timing of synthesis of α-glucosidase in various M strains of *S. cerevisiae*. The relative rate of enzyme synthesis is plotted as a function of the generation time. The data is derived from the second generation of the synchronous culture (data from Tauro, 1965).

For example, gene M_3 is located on chromosome III at close proximity to the centromere while gene M_1 is on chromosome VII at a greater but undetermined distance from the centromere. M_2 and M_4 have not yet been centromere linked and are therefore believed to be further from the centromere than M_3. Since gene M_3 functions much earlier in the generation than gene M_1, the transcription of the chromosome may be from the centromere to its extremities. Other models which would account for the reading of M_3 earlier than M_1 have not been ruled out. Experiments are in progress to examine the timing of expression of a series of genes on both arms of a single chromosome in yeast.

V. POSSIBLE MECHANISMS FOR CONTROLLING SEQUENTIAL TRANSCRIPTION

The study of synchronously growing cultures of cells has provided a means of examining periodic phenomena within each life cycle. It is

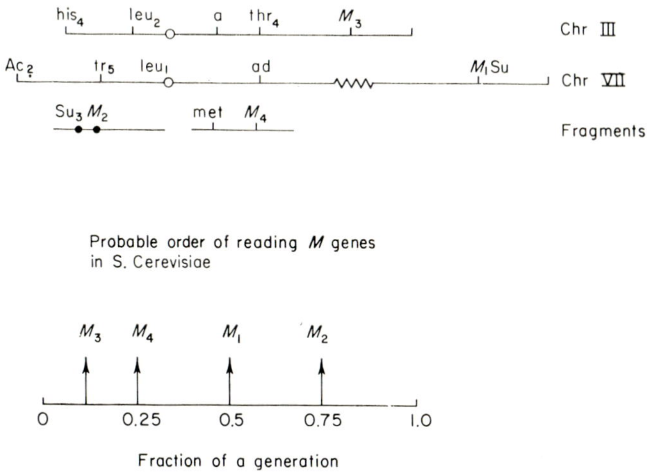

FIG. 8. The probable order of expression of the different *M* genes and the chromosome map of *Saccharomyces*. See Fig. 7 for timing of enzyme synthesis. This is plotted as a fraction of the generation time.

clear that the biosynthesis of discrete enzymes is not a random process occurring throughout the life cycle of the cell, but for each enzyme there is a particular period of expression whether its synthesis had been induced, repressed or occurred at a basal level. The molecular mechanisms by which such timing is accomplished are as yet unknown. The phenomenon has been observed both in germinating spores during periods when net DNA synthesis does not occur, in yeast cells during synchronous growth, and under conditions where DNA synthesis occupies the majority of the time during the cell cycle. Comparison of the time of expression in induced and basal enzyme production indicates that the process of induction does not alter the time at which expression occurs and, in fact, that induction can occur only over a very limited period of the cell cycle. Comparison with the genetic arrangement of the structural genes for several enzymes has shown that closely linked genes are expressed at nearly the same period and that for more distantly linked genes the time of expression during the generation appears correlated to the genetic mapping distance out from the centromere. The finding that the multiple alleles for a given enzyme which are subject to common metabolite control are expressed each at its own characteristic time implies that the time is not the result of periodic production of inducers or repressors within the cell.

The ordering process and timing of expression of these enzymes could be the result of a purely mechanical phenomenon whereby the enzyme which transcribes the DNA into mRNA is unable to leave the DNA once it has started on a round of readout. Either there are a small number of such enzymes, or if the number is large, all of them are clustered in limited regions on each chromosome and are transcribing cistrons that are closely linked compared to the total distance along the chromosome. It is well documented that the replication of DNA is an ordered process with one or a very small number of replicating points on any one chromosome. One can inquire whether the movement of the point of replication has a relation to the timing of the transcription and expression of the genetic information. For example, if the DNA at the point of replication or immediately after replication is more accessible to RNA polymerases, an ordering process could easily occur. We are in the process of directly testing this possibility by preparing at identical periods during a generation 5-BUdR-labeled DNA and radioactively labeled RNA. The 5-BrU-DNA segment is separable from those portions of the DNA which replicate at other characteristic periods within the generation. It is known that the region of DNA which replicates at any period in a generation is characteristic of that time in each generation. Thus, if the RNA made at a characteristic period in a generation is tested by hybrid formation against the heavy labeled DNA made in that same period one may determine whether replication and transcription occur at the same region in the same direction and in the same order. If indeed replication and transcription phenomena are occurring simultaneously at closely related regions in the chromosome, then there will be a strong selectivity for hybrid formation between any set of RNA and that DNA which had undergone replication during the same period of the cell cycle when the RNA was collected. The power of discrimination of this test becomes greater when shorter and shorter times of labeling are used and when the degree of synchrony of the cells nears perfection. Even with the current state of the art, this method can rule out many possibilities for the process of ordered gene expression. The studies on germinating spores demonstrate that DNA replication is not essential for an ordered transcription of the genetic information (Kobayashi *et al.*, 1965; Steinberg, 1965). Yet even in this situation the order of appearance of the various mRNA's during spore germination can be compared to the order of DNA replication in later generations after the cells have gone into vegetative growth. The studies of Cutler and Evans (1965) on hybrid formation of 5-BrU-DNA with RNA-P^{32} imply continuous transcription of all regions of DNA but at periodically increasing rates. The hybrid study which we independently had developed has progressed slowly until recently because reproducible isolation of DNA from yeast and from certain stages of bacillus spores was an undeveloped art. The models we favor predict strong relations between the order of replication and transcription and the experiments which could disprove our models are now in progress.

Several other models have been proposed by which readout of a given region of DNA is selected by cytoplasmic conditions within the cell. For example, gradation in the salt content or composition, the amount of polyamines, histones, or even amino acid pools are metabolic variables which could show a continuous drift throughout the cell cycle (Clever, 1964; Swift, 1964). If there are regions of DNA which are responsive to such components then they may be expressible only during a small fraction of the cell cycle. In this event, as well as in the event that cyclic productions of repressors or inducers is occurring, it is neither requisite nor likely that detailed correspondence will be found between genetic order or distances and time of expression of the corresponding genetic information.

The study of synchronous cultures of several cell types under varied metabolic states must be extended before the above findings can be generalized. The progress reported in this symposium indicates that such extensions as well as improvements in the scale and precision of synchronization are developing rapidly.

ACKNOWLEDGMENTS

This research was supported by research grants from the National Science Foundation (B-1750), the Public Health Service (E-1450), and the Air Force Office of Scientific Research (AF 49 (638) 314).

REFERENCES

Campbell, A. (1957). *Bacteriol. Rev.* **21,** 265.

Clever, U. (1964). *In* "Nucleohistones" (J. Bonner and P. Ts'o, eds.), p. 317. Holden Day, San Francisco, California.

Cutler, R. G., and Evans, J. E. (1965). *Abstr. 10th Meeting Biophys. Soc., San Francisco* p. 160.

Donachie, W. D. (1965). *Nature* **205,** 1084.

Engelberg, J. (1964). *Exptl. Cell Res.* **34,** 111.

Ganti, T. (1964). *Acta Microbiol. Polon.* **11,** 147.

Gilliland, R. B. (1954). *Nature* **173,** 404.

Gorman, J., Tauro, P., La Berge, M., and Halvorson, H. O. (1964). *Biochem. Biophys. Res. Commun.* **15,** 43.

Halvorson, H. O., Winderman, S., and Gorman, J. (1963). *Biochim. Biophys. Acta* **67,** 42.

Halvorson, H. O., Gorman, J., Tauro, P., Epstein, R., and La Berge, M. (1964). *Federation Proc.* **23,** 1002.

Hanawalt, P., and Wax, R. (1964). *Science* **145,** 1061.

Hayashibe, M., and Nose, Y. (1958). *Bull. Agr. Chem. Soc. Japan* **22,** 363.

Kobayashi, Y., Steinberg, W., Higa, A., Halvorson, H. O., and Levinthal, C. (1965). *In* "Spores" (L. L. Campbell and H. O. Halvorson, eds.), Vol. 3, p. 200. Am. Soc. Microbiol., Ann Arbor, Michigan.

Kuempel, P. L., Masters, M., and Pardee, A. B. (1965). *Biochem. Biophys. Res. Commun.* **18,** 858.

Lindgren, C., Lindgren, G., Shult, E., and Hwang, Y. C. (1962). *Nature* **194,** 260.

Maruyama, Y., and Lark, K. G. (1959). *Exptl. Cell Res.* **18,** 389.

Masters, M., Kuempel, P. L., and Pardee, A. B. (1964). *Biochem. Biophys. Res. Commun.* **15,** 38.

Mitchison, J. M. (1957). *Exptl. Cell Res.* **13,** 244.

Mitchison, J. M., and Cummins, J. E. (1964). *Exptl. Cell Res.* **35,** 394.

Mitchison, J. M., and Vincent, W. S. (1965). *Nature* **205,** 987.

Rudert, F., and Halvorson, H. O. (1963). *Bull. Res. Council Israel* **A11,** Part 4, 337.

Rudner, R., Porkop-Schneider, B., and Chargaff, E. (1964). *Nature* **203,** 479.

Sando, N. (1963). *J. Gen. Appl. Microbiol.* **9,** 233.

Scherbaum, O. H. (1964). *Exptl. Cell Res.* **33,** 89.

Scopes, A. W., and Williamson, D. H. (1964). *Exptl. Cell Res.* **35,** 361.

Steinberg, W. (1965). Master's Thesis, University of Wisconsin.

Swift, H. (1964). *In* "Nucleohistones" (J. Bonner and P. Ts'o, eds.), p. 169. Holden Day, San Francisco, California.

Sylvén, B., Tobias, C. A., Malmgren, M., Ottoson, R., and Thorell, B. (1959). *Exptl. Cell Res.* **16,** 75.

Tauro, P. (1965). Master's Thesis, University of Wisconsin.

von Borstel, R. C. (1963). Carbondale Yeast Genet. Conf. *Microbial Genet. Bull.* Suppl. 19 p. 21.

Williamson, D. H., and Scopes, A. W. (1960). *Exptl. Cell Res.* **20,** 338.

Winge, O., and Roberts, C. (1948). *Compt. Rend. Lab. Carlsberg* **24,** 263.

Winge, O., and Roberts, C. (1950). *Nature* **116,** 1114.

CHAPTER 7

Light-Induced Synchronous Sporulation of a Myxomycete—The Relation of Initial Metabolic Changes to the Establishment of a New Cell State

John W. Daniel

I. INTRODUCTION

The plasmodial myxomycete *Physarum polycephalum* is a highly syncytial organism, exhibiting rapid reciprocal protoplasmic streaming and synchronous mitosis during rapid growth. The occurrence of this natural highly synchronous mitosis, maintained by conditions favoring high growth rates, implies the operation of a sequence of synchronous processes of which mitosis is a visible manifestation. Although the origin of an extensive syncytial form such as *P. polycephalum* is completely unknown as distinguished from the classical unicellular types, it is of interest to note that multinuclearity is induced in mammalian cells by some viruses (Roizman, 1962) and has been attributed to a defect in membrane metabolism. In many cells during mitosis, a residue supposedly from the nuclear membrane has been suggested to participate in elaboration of the new cell partition. The nuclear membrane of *P. polycephalum* apparently remains intact during mitosis (Guttes *et al.*, 1961; Kessler, 1964). However, the protoplasmic extensiveness resulting from this lack of cytokinesis is endowed with a mechanism for rather rapid metabolic communication by the continuous oscillatory flow through the veinlike channels. This flow occurs uninterrupted except just prior to the formation of sclerotia or of sporangia. The protoplasmic continuum is then either fragmented to form much smaller, but multinucleate walled units (spherules) or, in the latter case, to form uninucleate sexual spores.

At low growth rates or in the absence of growth nutrients, the compact morphology characteristic of the rapidly growing plasmodium, and the associated mitotic synchrony, is lost. This state leads to another stage of the life cycle, sporulation (fruiting), where metabolic synchronization induced by exposure to light is manifest in simultaneous formation of sporangia, from a single plasmodium as well as from plasmodia prepared in parallel.

When deprived of organic nutrition and incubated on a salts medium containing niacin, the plasmodium acquires the ability to undergo this sexual morphogenesis when exposed to light (Gray, 1938; Daniel and Rusch, 1958; Daniel and Baldwin, 1964). The light-sensitive state is of particular interest since either growth or sporulation can be indued from it, by applying the proper stimulus. The stimuli, light or growth medium, are mutually antagonistic, although once induced, sporulation proceeds without alteration in the presence of optimal exogenous growth conditions. The mold is thus in a biochemically metastable condition, induced by and susceptible to environmental conditions. Furthermore, the sporulation process is actinomycin D-sensitive during the light-dependent induction period, being inhibited before but not after light induction is completed.

The myxomycete consequently offers a morphogenic system requiring specific environmental conditions to produce the sensitive inducible metabolic state, which upon receiving a specific and controlled stimulus, activates the required genetic information by which the metabolic transition is completed. It is this possibility of identifying the inducing reaction (photoreaction), the favorable metabolic environment (niacin-oriented metabolism) allowing the transmission of this signal, and its role in evoking genetic expression that recommends this system. Since light is an external source of metabolic energy, the problem is one of determining the pathway by which this light energy is absorbed and utilized. Information will be presented which suggests that in response to proper environmental conditions, light alters critically and specifically the pathways of energy metabolism upon which the successful evocation and translation of metabolic information appears to depend.

II. CHANGES IN PLASMODIAL MACROMOLECULAR COMPONENTS DURING THE DARK AND LIGHT PERIODS

A. Nucleic Acids and Protein

As *P. polycephalum* metabolizes on an inorganic salt medium, after removal from the growth medium, the protein and nucleic acids decline catabolically as shown in Fig. 1. Cultures incubated in the presence or absence of niacin, necessary for induction of light sensitivity (Daniel and Rusch, 1962a,c), behave differently, especially in the first 48 hours. At 7 days (4–5 days in other experiments), plasmodia are sensitive to light as

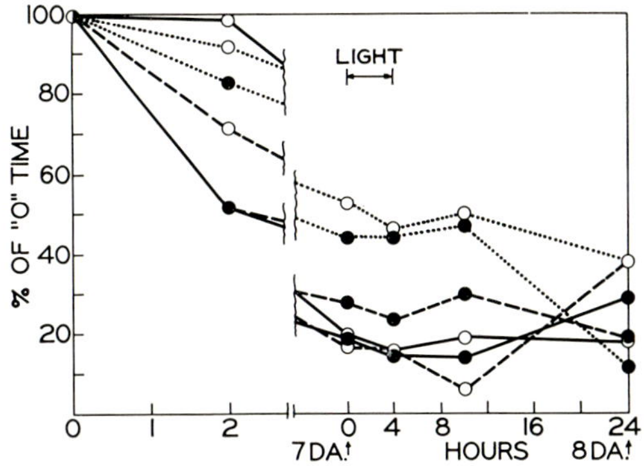

Fig. 1. Protein and nucleic acid changes during sporulation. Petri dish cultures prepared from submerged cultures (ca. 3 days old), M3c strain, and incubated with or without niacin at 22°C in the dark. After 7 days dishes illuminated for 4 hours and replaced in the dark for sporangial development, complete after approximately 16 hours. Values at 8 days for +niacin are for spores. Values in parentheses indicate quantities at "0" time (100%). ●····●, protein (20.9 mg) (+niacin); ○····○, protein (21.6 mg) (−niacin); ●----●, RNA (5.6 mg) (+niacin); ○----○, RNA (6. mg) (−niacin); ●——●, DNA (0.100 mg) (+niacin); ○——○, DNA (0.088 mg) (−niacin). Note that the 0–2 day values for RNA (+niacin) and DNA (+niacin) are close and indicated by a single line.

previously described. At this point the original plasmodium has declined by 56% in protein, 72% in RNA, and 81% in DNA. The precipitous drop in protein occurring near the end is associated with elaboration of sporangial structure.

Cultures incubated with niacin display 3 marked differences: (*a*) a relative maintenance of the RNA level before and during the illumination period, suggesting a possible conservation or nucleotide priming by niacin; (*b*) a rapid drop in alkali-soluble protein after illumination, associated with the elaboration of sporangial structure which does not occur without niacin; and (*c*) an approximate doubling of DNA during the same general period, the significance of which is not yet clear since a single meiotic division unaccompanied by mitosis has been observed late in this period (Guttes *et al.*, 1961; for a discussion, see Alexopoulos, 1963).

The feedback function of niacin on the synthetic pathway of the pyridine nucleotides, to cause an accumulation of nicotinamide ribotide in bacteria (Imsande and Pardee, 1962), and to stimulate transient pyridine nucleotide synthesis in liver (Kaplan *et al.*, 1956; Shuster and Goldin, 1959), may be involved in the increase of pyridine nucleotide during illumination (Daniel and Rusch, 1962b) as well as in the above RNA effect. It is of interest that pyridine nucleotide synthesis might be controlled in this way by niacin which was produced through the action of nucleotidases on unbound oxidized pyridine nucleotides and made available by the fluctuations in the rates of various dependent enzymes. These rates would in turn reflect local or overall anabolic-catabolic balance. Klingenberg (1963) has proposed, for example, a mechanism for the energy-dependent mitochondrial compartmentation of pyridine nucleotides in which a medium degree of pyridine nucleotide reduction favors maximum turnover rates. A high degree of reduction favors binding but inhibits oxidation. A high degree of oxidation favors nucleotide release which he suggests makes them available to nucleotidases, acting on the oxidized form of these nucleotides only.

B. Polysaccharide and Glucose

Plasmodial polysaccharide also changes dramatically during the dark and light periods. Fig. 2 shows that a glycogen-like polysaccharide (Daniel, 1964a) accumulated during the growth period, being very low at the onset of light sensitivity. Thus, as catabolic dissipation proceeds, protein as well as polysaccharide is oxidized to provide maintenance energy. Polysaccharide is used preferentially but is not completely exhausted at 5 days. However, when the plasmodium becomes light sensitive (4–5 days or 7 days for the nucleic acid experiments) polysaccharide has declined to a very low value. If now, the sporulation medium is replaced by

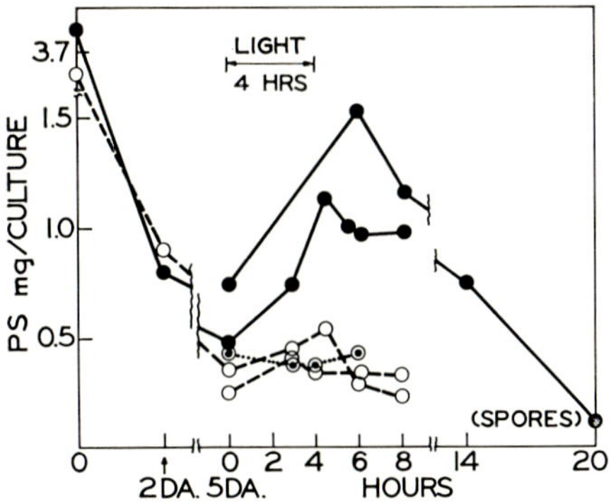

FIG. 2. Changes in plasmodial polysaccharide during sporulation. See Fig. 1 for culture preparation. Cultures illuminated at 5 days instead of 7 days. Plasmodia digested in 30% KOH, polysaccharide (PS) precipitated with hot ethanol and estimated by the anthrone reaction. ●——●, sporulating cultures (2 experiments); ○----○, nonsporulating cultures (2 experiments); ◉ · · · · ◉, cultures in Corning red "actinic" flasks (absorbs approximately 90% of light less than 500 mμ) exposed to normal light (no sporulation).

growth medium, growth, but not sporulation, occurs even after illumination (Daniel and Rusch, 1962a). The component essentially responsible for this inhibition of sporulation is glucose, which causes 50% inhibition at about 100 μg/ml (5×10^{-4} M). Growth follows in the presence of a complete medium. However, after light induction of sporulation is complete, glucose is no longer utilized or taken up (Daniel, 1964b). Thus, light, in effect, shuts down glucose transport. Reversible glucose transport can be demonstrated when glucose is added in the light (Fig. 3) and may resemble the glucose extrusion described by Hoffee and Englesberg (1962). Light, in effect, interferes with a component of the glucose "permease" system either by deletion or possibly by relocation, as, for example, under other conditions, the reversible binding of hexokinase to mitochondria (Rose and Warms, 1965).

Of particular interest is the effect of light on the polysaccharide content (Daniel, 1964a) summarized in Fig. 2. In the absence of exogenous glucose, a large net increase in the glycogen-like polysaccharide occurs near the end of the illumination period. The increase does not occur when niacin is omitted from the medium during the dark period, or when red light which does not induce sporulation is used. The polysaccharide disappears during the subsequent dark period. The obvious conclusion that a photodependent biosynthesis of this polysaccharide occurs from nonhexose precursors, (respiratory skeletons, carbon dioxide) must be tentative until the possibility of its formation by oligosaccharide interconversion has been eliminated. In occasional experiments, the "0" time poly-

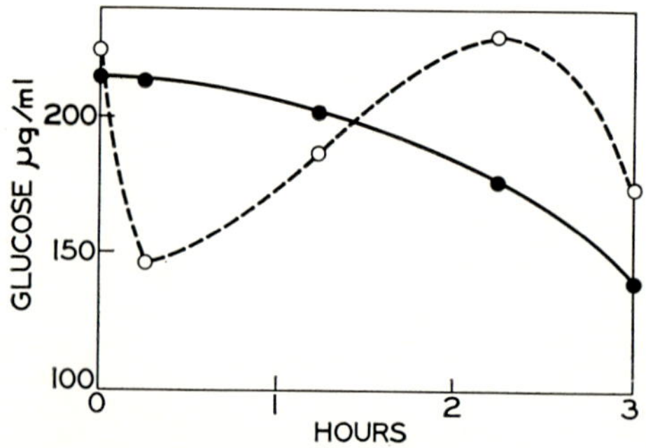

FIG. 3. The effect of light on glucose uptake. See Fig. 1 for culture preparation. Cultures illuminated at 5 days; glucose added at "0" and samples taken immediately and as indicated for estimation by glucose oxidase method (Glucostat); no glucose detected in medium before addition. Each dish contains 13 ml of medium and 4.5 mg plasmodial protein. ●——●, dark; ○----○, light.

saccharide level is about as high as that of the light-induced peak which then follows a transient drop.

Correlated with the course of the polysaccharide curve are the results from chromatography of extracted phosphate esters during this time. Qualitatively, hexose phosphates originally present in the dark strongly decrease or disappear during the first hour, followed by a strong increase in polyol phosphates (but not glyceraldehyde 3-phosphate). At 3 to 4 hours the approximate original pattern reappears.

Once glucose transport is reestablished, continued sporulation inhibition may also be effected by means of a catabolite repression mechanism very possibly in operation during polysaccharide depletion. The role of catabolite repression in bacterial sporulation has been recently discussed by Schaeffer *et al.* (1965).

Light appears to strongly reorient pathways of carbon and energy metabolism, which at once drive polysaccharide synthesis but inhibit glucose transport as a part of subsequent morphogenic transformation. The effect on glucose transport is particularly significant, since glucose is the only energy source supporting rapid growth. The effect of exogenous glucose on the plasmodial ATP levels in the dark and light, described in the following section, also supports this conclusion.

III. LIGHT-INDUCED CHANGES IN THE PLASMODIAL ATP LEVEL

Since marked increases in both reduced and oxidized pyridine nucleotides had been observed (Daniel and Rusch, 1962b) during the light induction period, and in conjunction with the above studies, changes in the ATP level of the whole plasmodium were also examined (Daniel, 1964a,b).

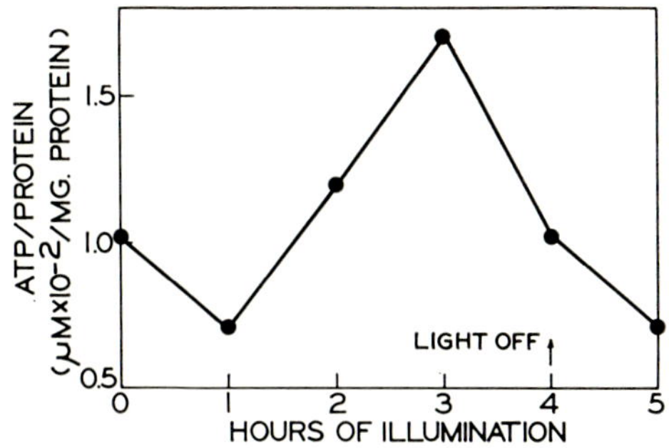

FIG. 4. The effect of light on the plasmodial ATP level. See Fig. 1 for culture preparation. Cultures illuminated at 5 days. Plasmodia extracted in pH 8.0 Tris buffer at 100°C for 1 minute and the ATP content estimated by a fluorometric firefly method using both independent and internal standards (other nucleoside triphosphates contributed less than 10%). Protein estimated by a modified Folin method.

Light produces a marked biphasic effect on the ATP level during the 4-hour illumination period, as shown in Fig. 4. The initial drop, maximal at about 1 hour, is followed by a rise, peaking at about 2.5 hours. The initial drop may vary from 25 to 50% of the initial dark level, while the peak may vary between 150 and 200% of that value.

The fluctuation is closely associated with the light-dependent induction period, since (*a*) it occurs only in the light, no change occurring in dark during this 4-hour period; (*b*) interruption of illumination after 2 hours results in a decline of the ATP level back to the dark level (as after the normal interruption at 4 hours); and (*c*) cultures incubated in the absence of niacin, a nutritional requirement, show at best an abortive rise during the light period. From a large number of experiments, a minimum light period of between 2 and 3 hours is required for sporulation frequency of 0.8 or greater, while 1 hour gives a very low frequency. This suggests that the appearance of the ATP peak corresponds approximately to the end of the light-dependent phase of metabolism, required to overcome an energetic barrier blocking the induction of sporulation. It also demonstrates the necessity for a substantial period of light metabolism, although the exact duration is dependent on the nutritional and cultural history of the organism.

The response of the ATP level indicates that light alters the supply-demand balance of the dark metabolism, either by inducing a new light-dependent pathway contributing to the ATP level, differing from but interacting with dark pathways, or, alternatively, by directly stimulating, existing, functioning dark pathways. The former choice might also result in at least partial inhibition of dark ATP synthesis. The first mechanism is also compatible with the effect of glucose on the ATP level. Added in

TABLE I

EFFECT OF ANTIMYCIN A AND NOHQO ON GROWTH[a]

Inhibitor concentration	Growth: protein, mg/ml	
	Antimycin A	NOHQO
1.0×10^{-4} *M*	—	2.77
3.0×10^{-5} *M*	3.28	2.45
1.0×10^{-5} *M*	3.23	2.49
1.0×10^{-6} *M*	3.62	—
None	3.29	2.83

[a] Submerged cultures with inoculum 0.13–0.17 mg protein/ml + inhibitor incubated, and growth determined at 65 hours (antimycin A) or at 67 hours (NOHQO) as Folin protein.

TABLE II

INHIBITION OF SPORULATION BY ANTIMYCIN A AND NOHQO[a]

Inhibitor concentration	Sporulation frequency[b]	
	Antimycin A	NOHQO
3×10^{-5} *M*	0/3	1/6
1×10^{-5} *M*	0/3	0/6
3×10^{-6} *M*	1/3	4/6[c]
1×10^{-6} *M*	3/3	2/3
None	3/3	6/6

[a] Inhibitor added 15 minutes before illumination at 5 days dark incubation.
[b] Number of cultures sporulating per number tested.
[c] One culture produced abnormal sporangia.

the dark, glucose (10^{-3} *M*) substantially raises this level which is suppressed by light but again rises in the dark.

More convincing support is found in the differential effect of antimycin A and *n*-nonyl 4-hydroxyquinoline-*N*-oxide (NOHQO) on growth and sporulation shown in Tables I and II. Although at high levels neither inhibitor is effective against growth, both are effective against sporulation. Antimycin A, at 3×10^{-5} *M*, causes no more than 15% inhibition of respiration (oxygen uptake) of either growth or sporulation plasmodia, indicating that the sensitive reaction is a part of the light-dependent metabolism.

Table III shows the strong effect of these inhibitors in suppressing the light-dependent rise in ATP level of sporulating plasmodia. NOHQO effects a slow inhibition in the dark. The similar pattern in the light suggests that the light-dependent ATP increase has been preferentially blocked.

Since antimycin A inhibits the respiration in mammalian mitochon-

TABLE III

EFFECT OF ANTIMYCIN A AND *n*-NONYL-4-HYDROXYQUINOLINE *N*-OXIDE ON PLASMODIAL ATP LEVELS[a]

Illumination period (hours of light)	Control: no addition (10^{-2} μM/mg protein)	Antimycin A (9×10^{-6} M) (10^{-2} μM/mg protein)	% of control at "0"	Control: no addition (10^{-2} μM/mg protein)	NOHQO (3×10^{-6} M) (10^{-2} μM/mg protein)	% of control at "0"
0	1.07	—	—	0.64	—	—
1	0.68	0.63	59	0.52	0.50	78
					0.45 (Dark)	70
2.5	—	—	—	0.90	0.40	45
					0.42 (Dark)	59
3	1.18	0.74	69	—	—	—
4	1.10	0.69	64	0.67	0.38	73

[a] Inhibitor added 15 minutes before illumination; at above concentration, inhibitors completely suppressed sporulation but were not visibly toxic to remaining plasmodia. Average plasmodial protein content was 6.7 mg (antimycin A) or 11 mg (NOHQO) per petri dish culture (analyzed in duplicate) containing 13 ml medum.

dria by acting apparently between cytochromes b and c_1 (Tyler *et al.*, 1965), the participation of this site in the light-dependent pathway is implied if a common inhibition mechanism is involved.

Cyanide (10^{-4} M) causes 62% inhibition, and 2,4-dinitrophenol (DNP) (10^{-4} M), 80% inhibition of the ATP peak level after 3 hours of illumination. Both inhibit sporulation and growth strongly. The sensitivity of both the growing and sporulating mold to cyanide indicates the importance of mitochondrial terminal respiration in maintaining the plasmodial ATP level. Light would thus appear to effect an alternative electron transport and phosphorylation pathway through an antimycin-sensitive site, apparently not a part of the dark respiration but whose function inhibits this respiration.

The normally observed transient peak is not considered as an "energy reservoir" but more likely as the resultant of synthesis and utilization rates, or as a progressive acceleration in light reaction rate. The ATP level may act, for example, through the ATP/ADP to regulate the subsequent metabolic pattern. This conclusion is strongly supported by the subsequent observation (Daniel, 1965) discussed in the following section, that light strongly inhibits respiration of both growing and sporulating plasmodia.

IV. INHIBITION OF RESPIRATION BY LIGHT

The most unusual biosynthetic deficiency of *P. polycephalum* is its inability to synthesize hematin (Daniel *et al.*, 1962d). Although growth and sporulation are strictly aerobic, the plasmodium survives and maintains motility, anaerobically or aerobically, without exogenous hematin for long periods of time. Since sporulation, growth, and respiration are quite sensitive to azide and cyanide (Kamiya, 1960; Ohta, 1954; Daniel, unpublished data, 1960; and Daniel and Rusch, 1962b), and mitochondria are a prominent cytoplasmic component, the tentative conclusion was that a mitochondrial respiratory pathway furnished aerobic energy. The conclusion is strengthened by the study of respiration of isolated mitochondria (Daniel, 1965) and the observation of a cytochrome series by low-temperature split-beam spectrophotometry of isolated mitochondria.*

These findings are of particular interest in view of the observation (Daniel, 1964c) of a marked differential sensitivity between growth and sporulation to antimycin A. Table II compares the effect of both antimycin A and NOHQO.

The operation of a differential electron transport pathway during morphogenesis was thus indicated. Measurement of oxygen uptake (respiration), measured polarographically, confirmed that growing plasmodia or mitochondria derived from them were not inhibited by antimycin A at 2×10^{-5} M by more than 15%, although respiration of plasmodia prepared for sporulation was also resistant.

The effect of light on respiration, however, provided at least a provisional explanation, suggested by its effect on plasmodial levels of ATP and reduced pyridine nucleotides (PNH).

Figure 5 shows this effect when a submerged culture suspended in a salts or growth medium was illuminated with an unfiltered tungsten lamp. An altered respiration rate was established rapidly within 15 seconds. The Clark electrode is sensitive to portions of the visible spectrum and was protected with aluminum foil. The spike effect is an electrode artifact from the light source. When the light was turned off, a partial or complete restoration of the dark rate occurred. In some experiments with growing plasmodia depending on the time schedule of intensity increase, a tendency to undergo damped oscillations was observed as the inhibited state was approached.

The normal dark respiration (1.0–2.5 mg protein/ml) is inhibited approximately 50% at 10^{-4} M cyanide and completely at 10^{-3} M, but not by 10^{-5} M antimycin A. Antimycin A (10^{-5} M), added prior to or during illumination, allows partial but usually not complete reversal of inhibition with sporulation plasmodia.

Light also inhibits the endogenous respiration of plasmodia prepared for sporulation (Fig. 6). Inhibition is again rapid. At approximately

* The author is greatly indebted to Dr. Walter Bonner for this spectrum.

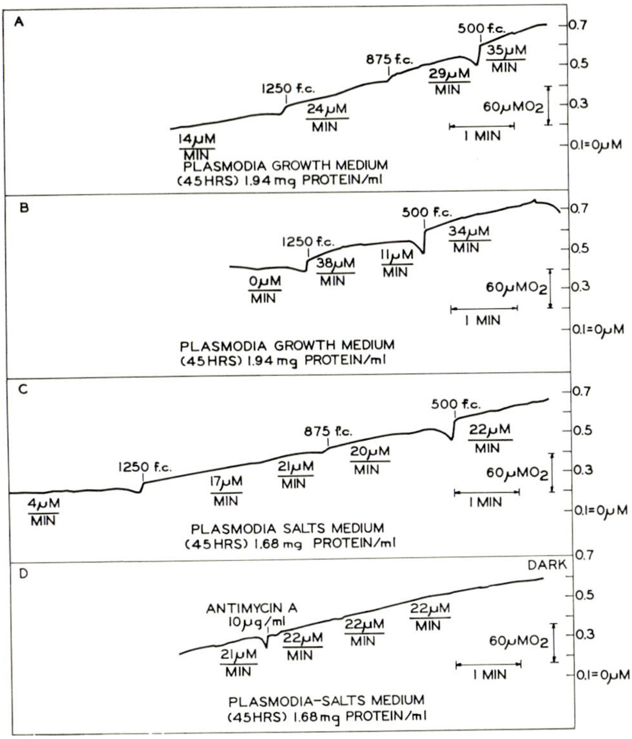

FIG. 5. Inhibition of respiration by light: growing plasmodia. Growing plasmodia were resuspended either in fresh growth medium (1% glucose), or its salts base and 2.5 ml aliquots were removed to the polarigraph cell. Respiration was measured with the Clark electrode at 25°C. The electrode stem was shielded from light with aluminum foil. Suspension was maintained with a magnetic stirrer. Light source was a 75 watt tungsten lamp.

1440 ft-cd or less, inhibition is complete in most preparations and in experiments lasting 5 minutes or less, it is at least partially reversible in the dark. This agrees with the observation (Fig. 6b) that plasmodia, exposed to light for the complete 4-hour induction period, regain approximately the original respiration rate when returned to the dark and have about the same sensitivity to light, although after 4 hours some decline is occasionally found.

Although respiratory rates measured in the polarograph chamber cannot be precisely applied to the plasmodia *in situ,* it appears that light controls the respiratory rate in a reversible manner, implying that either the inhibition per se, or a light-dependent alternative pathway interacting with the respiratory apparatus, participates in the morphogenic shift in metabolism. The latter possibility is compatible with the ob-

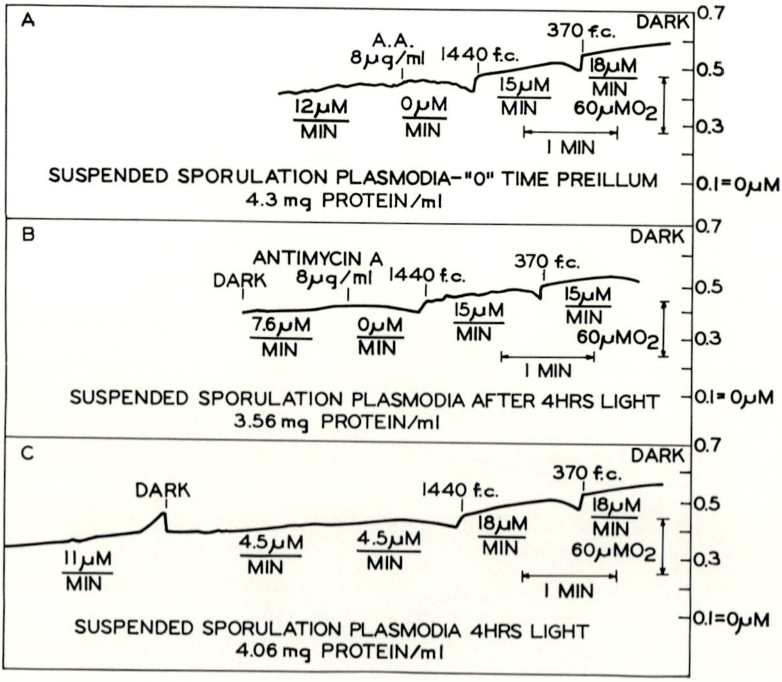

FIG. 6. Inhibition of respiration by light: sporulating plasmodia. Surface plasmodia prepared for sporulation (5 days old) (see also Fig. 1) were removed from the filter paper support and resuspended in 2.5 ml fresh sporulation medium in the polarigraph cell at 25°C and respiration measured as in Fig. 5. A, plasmodia measured before illumination; B and C, plasmodia measured after 4 hours of illumination.

served ability of antimycin A to at least partially reverse the light effect. Either effect would explain inhibition of growth.

The effect is also obtained with mitochondria isolated from growing plasmodia, as shown in Fig. 7. Mitochondrial respiration of malate + glutamate is strongly inhibited by light at an intensity approaching the minimum (200 ft-cd) for 4-hour induction. Inhibition also occurs in the absence of added ADP. The mitochondrial inhibition occurs rapidly and is at least partially reversible in the dark. Also, particularly interesting is the ability of DPNH to reverse the light inhibition.

Although the dark mitochondrial oxidation of succinate, DPNH, or malate + glutamate is antimycin-insensitive (10^{-5} M), its sensitivity to light is diminished by antimycin A. This agrees with the fact that the sensitivity of sporulation but not growth, is sensitive to antimycin A and NOHQO. On the basis of this evidence, the site of antimycin A

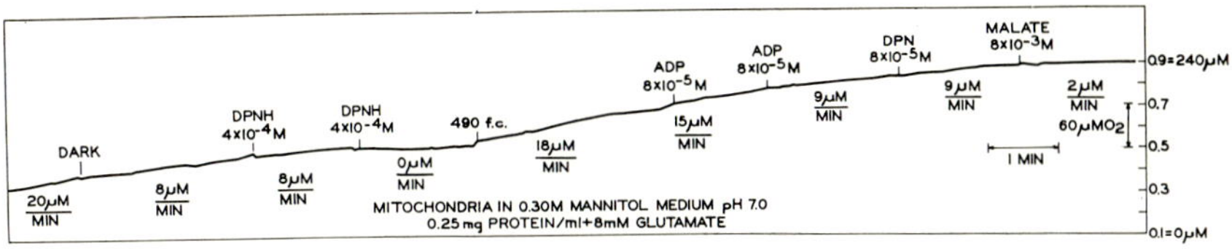

FIG. 7. Inhibition of respiration by light: mitochondria. Mitochondria were isolated from growing plasmodia in 0.3 *M* mannitol—5 m*M* ascorbate—5 m*M* EDTA—0.1% bovine serum albumin V, pH 7.2; for measuring respiration mitochondria were suspended in 0.3 *M* mannitol—20 m*M* KPO_4, 5 m*M* $MgCl_2$—20 m*M* KCl. Other conditions are as in Fig. 5.

action after mitochondrial cytochrome b and before c_1 in mammalian tissue (Tyler *et al.*, 1965; Estabrook, 1962), would seem not to participate in the main *P. polycephalum* respiratory chain but would function in the light-dependent metabolism, possibly forming a branch pathway or short circuit from the main chain. The ability of antimycin A to antagonize light inhibition of growth also suggests that the initial light reaction is present in both the growing and presporulation states, but that the metabolic consequences of this function are quite different in the two environments (as well as their apparent ability to transmit this energetic information). It also implies similar action spectra for these two light effects.

Since respiration appears to be reinstituted rapidly in the dark (Fig. 6), interruption by light may occur by diversion of the hydrogen or electron flow into the alternative pathway, the function of which might further alter the dark respiration through metabolites common to the two pathways. The identity and location of the photoreceptor and its role in respiratory inhibition can more nearly be defined when an action spectrum and the dark pathways of mitochondrial respiration have been exactly defined. The direct effect of light on mitochondrial preparations indicates that the mitochondria are the probable site of the light reaction, the effect of which is then transmitted to react with extramitochondrial systems including the yellow pigment (Daniel, 1964c). Although actinomycin D does not affect the mitochondrial inhibition, it does affect subsequent events during the light period. The mitochondrial preparations which contain little or no yellow pigment do contain 6 to 8 μg DNA/mg mitochondrial protein (see also Guttes and Guttes, 1964) and approximately 7 μg RNA/mg mitochondrial protein. External evidence of these reactions is reflected by the pH of the medium which increases as illumination proceeds (Table IV), but is stationary in the following dark period during actual sporangial development. Rakoczy (1965) has recently determined the sporulation action spectrum of a related myxomycete.

TABLE IV

INCREASE IN pH OF CULTURE MEDIUM DURING ILLUMINATION[a]

Illumination period (hours)	pH change during illumination	pH increase
0	—	0.0
2	6.20 → 6.35	0.15
3	6.05 → 6.40	0.35
4	6.30 → 7.00	0.70

[a] Cultures 4 days old at "0" hours illumination; Light intensity ≅ 1800 ft-cd using GE "cool white" 40 watt fluorescent lamp. 3/3 cultures sporulated; 0/3 cultures illuminated in Corning red "actinic" flasks sporulated, but no change in medium pH. Average medium pH when prepared (0 day) = 5.3.

Gordon and Surrey (1960) have reported the effects of red and far-red light on rat liver mitochondrial phosphorylation; however recently Ikuma and Bonner (1964) were unable to verify these observations.

V. EVIDENCE OF LIGHT-DEPENDENT OXIDATION-REDUCTION REACTIONS

Two reactions have been observed which may indicate the presence and nature of electron transport reactions arising from the light inhibition of respiration.

The first is the reduction of inorganic iron. α,α'-Dipyridyl (AAP) (10^{-3} *M*) and *o*-phenanthroline (3×10^{-4} *M*) inhibit sporulation when added to the culture medium just prior to illumination (Table V). AAP allows the light-specific accumulation of the AAP·Fe^{++} complex in both the medium and in the plasmodia of illuminated cultures (Daniel, 1964c), indicating a light-dependent reduction of Fe^{+++} (Table VI).

In addition, AAP affects the plasmodial ATP level in a manner quite similar to that of antimycin A or NOHQO (Table VII), indicating a possible link between the above Fe^{+++} reduction and the light-

TABLE V

SPORULATION INHIBITION BY IRON CHELATORS[a]

Inhibitor concentration	Sporulation frequency	
	α,α'-Dipyridyl	*o*-Phenanthroline
1.0×10^{-3} *M*	0/6	0/6
3.0×10^{-4} *M*	3/6	1/6
1.0×10^{-4} *M*	5/6	3/5
None	6/6	5/6

[a] Inhibitor added 15 minutes before illumination at 5 days; plasmodia grown and sporulated on media containing 3×10^{-4} *M* Fe^{++}.

TABLE VI

FORMATION OF α,α'-DIPYRIDYL-Fe^{++} COMPLEX IN THE MEDIUM OF ILLUMINATED CULTURES[a]

Time (hours)	Light (OD 520 mμ)	Dark (OD 520 mμ)	Light[b] (−Fe medium) (OD 520 mμ)
0.0	0.000	0.00	0.000
0.5	—	—	0.144
1.0	0.106	—	0.212
1.5	0.145	—	—
2.0	0.183	0.00	0.269
2.5	0.216	—	—
3.0	0.216	—	—
3.7	0.300	—	—
4.0[c]	0.317	0.005	0.377
	(3.4×10^{-4} *M* Fe)[d]		(4.0×10^{-4} *M* Fe)[d]
5.0	0.276	—	—

[a] Cultures (2.5 mg protein per dish) 5 days old at "0" hour when 10^{-3} *M* AAP was added.

[b] Cultures on −Fe medium transferred, after washing from normal medium (3×10^{-4} *M* Fe when cultures prepared), to medium free of added Fe containing 2×10^{-3} *M* AAP protein per dish = 3.3.

[c] Light off at 4.0 hours.

[d] Calculated from E_M for AAP·Fe^{++} = 9.4×10^3.

TABLE VII

EFFECT OF α,α'-DIPYRIDYL ON PLASMODIAL ATP LEVELS[a]

Illumination period (hours of light)	Control: no addition (10^{-2} μ*M*/mg protein)	α,α'-Dipyridyl (2×10^{-3} *M*) (10^{-2} μ*M*/mg protein)	% of control at "0"
0.0	0.46	—	—
0.5	0.41	0.34	73
1.5	0.47	0.23	48
3.0	0.89	0.35	77
4.0	0.76	0.086	19

[a] Inhibitor added 15 minutes before illumination; 2×10^{-3} *M* AAP completely inhibits sporulation. Plasmodia plus the inhibitor showed gross toxicity in the light but not in the dark after 4 hours. Average plasmodial protein content was 4.8 mg per petri dish culture (analyzed in duplicate) containing 13 ml medium.

dependent ATP increase. The function of iron in the light-driven metabolism is further indicated by the dependence of sporulation on exogenous iron seen in Table VIII.

Photoreduction of Fe^{+++}, especially in the presence of sulfhydryl (SH) compounds, is strongly stimulated by nonenzymatic preparations of yellow

TABLE VIII

IRON REQUIREMENT FOR SPORULATION[a]

Cultures transferred to no iron sporulation medium at:	Sporulation frequency
Harvest time (0 days)	7/15
5 days (in dark) (15 minutes before illumination)	3/12
Control (+Fe at 0 day)	12/15

[a] Plasmodia grown and sporulated on medium containing 3.0×10^{-4} *M* Fe^{++}. No iron medium prepared without added Fe^{++}.

pigment which are bleached in the dark by the Fe^{++} produced (Daniel and Rusch, 1962b). The light-dependent formation of AAP·Fe^{++} (Daniel, 1964c) by plasmodia suggests that Fe^{+++} reduction also participates in the *in vivo* light reactions.

Although the yellow pigments appear able to mediate Fe^{+++} reduction by a nonenzymatic light reaction, *in vivo* Fe^{++} may be formed not at the pigment particle site where bleaching presumably occurs, but for example, in the mitochondria or at another cytoplasmic site. Thus *in vivo* Fe^{+++} reduction may lead to pigment bleaching without being pigment mediated, or less likely if bleaching is reversible, it may be effected by reverse operation of the pigment mediated nonenzymatic reaction. The latter possibility now appears to be unlikely.

The second reaction suggesting the occurrence of light-dependent electron transport involves the formation of sulfhydryl (SH) groups. It has been recently observed (Daniel, 1963; Daniel, 1964c) that *N*-ethylmaleimide (NEM) but not other SH reagents, such as *p*-chloromecuriphenylsulfonate (PCMPS), Cd^{++} or arsenite, causes relatively specific and rapid release of the yellow pigments, indicating a function of SH metabolism in pigment binding. The other SH reagents do inhibit sporulation, usually causing delayed sporangial pigmentation. In light, the release by NEM is accelerated considerably (Table IX). Since good recovery of total pigment is obtained, pigment bleaching which normally occurs during illumination is strongly inhibited by NEM. This is confirmed by the high pigment recoveries obtained when NEM is added at later times during the light period. Whether or not the pigments are primarily involved as photoreceptors, these observations point out that (*a*) SH* groups are involved in binding of the pigments to particulates, and (*b*) light accelerates formation or turnover of SH groups, and in doing

* NEM has been shown to react with groups other than SH (Smyth *et al.*, 1960; Flavin, 1965), and therefore the NEM pigment release may also involve other groups. It must also be remembered that specific SH groups can show reaction specificity toward particular SH reagents.

TABLE IX

Release of Yellow Pigments from Plasmodia by *N*-Ethylmaleimide[a]

Time (minutes)	Yellow pigment released (total OD/culture)	
	Dark	Light
0	0.0	0.0
15	15.3	18.9
30	36.3	51.3
60	65.7	59.1
120	77.0	63.9
240[b]	83.1	66.9

[a] 10^{-3} *M* NEM added to dish culture at "0" minutes (5-day-old culture) and medium samples removed at indicated times, diluted in 4% TCA in 50% aqueous acetone and read at 415 mμ; total OD represents pigment content of medium for 1 culture of average 3.9 mg protein. Other culture conditions as in Fig. 1. At "0" minute pigment OD per culture = 92.3. Pigment recovery (4 hours) (medium + plasmodium) = 96% light; 131% dark; *in vivo* pigment bleaching by light = 56% at 4 hours.

[b] Lights out at 240 minutes.

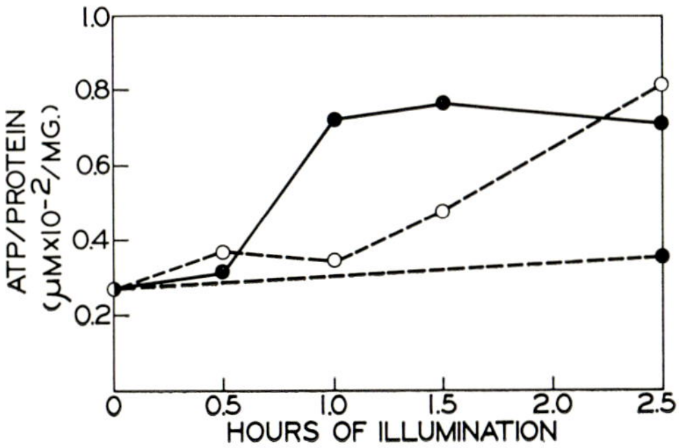

Fig. 8. Glutathione enhancement of the light-dependent rise in plasmodial ATP level. See Fig. 4 for culture preparation and experimental details. Reduced glutathione (3×10^{-3} *M*) added as indicated at "0" time. Experiment covers only the first 2.5 hours of the normal 4 hour illumination period (Fig. 4). ○----○, light control; ●——●, light + glutathione; ●----●, dark control.

so promotes photobleaching. Both conclusions are in agreement with the nonenzymatic SH-stimulated bleaching.

Direct SH involvement is further indicated through the specific enhancement of the light-dependent ATP level by exogenous reduced glutathione, shown in Fig. 8. In the presence of reduced glutathione, the maximum ATP level is attained much more rapidly during the "lag" period of the control. In the dark, reduced glutathione has little effect at 2.5 hours, or at other times tested in similar experiments. Exogenous, reduced glutathione specifically hastens the appearance of the light-dependent ATP peak without affecting the dark level. NEM (10^{-4} *M*) strongly inhibits the control pattern. Reduced glutathione may function directly as a hydrogen donor, or indirectly by modifying enzyme activity necessary for the expression of the ATP peak. In either case the result supports the conclusion from the light stimulation of NEM pigment release, that a photooxidation or light-accelerated turnover of SH groups occurs. A net increase in SH concentration occurs at least in the presence of NEM. The effect may be related to that previously observed (Petrack and Lipman, 1961; Petrack *et al.*, 1965), in which depending on the concentration, reduced glutathione, cysteine or lipoic acid stimulated either photophosphorylation or ATP photohydrolysis.

The function of the new light pathway may be to limit the energy available for function of the dark endogenous state, by inhibiting dark respiration while simultaneously furnishing a new and specific energetic pathway supporting transition to the new spore state. The consequences of such a shift, if protein SH groups are involved, may also be the critical alteration of mitochondrial and other structure (protein subunits) to effect changes in enzyme patterns and even possibly to effect changes in repressor function (see also Section VII).

The evidence so far obtained is compatible with a light-induced reduction of —S—S— groups which affect pigment binding and bleaching, and which may also participate in an electron transport pathway sensitive to antimycin A and NOHQO, the function of which is possibly coupled to ATP formation as evidenced by reduced glutathione light-dependent stimulation of the plasmodial ATP level. Functions of such a pathway would oppose or compete with the normal dark pathways of respiration.

Although the above observations indicate that iron and SH function in the transient antimycin-sensitive light-dependent ATP increase, definition of the complete pathway, its location and exact role in morphogenesis, will depend upon extension of these observations to mitochondria and to recombined cell fractions.

The recent observations of Butow and Racker (1965), implicating the function of an inorganic iron sulfide component with mitochondrial oxidative phosphorylation in the DPNH-flavoprotein region, may be pertinent to the preceding observations.

VI. METABOLISM AND NATURE OF THE YELLOW PIGMENT

A complex of apparently related yellow pigments, essentially if not completely localized in individual cytoplasmic particulates, occurs in all stages of the life cycle except the spores. There is some evidence that this may function as a photoreceptor (Gray, 1953; Daniel and Rusch, 1962a), though the present effect of light on respiration modifies this possibility. During the photoinduction period, considerable pigment is bleached to unidentified products. Paper chromatography of pigment extracts reveals 3 distinct normal pigment components. Although all 3 undergo photobleaching during the illumination period, only one is preferentially and almost completely lost. However, cultures that have lost the ability to sporulate, or nearly so because of their nutritional history, exhibit much greater, more rapid, and less component-specific bleaching.

Later in sporogenesis the remaining yellow pigments are lost, although the particulate structures of origin can be identified in electron micrographs of the spores. Photobleaching of these extracted pigments also occurs nonenzymatically (Daniel and Rusch, 1962b) but requires Fe^{+++} and a source of SH or, in the dark, Fe^{++} (see Section V).

The chemical nature of these pigments until recently was completely unknown. Recent work (Brewer, 1965) indicates that 3 isolated components are conjugated polyenes containing nitrogen and a carboxyl group. One component contains a strongly basic group. None contain peptide or carbohydrate components. One component, on catalytic hydrogenation, yielded a crystalline derivative, the first reported for this pigment group.

Observations in this laboratory (Daniel, unpublished data, 1965) indicate the presence of a carbonyl group as an essential part of the chromophore. A pigment component absorbing maximally in acid solution at 380 to 390 mμ and in alkaline solution at 415 mμ reacts, as measured by loss of light absorption in the 390 and 415 mμ regions, with hydroxylamine, 2,4-dinitrophenylhydrazine and semicarbazide. The reaction occurs slowly with semicarbazide, but goes to completion with a single pigment component isolated chromatographically. Semicarbazide reacts with Schiff bases, especially in the presence of certain free amines (Cordes and Jencks, 1962).

The pigment is also reduced (loss of 380–390, 415 mμ absorption) by $NaBH_4$, sulfite, dithionite, and Fe^{++}-mercaptoethanol. The tentative conclusion is that destruction of the carbonyl group results in loss of the characteristic absorption.

It is of interest that the indicator yellow derivative, obtained by the action of light on rhodopsin (Lythgoe, 1937; Pitt *et al.*, 1955), and related retinene Schiff base derivatives show a number of similar spectral and chemical properties although molar absorptions and solubility characteristics differ. Similar spectral and pH-dependent shifts are also characteristic of the Schiff bases of some pyridoxal derivatives and enzymes (Snell,

1962). The quaternary form of a Schiff base would explain the strong basic nitrogen function observed (Brewer, 1965).

The isolated pigment components give strong typical color reactions with concentrated sulfuric acid, indicative of conjugated polyenes (Brewer, 1965). We have recently observed a small but definite bathochromic shift from 416 mμ to 430 mμ when one purified pigment component reacts with $SbCl_3$-CH_3OH.

The characterization by Brewer (1965) in Strong's laboratory and the additional characteristics summarized above suggest that the yellow pigments are polyenes with at least some components containing a marked or inhibited carbonyl group, very possibly a Schiff base. The complete characterization of these pigments and their role in the light-dependent phenomena of this organism will be of great interest.

VII. MELANOGENESIS

Another but delayed effect of NEM is activation of a polyphenol oxidase (Daniel, 1963). Plasmodia prepared for sporulation but treated with NEM, first lose pigment, a phenolic component associated with pigment, and then after several hours begin to darken and lose viability. No other morphological or cytological indication of sporangial formation occurs. The effect is intensified by, but not completely dependent on exposure to light. It is, however, dependent on the nutritional conditions, since plasmodia growing in submerged cultures readily lose pigment and the phenol, upon NEM addition, but do not melanize even when incubated in a nonnutrient medium before NEM treatment.

These induced cultures will utilize a variety of added mono- and dihydric-phenols for synthesis of melanin-like pigments, including tyrosine, 3,4-dihydroxyphenylalanine (dopa), catechol, *p*-cresol, tyramine, or arterenol. None of these compounds, including 3-hydroxykyneurenine or indoles, can be detected chromatographically in various plasmodial extracts. Tyrosine can be detected in extracts of growing plasmodia. The unidentified plasmodial phenol, associating with the yellow pigment during chromatography and present in relatively large amounts, decreases just prior to and disappears during the normal melanization of sporangia. The phenol separated by paper chromatography is a substrate for purified (4200 units/mg) mushroom polyphenol oxidase and forms an intermediate with a spectrum maximum at 470 mμ, quite similar to dopachrome.

Phenolase activity can readily be detected in the NEM-treated mold prepared for sporulation. However it can not be detected, except during normal sporangial development, in untreated plasmodia or their homogenates at any other stage of the life cycle, by use of the chronometric assay (Miller *et al.*, 1944). It is of particular interest that with homogenates of untreated plasmodia, NEM alone is not sufficient to induce the phenolase, but, in addition, simultaneous incubation with 5 *M* urea (Guest and Horowitz, 1958) is required. The supernatants from such crude preparations act enzymatically on dopa or tyrosine to produce a product whose spectrum closely resembles that of dopachrome.

Finally, the spectrum of solubilized, reprecipitated, melanin pigment from spore wall preparations, having maxima at about 280 mμ and 470 mμ, with a strong shoulder at 340 mμ, bears only slight resemblance to the pigment derived from the NEM-treated plasmodium, utilizing exogenous L-tyrosine or endogenous substrates only. According to Yasunobu *et al.* (1959) the spore pigment spectrum would indicate the presence of phenolase-reacted tyrosine or phenol groups, in protein or polypeptide chains, (340 mμ shoulder) and also in N-terminal positions (470 mμ absorption). Thus, the spore pigment is probably a melanoprotein derived from both free phenols and protein tyrosine. This is compatible with the melanization of plasmodial protein observed in NEM-treated plasmodia which have lost free phenol, the reactivity of the plasmodial phenol with mushroom polyphenol oxidase, and the presence of appreciable biuret but low Folin protein in purified spore-pigment preparations. The extensive participation of plasmodial protein in the formation of the spore walls and other sporangial structures, is also indicated by the rapid fall in Folin protein (Fig. 1) during the final stages of sporangial formation. Protein determined by the biuret reaction does not show this rapid drop.

The effect of NEM, which provides evidence for the important role of SH participation in the light reaction and for its role in pigment bleaching as affected by light (*in vivo* and *in vitro*), connects this oxidation-reduction action with a subsequent and late morphogenic event, melanogenesis, which occurs in the maturing spore walls as a kind of "wrapping up" or enclosure reaction delimiting new mononucleate units of protoplasm, the maturing spores.

If light serves as a means for accelerating or driving an SH-mediated, oxidation-reduction system, exhausting a metabolic electron source required for a metabolic poising action dependent on respiratory electron transport, a controlled shutdown of this system could result in release of yellow pigment and activation of an enzyme activity, phenolase, the substrate of which had previously followed an alternative pathway. In this way the effectiveness of the primary event, the initial light reaction, would be dependent on the nutritional history for induction potential, e.g., by reducing the specific endogenous energy resources, and would control a late step involving activation of a new enzyme, very possibly in this particular case, without the mediation of a new protein synthesis (Horowitz *et al.*, 1961). The requirement of urea and NEM for *in vitro* phenolase "activation," suggests the participation of protein subunits (e.g., Grossberg *et al.*, 1962) in this process, either by inactivation of an inhibitor, or by the actual production of a new activity.

Horowitz (Horowitz *et al.*, 1961; Horowitz and Metzenberg, 1965) has suggested the role of a protein repressor in the control of tyrosinase expression during sexual morphogenesis of *Neurospora*. The repressor stability is postulated to decrease under starvation conditions and in the presence of protein synthesis inhibitors, conditions which allow expression of the new tyrosinase activity. The action of NEM mimics this sequence, causing premature and lethal expression of this new enzyme activity.* Energy source starvation, required for induction of catechol oxidase (Imamoto *et al.*, 1964), appears to result in the ability to form the mRNA necessary for this sequence of enzymes.

Another example of light-dependent alteration of enzyme activity was seen in its effect on glucose transport, described in a previous section. This can be legitimately regarded as inactivation or loss of an enzyme activity of the glucose permease system, an effect maximal within 15 minutes of illumination. Thus the role of enzyme activity or function, but not necessarily of specific protein loss (Wright, 1963), as a direct consequence of light action, is unavoidable. The same result of decrease in activity was seen even more clearly in the earlier section on respiration. At the same time it must be emphasized that informational metabolism and new protein synthesis, does seem to play an obligatory role in the complete expression of this morphogenic sequence, since although actinomycin D does not directly affect the 2 above events, it does block morphological expression evoked by the primary light events, and required for normal completion of sporulation (Rusch *et al.*, 1964; Daniel, 1964c). This role may well be important during the dark incubation prior to illumination.

Recently Horowitz and Metzenberg (1965) (see also Grobstein, 1963), while emphasizing the great impact of the Jacob-Monod hypotheses, called attention to the present limitations of these postulations (Monod and Jacob, 1961b), particularly as they apply to morphogenesis and the mRNA concept (Monod and Jacob, 1961a; Jacob and Monod, 1963).

Although many or even most processes, termed differentiative or morphogenic, may well involve synthesis (by way of an mRNA) of at least one new protein, it is by no means certain, or even likely, that such synthesis is the only important means of obtaining new catalytic activities. Although such new protein synthesis may remain an essential step

* NEM is not necessarily lethal, since simultaneous addition of diethyl dithiocarbamate will allow survival of a bleached (colorless) motile plasmodium.

of such processes, it can not always be regarded as the definitive event described by Jacob and Monod (1963). Rather, new protein formation would seem to be where applicable, the second of three general stages; the first being the occurrence of environmental changes allowing new information transcription (repressor-inducer function), the second, the actual information transcription with subsequent but possibly delayed expression, and third, the consequential metabolism accruing from this initial expression. The above three stages constitute the differentiating or transient state metabolism, leading from one differentiated steady state to another.

It must also be noted that although the nuclear genome is a central and possibly ultimate repository for the metabolic information of the cells, it is not the only functional location and may be supplemented by other loci to constitute a collective genome. Such loci occur particularly at centers of energy metabolism (Luck and Reich, 1964; Sager and Ishida, 1963).

The coordinate if not pivotal role of specific energy metabolism, as evidenced for example, by the rapid effect of light on respiration in switching or altering major pathways of metabolic energy, acts as an energetic barrier limiting or promoting the transitions in enzymatic patterns required for mobility between differentiated states. This results most likely from transient instability (oscillation) of such pathways, occasioned by a fluctuating environment, setting the metabolic stage for resolution into a new steady state. Although such changes involve fluctuation in enzyme activities they do not necessarily involve changes in enzyme concentration (Wright, 1963). Ample evidence of the variety and the subtlety of energetic metabolism, and its ability to modulate the flow of substrate-available energy into useful metabolic channels is available (Chance, 1963; Mitchell, 1963; George and Rutman, 1960; complete reference cited in Butow and Racker, 1965). The concept often implied or involuntarily invoked, that of a central rather ill-defined energy reservoir or source of ATP, cannot be constructively invoked. Recent studies (Fan *et al.*, 1964; Nakada and Fan, 1964; Soffer and Gros, 1964), suggest some of the possible effects that alteration or limitations in energy metabolism may have on information transcription and expression and draw attention to the regulatory role of energetic barriers as elements of information.

An interesting application of this point of view can be made in relation to the mechanism of mitosis and the suggestion (e.g. Lark, 1963) that mitosis induction occurs when at least partial single strandedness of DNA occurs. The observations of Peller and Barnett (1962) concerning the thermodynamic instability of the large DNA molecules indicates that both the double strandedness and the integrity of the constituent DNA molecules require continuous energetic support *in vivo* which if interrupted or diminished could initiate fragmentation.

It is also of considerable importance that a presumably important resultant of the sporulation process is the occurrence of meiosis, the formation of sexual unicellular spores from an otherwise coenocytic organism. Since mitosis does occur during the dark preillumination period, though more slowly (Guttes *et al.*, 1961), light may effect a kind of differentiation sequence comparable in biochemical function to a single mitotic growth cycle, but in which net macromolecular synthesis is at least partly replaced by endogenous reutilization (Daniel and Baldwin, 1964) for morphogenic function, blocking DNA synthesis, and terminating in karyokinesis without DNA synthesis.

VIII. SOME OBSERVATIONS ON ENERGY-CONTROLLED METABOLISM RELATING TO MORPHOGENESIS

A. Flow of Energy

The environment of a living organism is generally considered, almost by necessity, as the ultimate source of change. The transformation (biosynthesis) of environmental matter into living cells, and their subsequent material flux is supported by free energy supply which is also supplied by the environment (including other cells). The material flux is thus accompanied by one of energy. The energy source may be other matter, organic or inorganic, light and heat. A given vital cellular function will depend ultimately for its regulation and existence upon fluctuations in these critical environmental quantities.

This flow of energy entailing a dismutation of free energy to support biosynthesis with its concomitant return to the environment in the degraded form of products higher in entropy content (Lumry and Eyring, 1954; Prigogine, 1961; Lumry and Spikes, 1960), is regulated by the cell's ability to couple available free energy to synthetic function. The immediate source of the information for this function, is considered to be the collective cell genome which in response to environmental change regulates material and energy exchange between cell and environment. Simplified, there are two controlling factors: the cell's ability to recognize a specific component (gene availability and transcription), and recognize a specific component (gene availability and transcription), and its subsequent ability to provide the energetic capacity or capability for the expression of this information as the synthesis of single or series of specific enzymes and other proteins. The enzyme's subsequent utility is then governed by its kinetic performance (rate), and its coupling function in utilizing free energy flux for cell function. The flux rate of free energy through the cell, and the efficiency of its utilization for cell fabrication are two factors which in the evolution of cell economy apparently have been coupled for increased survival value (Lotka, 1922).

A shift from one metabolic pattern to another, potential (recognition) or actual (expression), depends on a prerequisite source of free energy. Any number of shifts might be allowed within a given energetic framework, but others would require a characteristically different one. Thus certain possibly large groups of enzymes corresponding to a characteristic metabolic pattern would be dependent on a given source of free energy, distributed and modulated through metabolic networks (e.g., Bradley and Calvin, 1956). Any two sources would support a large number of common components, but in addition, each would forbid or allow other components. Such components would characterize the system by virtue of their ability to act as or furnish repressor-inducer metabolites controlling the induction of enzyme sequences or the required synthesis or the availability of coenzymes. Utilization of a given primary exogenous source of free energy would restrict cellular response to others, for example diauxie phenomena and catabolite repression (Mandelstam, 1962; Nakada and Magasanik, 1964). The solicitation of the genome would then appear as energy controlled. The source of energy could be radiant or organic, of exogenous or endogenous origin, and may or may not support replication of the organism, that is "symmetrical" growth.

An interesting example of the effect of shifting external free energy sources has recently appeared (Imamoto *et al.*, 1964). In this case, formation of mRNA apparently for catechol oxygenase and subsequent enzymes, is dependent on exhaustion of an exogenous energy source, other added sources suppressing this response. Any of the inhibiting compounds as well as catechol, could support growth. This may be particularly pertinent to the appearance of a phenolase during *Physarum* sporulation, as well as its formation induced prematurely by NEM. In accord with this example (see also Section VII), energy dependence of genome operation may lie in function of the energy source as the inducer of sequential enzyme synthesis making metabolically available a new or "secondary" source of free energy having a different specificity. In addition, catabolic or differential anabolic exhaustion of e.g., stored energy sources may be equated with exhaustion of repressor molecules to allow response to inducer. In turn this may depend on the availability of key coenzymes (ATP, TPN, DPN, etc.) regulated by competition between pathways as well as compartments. Although in the case of net catabolism such a mechanism must operate by competitive redistribution of coenzymes, for a system undergoing symmetrical or differential growth, a similar shift would necessitate the net synthesis of coenzymes. The photoreactivation and dark recovery of genetic damage may be direct examples (Setlow, 1964; Muhammed, 1965). In a somewhat different way, for example, the operation of the nucleotide reductase for DNA synthesis (Cohen and Barner, 1962; Holmgren *et al.*, 1965) may likewise be affected. The frequency and therefore the effective existence (probability) of such events could be regulated through networks of variable capacities driven by a primary effectively noncyclic pathway, such as a respiratory chain terminal oxidase, photochemical reaction or glycolysis. It is also of interest that respiration and oxidative phosphorylation resembling that in mitochondria have been observed in nuclei (Betel and Klouwen, 1964).

The energy source would then assume the significance of an information factor, and the genome response (information transcription) one of modulating the energy flux between the cell and the environment. In this sense, the genetic content would appear as a memory (by mutation and selection) of cyclic environmental changes favoring survival.

When a recognizable energy source occurs in excess (maximum flux), reduplication of the cell unit would be favored under otherwise ideal and constant conditions, at a maximal rate for its minimal components (Daniel and Baldwin, 1964). Such conditions may not necessarily result in efficient free energy utilization but might result from associated metabolic flux requirements, (Chance and Williams, 1956) and possibly have survival value by allowing maximum production of cell units with minimal entropy content. The tightness of coupling of electron transport to phosphorylation, for example, may vary with functional cell state (Chance and Williams, 1956) demanded by the environment. Under growth-rate-limiting conditions, which normally prevail, and in the extreme case where the cell is totally dependent on internal or endogenous resources, the surviving or most "fit" system would shift metabolism to minimize the conversion of cell substance for energetic use. Such would appear to be the conditions preceding the development of light sensitivity for sporulation. This would result in a tendency to minimize the overall metabolic rate or intensity of metabolism, either by increasing the efficiency of energy yielding processes, or by metabolite recycling (turnover) efficiency. Diversity of cell function may in this way be preferentially curtailed with concomitant increased metabolic specialization. This tendency has been equated thermodynamically by Prigogine (1961), with an approach to a state of minimum entropy production, described as a characteristic of stable evolution. Fisher (1958) (see also Wright, 1964) has compared the rate of increase in genetic fitness of an organism with the tendency to decrease or minimize the rate of entropy flow. Such may be the state approached during the transition from the growing to the spore state.

This tendency taken to its limit would describe viruslike entities, characterized by the virtual lack of catalytic metabolism, but well suited either to inducing rapid replication in a suitable environment (cell) or to survival in others. Virus metabolism per se so far detected is that initiated by its interaction with host cell interfacial structures as, for example, with penetration of the T-even phages of *E. coli* (Kozloff and Lute, 1959, 1960; Wahl and Kozloff, 1962), and enclosure of virus during its biosynthesis at the cell membrane (Novikoff *et al.*, 1962). In the former case, a complement of virus-bound purine nucleoside triphosphates serves as substrate for the ATPase activity of the contractile protein participating in penetration of the host. In both cases, attention is called to the nature of the membrane and its special role in maintaining viral integrity and in host recognition.

B. Flow of Matter

When the cell is considered as a heterogeneous system maintained by membrane barriers, an important mode of energy dissipation is through molecular diffusion. Maintenance of this heterogeneity, through control of distribution of components between compartments, requires energetic support and applies to molecules of all sizes (Garfinkel, 1963) and inorganic ions. The source of metabolic energy will have a specificity appropriate to the given component or system. The function of most energetic pathways (of energy dispersal) must be such that by localized and oriented function on membranes, communication between metabolic compartments (membrane-bound spaces) is regulated, i.e., flows, gradients, and chemical potentials of water, substrates, coenzymes, and even catalysts appropriate to existing biosynthetic capacities. Membranes would thus appear to be the major sites of control for free energy flow and dissipation, and accordingly closely connected with primary sites of metabolic energy generation. Whatever this primary mechanism for the flux of a given component (species), this and other fluxes may also be coupled by diffusional interaction, the biological significance of which has been emphasized by Hearon (1950).

Hearon (1950, 1951) and Katchalsky and Kedem (1961) have emphasized the role of free energy dissipation by diffusion and the importance of this factor, especially as described by the Onsager reciprocal relation, for transport regulation of metabolism both for reciprocal transport by metabolites *not* directly related, and for the possible redistribution of catalytic activities ensuing from this in a heterogeneous system. Such pathways serve to energize compartmental communication and maintenance, particularly by governing the flux of primary free energy sources and therefore maintaining heterogeneity. A mechanism of this kind would also minimize the rate of entropy production due to diffusion.

Of particular importance in maintaining constant metabolic rates are the factors which govern the stability and mobility of catalytic activities. When protein or catalytic components are localized in a given metabolic or membrane-defined compartment, then the nature of the binding forces responsible for this localization and for transport of these components to their respective sites of function governs the stability of this state. When the diffusible small molecules serving as substrates for catalytic activity and resynthesis are not limiting, and their flows to these sites are maintained, the steady state will obtain. If the binding forces maintaining macromolecular orientation are also maintained, the catalytic pattern is stable and therefore the dependent metabolism would fluctuate with the substrate supply for catalytic activity, even if precursors and the machinery for synthesis of enzymes decline. If however, fluctuation of any one of these factors allows the enzyme itself (or coenzyme) to dissociate appreciably or to become a substrate of another system at a sufficient rate, a steady mobility and/or loss of this component will occur. The relocation or loss of the activity would disrupt the steady state and initiate a transition state from which a new steady state emerges, characterized by a rebalance of remaining activities accompanied in many cases by the appearance of one or more new catalytic activities. Since on one hand a high rate of net synthesis requires a high energy flux, and on the other, the high rate of catabolism furnishes a corresponding flux of carbon skeletons, also intermediate products of a high energy flux, the sites for generation of metabolic energy, for example, mitochondria in higher aerobes, become focal points for regulation.

Two recent examples emphasize this possibility. Lee *et al.* (1965) have reported that the thyroxine-dependent increase in α-glycerol phosphate oxidation appears to be due to the synthesis of new dehydrogenase. Similarly, the dependence of protein synthesis in isolated mitochondria on the state of mitochondrial respiration as it affects the synthesis of amino acids has been noted by Kroon (1963) (see also Slater and Tager, 1963).

The hydrolytic steps of such catabolism, for example, may well depend on changes in the configuration or environment constituting or neighboring the enzymatic sites catalyzing the corresponding synthetic reactions. The balance between synthetic and hydrolytic function also may be governed by availability of acceptor molecules for synthesis. Thus changes causing mobility of an enzyme may simultaneously lead to alteration in its function, for example as above, and give rise to centers for enzyme, coenzyme, and metabolite turnover, and catabolism, providing for the mobility of cellular components through "pools" of small molecules. This autodigestive facet of metabolism (Swift and Hruban, 1964) has many ultrastructural similarities to vacuolar ingestive processes, recently observed in certain slime molds (Hohl, 1965). The reutilization or cycling of cellular material under limiting nutritional or otherwise adverse conditions and the primary ingestion of complex substrates, have in common the digestion and retransport of substrates to sites of biosynthesis and respiration. The formation (and function) of membranes associated with these processes and dependent on net flux of substrates through cellular polymers may provide a source for renewal or net synthesis of membrane, destined for various cell organelles. These and other observations, infer a continuous and possibly obligatory fabrication of membrane during both the growing and the endogenous cell state. At the time of formation, such membranes would appear to be loci for hydrolytic enzymes and might represent a more efficient and versatile mechanism for substrate and cellular conservation than less efficient digestive mechanisms, involving the excretion of hydrolases. Van Lancker (1964) has briefly discussed the origin and localization of hydrolases in mammalian cells. The mechanism of catabolism is of particular interest in its relation to the conservation of limiting metabolites, i.e., those the cell can not synthesize.

In much of the foregoing discussion, the control of water availability and movement is implied. The organization of cellular water, certainly as heterogeneous as other cell components, is of equally great interest. External conditions affecting intracellular water content impose osmotic work loads on the cell and may therefore, under otherwise favorable conditions, limit cell function and growth by critically diverting the cell's energetic resources. Evidence for the ordered nature of cellular water has been discussed (Klotz, 1962; Trintscher, 1964).

At a molecular level, the interaction of the hydrophobic sidechains of proteins plays a central role in protein folding (Kauzmann, 1959; Klotz, 1960). The driving force provided by this interaction (e.g. Whitney and Tanford, 1962) and hydrogen bonding allows the folding of polypeptide chains and the ordering of adjacent water molecules. The protein molecules are folded in such a way that water does not have equal access to all molecular sites. *In vivo,* such a property may be of prime importance in protecting synthetic reactions having water as a possible reactant or product. Conversely, uncontrolled access of the active sites to water would favor the hydrolytic or reverse reaction. Thus by altering the protein environment at this site, metabolic direction between synthesis and hydrolysis could be regulated, as well as the local water concentration. Recently, more and more attention has been focused on the importance of such hydrophobic sites *in situ.* Mitchell (1961) has emphasized the likely role of this factor in the mechanism of ATP synthesis and hydrolysis. Sanadi (1965) has also drawn attention to the importance of this factor as evidenced by the action and specificity of inhibitors.

The affinity of the catalytic and structural environment for water thus focuses attention on the role of directed diffusion and transport of water, especially at or near the synthetic-hydrolytic centers, and thus the regulation of both the transport of water and hydrophilic molecules. The movement and organization of cellular water are obviously of great importance and participate strongly in controlling the rate and direction of diffusion into and within the cell. It follows that factors (metabolites, temperature, etc.) regulating protein and protein-lipid complex conformation within the limits of the established ultrastructure could also control mobility of associated water. Concentration-dependent binding of various ions and metabolites to these complexes would provide a means of regulating interaction with water, and therefore their own diffusion, possibly linked in this way with anisotropic catalytic activity. The diffusion rate of protons is particularly affected by the state of water (e.g. Klotz, 1962) which may well provide a means for selectively controlling the cellular diffusion and transport of this important metabolite.

Mitchell (1961, 1963) has offered a mechanism for membrane-localized energetically coupled transport, or vectorial metabolism. Specificity and direction were incorporated by virtue of the substrate dehydrogenases located on the membranes, the driving force arising from the anisotropicity of membranes to H^+, from respiration, and OH^- ions to effect the phosphorylation of ADP by dehydroxylation of inorganic phosphate. The theory of Mitchell emphasizes the coupling of proton transport with membrane-localized coupling of electron transport and phosphorylation. Klingenberg (1963) has recently extended this point of view for mitochondria as has Mitchell (1965). Observations recently presented by Chance (1965), however, do not support this view of proton transport.

Control of the cation and associated water fluxes in mitochondria does appear to be closely dependent on the availability of high energy intermediates (Packer *et al.*, 1963; Rasmussen *et al.*, 1965). Pressman (1965) has recently focused attention on the association between potassium, proton, and water transport during oscillatory contractions of mitochondria. Just how transport dependent on mitochondrial energy is mediated from the environment into the cell is not yet clear.

Recent observations have extended, at least cytologically, the close biochemical and structural association of oxidative with phosphorylative capacity in mitochondria to bacterial membranes (Abram, 1965; Van Iterson, 1965) and to the photosynthetic pigment of *Rhodospirillum rubrum* localized on invaginated extensions of a surface membrane (Cohen-Bazire and Kunisawa, 1963; Holt and Marr, 1965). These observations further emphasize the location of centers of energy metabolism at the sites of metabolite transport.

The development of membrane systems such as those studied during mitochondrial genesis in yeasts undergoing aerobic adaptation (Schatz, 1963; Schatz, 1965; Linnane and Wallace, 1964) very possibly offers a good example of energy-dependent differentiation induced by oxygen especially in relation to the suggested fission proliferation of mitochondria (Luck, 1963; Luck and Reich, 1964). Light-induced chloroplast development is another such example.

A particularly interesting example is the early changes in energy metabolism accompanying fertilization of sea urchin eggs (Epel, 1964a, b). The rapid increase in reduced pyridine nucleotides dependent on the initial conversion of DPN to TPN by DPN kinase, the induction of ion fluxes and changes in adenine nucleotides all prepare the egg for the intensification of synthetic activity of the developing embryo, preceded and dependent on these prior changes in energy metabolism. In a number of ways, the initiation of this developing system appears to require the mediation of many of the same pathways apparently functioning during the light induction of *Physarum* sporulation, but with the end result that the cell is activated from a relatively dormant state in contrast to the attenuation of a relatively active system in the case of *Physarum*. Thus, for two oppositely directed morphogenic systems the same basic pathways may mediate the development of these very different new metabolic directions most likely determined by the environmental resources available at the time the initial stimulus is applied. Both transitions depend on quite early and rapid changes in energetic metabolism.

Thus it is of interest that as the energetic resources available to the cell fluctuate in kind and quantity, the cell responds by elaboration of specific membrane structures (increased compartmentation) to initiate and direct the cellular flux of a new or altered energy source. When exogenous sources are exhausted, the same response may occur as the cell becomes dependent on catabolism for energetic support. The cell's ability to metabolically distribute and modulate this primary flow across membranes through energy-dependent substrate networks (e.g. Bradley and Calvin, 1956) must regulate the division of labor (kind and rate of work performed) depending on the substrates available, varying from most efficient replication of a "minimal" cell unit, to most efficient function for specialized biosynthesis or function (differentiation)—i.e., conditions different from those for maximum and most efficient duplication rates allowed by a given environment.

One of the striking changes during the sporulation process for *P. polycephalum* is the apparent dehydration or preferential loss of cellular water since the plasmodium is about 80% water and the spores (not including sporangial structure) approximately 80% dry weight. The rapid inhibition of mitochondrial respiration by light and the subsequent relatively rapid change in glucose permeability and the increase of the pH in culture medium, suggest the alterations of membrane function or synthesis as a possibly preferential consequence of the initial rapid but partial respiratory inhibition. These initial changes are reflected by the inability of the plasmodium to maintain its morphological unity and water content as demonstrated by the stepwise fragmentation first into globular sporangial primordia and finally into mononucleate spore units. This final fragmentation occurs when the oscillatory cytoplasmic flow ceases after sporangial erection is complete. The newly delimited cytoplasmic units are then enclosed in melanoprotein spore walls which functionally replace the original bounding membrane of the now disasembled plasmodium. The spores remain as highly condensed (contracted ?) protoplasmic units until favorable conditions for germination occur. Some evidence for the existence of the bacterial spore as a highly condensed protoplasmic network spatially but not metabolically in equilibrium with external water has been presented (Black and Gerhardt, 1962).

It is of interest that reactivation of this inactive state in the case of bacteria may require a specific energy source for germination (Freese *et al.*, 1964).

IX. SUMMARY

During the transition from the rapidly growing to the light-sensitive state on a nonnutrient medium plasmodia of *P. polycephalum* undergo selective niacin-controlled depletion of endogenous polysaccharides, proteins, and nucleic acids. The absence of exogenous nutrition prevents net growth and as catabolism continues to provide energy for maintenance function, the respiratory rate is also limited. Under these conditions, the conversion of protoplasmic substance for energetic needs becomes a survival factor and most likely favors a more selective metabolic operation by the plasmodium.

Light intervenes to induce sporulation by supplying a specific energy source to support and direct the controlled shutdown of energetic and synthetic metabolism. The final morphogenic product, unicellular meiotic spores, lacking motility and having greatly reduced transport, osmotic, and synthetic functions, is enabled to exist in this state characterized by greatly decreased energy demands and a correspondingly low respiration rate.

Light appears to act by inhibiting mitochondrial respiration thereby strongly reorienting energy metabolism. This rapid but partial inhibition, released in the dark, appears to govern subsequent light-dependent events during this induction period. These include reduction of nonheme iron, a rapid inhibition of exogenous glucose uptake and transient increases in plasmodial ATP, and glucose polysaccharide. Glucose strongly inhibits sporulation when added before but not after light induction. SH groups may participate in the ATP increase. Bleaching of the yellow pigments also occurs but is probably not a part of the initial photoreception.

As sporangial structure develops, prior to spore pigmentation, a large fraction of plasmodial protein appears to be diverted to their formation. A very late morphogenic reaction, the activation of a phenolase required for spore wall melanogenesis, can be induced prematurely by NEM without the formation of sporangial structures.

Experimentally, the questions are what the photoreceptor is, how the information is metabolically transmitted from this light reaction (actinomycin D-insensitive) and how it is translated (actinomycin D-sensitive) into the presumed formation of new proteins required for the implementation of this morphogenic sequence.

Assuming a specificity of light absorption, transmission specificity may depend on the essential depletion or suppression of alternative pathways or networks during the catabolic dark period, which under other environmental conditions would survive to act as nonspecific transmittors, or traps for the series of essential products arising from photometabolism. In this way the specific information content of the stimulus would be greatly enhanced by this prior deletion of alternative competing pathways to select the relevant one, very possibly as a function of the energetic limitations of the environment. Such a mechanism would also control sensitivity to the stimulus since the presence and capacities of the competing pathways would regulate the stimulus-driven flux required, for example, to attain and maintain the threshold concentration of the terminal gene-reacting product.

This complex morphogenic response to change of environmental conditions is accomplished by a coordinated shift in metabolism and structure initiated through alteration of energy metabolism and mediated through specific actinomycin D-sensitive genetic pathways. Since light also inhibits growth, both this effect and the induction of sporulation may be alternative expressions of the same photometabolism as it occurs in different metabolic environments.

REFERENCES

Abram, D. (1965). *J. Bacteriol.* **89,** 855.

Alexopoulos, C. (1963). *Botan. Rev.* **29,** 1.

Betel, I., and Klouwen, H. M. (1964). *Abstr. 6th Intern. Congr. Biochem.,* New York, 1964 p. 774. Pergamon Press, Oxford.

Black, S. H., and Gerhardt, P. (1962). *J. Bacteriol.* **83,** 960.

Bradley, D. F., and Calvin, M. (1956). *In* "General Systems Yearbook" (L. von Bertalanffy and A. Rapoport, eds.), Vol. 1, pp. 56–65. Braun-Brumfield, Ann Arbor, Michigan.

Brewer, E. N. (1965). Doctoral Thesis, Dept. Biochem., University of Wisconsin.

Butow, R., and Racker, E. (1965). *In* "Non-Heme Iron Proteins" (A. San Pietro, ed.), pp. 383–392. Antioch Press, Yellow Springs, Ohio.

Chance, B., ed. (1963). "Energy-linked Functions of Mitochondria." Academic Press, New York.

Chance, B. (1965). *Abstr. 5th Ann. Meeting Am. Soc. Cell Biol.*

Chance, B., and Williams, G. (1956). *Advan. Enzymol.* **17,** 65.

Cohen, S. S., and Barner, H. D. (1962). *J. Biol. Chem.* **237,** PC1376.

Cohen-Bazire, G., and Kunisawa, R. (1963). *J. Cell Biol.* **16,** 401.

Cordes, E. H., and Jencks, W. P. (1962). *Biochemistry* **1,** 773.

Daniel, J. W. (1963). *J. Cell Biol.* **19,** 18A.

Daniel, J. W. (1964a). *Federation Proc.* **23,** 320.

Daniel, J. W. (1964b). *Bacteriol. Proc.* p. 114.

Daniel, J. W. (1964c). *J. Cell Biol.* **23,** 23A.

Daniel, J. W. (1965). *J. Cell Biol.* **27,** 23A.

Daniel, J. W., and Baldwin, H. H. (1964). *Methods Cell Physiol.* **1,** 9.

Daniel, J. W., and Rusch, H. P. (1958). *Federation Proc.* **17,** 434.

Daniel, J. W., and Rusch, H. P. (1962a). *J. Bacteriol.* **83,** 234.

Daniel, J. W., and Rusch, H. P. (1962b). *Abstr. Am. Soc. Cell Biol.* p. 38.

Daniel, J. W., and Rusch, H. P. (1962c). *J. Bacteriol.* **83,** 1344.

Daniel, J. W., Kelly, J., and Rusch, H. P. (1962d). *J. Bacteriol.* **84,** 1104.

Epel, D. (1964a). *Biochem. Biophys. Res. Commun.* **17,** 62.

Epel, D. (1964b). *Biochem. Biophys. Res. Commun.* **17,** 69.

Estabrook, R. W. (1962). *Biochim. Biophys. Acta* **60,** 236.

Fan, D. P., Higa, A., and Levinthal, C. (1964). *J. Mol. Biol.* **8,** 210.

Fisher, R. (1958). "The Genetical Theory of Natural Selection" (2nd rev. ed.), pp. 39–40. Dover, New York.

Flavin, M. (1965). *Federation Proc.* **24,** 218.

Freese, E., Park, S. W., and Cashel, M. (1964). *Proc. Natl. Acad. Sci. U.S.* **51,** 1164.

Garfinkel, D. (1963). *J. Biol. Chem.* **238,** 2435.

George, P., and Rutman, R. J. (1960). *Progr. Biophys. Biophys. Chem.* **10,** 2.

Gordon, S., and Surrey, K. (1960). *Radiation Res.* **12,** 325.

Gray, W. D. (1938). *Am. J. Botany* **25,** 511.

Gray, W. D. (1953). *Mycologia* **45,** 817.

Grobstein, C. (1963). *In* "Cytodifferentiation and Macromolecular Synthesis" (M. Locke, ed.), pp. 1–14. Academic Press, New York.

Grossberg, A. L., Stelos, P., and Pressman, D. (1962). *Proc. Natl. Acad. Sci. U.S.* **48,** 1203.

Guest, H., and Horowitz, N. H. (1958). *J. Gen. Microbiol.* **18,** 64.

Guttes, E., and Guttes, S. (1964). *Science* **145,** 1057.

Guttes, E., Guttes, S., and Rusch, H. P. (1961). *Develop. Biol.* **3,** 588.

Hearon, J. (1950). *Bull. Math. Biophys.* **12,** 135.
Hearon, J. (1951). *Federation Proc.* **10,** 602.
Hoffee, P., and Englesberg, E. (1962). *Proc. Natl. Acad. Sci. U.S.* **48,** 1759.
Hohl, H. R. (1965). *J. Bacteriol.* **90,** 755.
Holmgren, A., Reichard, P., and Thelander, L. (1965). *Proc. Natl. Acad. Sci. U.S.* **54,** 830.
Holt, S. C., and Marr, A. G. (1965). *J. Bacteriol.* **89,** 1402.
Horowitz, N. H., and Metzenberg, R. L. (1965). *Ann. Rev. Biochem.* **34,** 527.
Horowitz, N. H., Fling, M., Macleod, H., and Wantanabe, Y. (1961). *Cold Spring Harbor Symp. Quant. Biol.* **26,** 233.
Horowitz, N. H., Fling, M., and Asano, K. (1964). *Abstr. 6th Intern. Congr. Biochem., New York, 1964* p. 721, Pergamon Press, Oxford.
Ikuma, H., and Bonner, W. D., Jr. (1964). *Abstr. 6th Intern. Congr. Biochem., New York, 1964* p. 780. Pergamon Press, Oxford.
Imamoto, F., Yamagishi, H., and Nozu, K. (1964). *J. Biochem.* (*Tokyo*) **55,** 303.
Imsande, J., and Pardee, A. (1962). *J. Biol. Chem.* **237,** 1305.
Jacob, F., and Monod, J. (1963). *In* "Cytodifferentiation and Macromolecular Synthesis" (M. Locke, ed.), pp. 30–64. Academic Press, New York.
Kamiya, N. (1960). *Ann. Rev. Plant Physiol.* **11,** 323.
Kaplan, N., Goldin, A., Humphreys, S., Ciotti, M., and Stolzenbach, F. (1956). *J. Biol. Chem.* **219,** 287.
Katchalsky, A., and Kedem, O. (1961). *Biophys. J.* **2,** 53.
Kauzmann, W. (1959). *Advan. Protein Chem.* **14,** 1.
Kessler, D. (1964). Doctoral Thesis, Dept. Zool., University of Wisconsin.
Klingenberg, M. (1963). *In* "Energy-linked Functions of Mitochondria" (B. Chance, ed.), pp. 121–142. Academic Press, New York.
Klotz, I. M. (1960). *Brookhaven Symp. Biol.* **25,** 25.
Klotz, I. M. (1962). *In* "Horizons in Biochemistry" (M. Kasha and B. Pullman, eds.), pp. 523–550. Academic Press, New York.
Kozloff, L., and Lute, M. (1959). *J. Biol. Chem.* **234,** 539.
Kozloff, L., and Lute, M. (1960). *Biochim. Biophys. Acta* **37,** 420.
Kroon, A. M. (1963). *Biochim. Biophys. Acta* **76,** 165.
Lark, K. G. (1963). *In* "Molecular Genetics" (Taylor, J. H., ed.), Part 1, p. 153. Academic Press, New York.
Lee, Y. P., Mochizuki, A., and Takahashi, T. (1965). *Federation Proc.* **24,** 188.
Linnane, A., and Wallace, P. (1964). *Nature* **201,** 1191.
Lotka, A. J. (1922). *Proc. Natl. Acad. Sci. U.S.* **8,** 147.
Luck, D. J. L. (1963). *Proc. Natl. Acad. Sci. U.S.* **49,** 233.
Luck, D. J. L., and Reich, E. (1964). *Proc. Natl. Acad. Sci. U.S.* **52,** 931.
Lumry, R., and Eyring, H. (1954). *In* "The Present State of Physics," pp. 189–210. Am. Assoc. Advance. Sci., Washington, D. C.
Lumry, R., and Spikes, J. D. (1960). *Radiation Res.* Suppl. 2, 539.
Lythgoe, R. J. (1937). *J. Physiol.* (*London*) **89,** 331.
Mandelstam, J. (1962). *Biochem. J.* **82,** 489.
Miller, W. H., Mallette, M. F., Roth, L. J., and Dawson, C. R. (1944). *J. Am. Chem. Soc.* **66,** 514.
Mitchell, P. (1961). *Nature* **191,** 144.
Mitchell, P. (1963). *Biochem. Soc. Symp.* (*Cambridge, Engl.*) **22,** 142.
Mitchell, P. (1965). *J. Cell Biol.* **27,** 67A.
Monod, J., and Jacob, F. (1961a). *Cold Spring Harbor Symp. Quant. Biol.* **26,** 389.
Monod, J., and Jacob, F. (1961b). *Cold Spring Harbor Symp. Quant. Biol.* **26,** 193.

Muhammed, A. (1965). *Abstr. 9th Meeting Biophys. Soc., San Francisco* p. **112**.
Nakada, D., and Fan, D. P. (1964). *J. Mol. Biol.* **8,** 223.
Nakada, D., and Magasanik, B. (1964). *J. Mol. Biol.* **8,** 105.
Novikoff, A. B., de Thé G., Beard, D., and Beard, J. W. (1962). *J. Cell Biol.* **15,** 451.
Ohta, J. (1954). *J. Biochem.* (*Tokyo*) **41,** 489.
Packer, L., Marchant, R. H., and Mukohata, Y. (1963). *In* "Energy-linked Functions of Mitochondria" (B. Chance, ed.), pp. 51–73. Academic Press, New York.
Peller, L., and Barnett, L. (1962). *J. Phys. Chem.* **66,** 680.
Petrack, B., and Lipman, F. (1961). *In* "Light and Life" (W. D. McElroy and B. Glass, eds.), pp. 621–630. Johns Hopkins Press, Baltimore, Maryland.
Petrack, B., Craston, A., Sheppy, F., and Farron, F. (1965). *J. Biol. Chem.* **240,** 906.
Pitt, G. A. J., Collins, F. D., Morton, R. A., and Stok, P. (1955). *Biochem. J.* **59,** 122.
Pressman, B. C. (1965). *Proc. Natl. Acad. Sci. U.S.* **53,** 1076.
Prigogine, I. (1961). "Introduction to Thermodynamics of Irreversible Processes" (2nd rev. ed.), p. 85. Wiley, New York.
Rakoczy, L. (1965). *Acta Soc. Botan. Polon.* **34,** 97.
Rasmussen, H., Chance, B., and Ogata, E. (1965). *Proc. Natl. Acad. Sci. U.S.* **53,** 1069.
Roizman, B. (1962). *Cold Spring Harbor Symp. Quant. Biol.* **27,** 327.
Rose, I. A. and Warms, J. V. B. (1965). *Federation Proc.* **24,** 297.
Rusch, H. P., Braun, R., Daniel, J. W., Mittermayer, C., and Sachsenmaier, W. (1964). *In* "Cellular Control Mechanisms and Cancer" (P. Emmelot and O. Mühlbock, eds.), pp. 80–85. Elsevier, Amsterdam.
Sager, R., and Ishida, M. R. (1963). *Proc. Natl. Acad. Sci. U.S.* **50,** 725.
Sanadi, D. R. (1965). *Ann. Rev. Biochem.* **34,** 31.
Schaeffer, P., Millet, J., and Aubert, J. P. (1965). *Proc. Natl. Acad. Sci. U.S.* **54,** 704.
Schatz, G. (1963). *Biochim. Biophys. Res. Commun.* **12,** 448.
Schatz, G. (1965). *Biochim. Biophys. Acta* **96,** 342.
Setlow, R. B. (1964). *J. Cellular Comp. Physiol.* **64,** Suppl. 1, 51.
Shuster, L., and Goldin, A. (1959). *J. Biol. Chem.* **234,** 129.
Slater, E. C., and Tager, J. M. (1963). *In* "Energy-linked Functions of Mitochondria" (B. Chance, ed.), p. 97. Academic Press, New York.
Smyth, D. G., Nagamatsu, A., and Fruton, J. G. (1960). *J. Am. Chem. Soc.* **82,** 4600.
Snell, E. E. (1962). *Brookhaven Symp. Biol.,* **15,** 32.
Soffer, R. L., and Gros, F. (1964). *Biochim. Biophys. Acta* **87,** 423.
Swift, H., and Hruban, Z. (1964). *Federation Proc.* **23,** 1026.
Trintscher, K. S. (1964). *Biochim. Biophys. Acta* **88,** 438.
Tyler, D. D., Estabrook, R. W., and Sanadi, D. R. (1965). *Biochim. Biophys. Res. Commun.* **18,** 264.
Van Iterson, W. (1965). *Bacteriol. Rev.* **29,** 299.
Van Lancker, J. L. (1964). *Federation Proc.* **23,** 1050.
Wahl, R., and Kozloff, L. (1962). *J. Biol. Chem.* **237,** 1953.
Whitney, P. L., and Tanford, C. (1962). *J. Biol. Chem.* **237,** PC1735.
Wright, B. (1963). *Bacteriol. Rev.* **27,** 273.
Wright, S. (1964). *In* "Stochastic Models in Medicine and Biology" (J. Gurland, ed.), pp. 199–244. Univ. of Wisconsin Press, Madison, Wisconsin.
Yasunobu, K. T., Peterson, E. W., and Mason, H. S. (1959). *J. Biol. Chem.* **234,** 3291.

CHAPTER 8

Studies on Chloroplast Replication in Synchronized *Euglena*

J. R. Cook

I. INTRODUCTION

The origin and replication of structure and function in subcellular organelles has been a subject of increasing interest in recent years. Chloroplasts and mitochondria in particular have received considerable study in this respect, in part perhaps because the presence of DNA in these organelles presents the interesting possibility that they are endowed with a certain independence from nuclear control in their self-replication. Synchronized populations of cells can prove useful in analysis of chloroplast development and replication, since it has been shown (Gibor and Granick, 1964) that the several proplastids of dark-grown *Euglena* divide synchronously at the time of cell division. It will be shown below that mature chloroplasts of *Euglena* also divide synchronously at the time of cell division in populations of cells synchronized with respect to cell division. It will be the purpose of this chapter to review some of the physiological and biochemical events occurring over the life cycle of chloroplasts in synchronized *Euglena*.

A. Historical

The origin, differentiation, structure, and function of chloroplasts have been recently the subjects of excellent review articles (Gibor and Granick, 1964; Granick, 1961; Grenson, 1964). The photosynthetic protozoan flagellate *Euglena gracilis* has proved to be exceptionally good material in such studies. These cells contain no chlorophyll or chloroplasts when grown heterotrophically for long periods in the dark; however, they do contain tiny colorless proplastids (Granick, 1961) which, on exposure to light, develop into mature chloroplasts, about $1 \times 6\ \mu$ in size. Epstein and Schiff (1961) have studied the greening process by fluorescence and electron microscopy; under their conditions maturation from proplastid to the mature chloroplast, containing about a dozen lamellae, requires about 13 days.

In a series of publications Brawerman and his colleagues (Brawerman and Chargaff, 1959; Pogo *et al.*, 1962) have demonstrated that greening of *Euglena* is associated with a massive synthesis of protein and RNA, the latter having a base composition different from that of cellular RNA from nongreen cells. Ribosomes isolated from chloroplasts have a base composition which differs from that of the microsome fraction of the cells (Brawerman, 1963).

Euglena gracilis irreversibly loses the ability to synthesize chlorophyll after treatment by several physical and chemical agents. Among these are extremely low doses of ultraviolet light (Pringsheim, 1958) which leaves the cells perfectly viable as long as an exogenous carbon source is provided. Lyman *et al.* (1959, 1961) have studied this bleaching phenomenon in considerable detail. The action spectrum for inactivation implicates nucleoprotein, presumably DNA, as the chromophore.

Sagan *et al.* (1964) have demonstrated the incorporation of tritiated DNA precursors into the cytoplasm of *Euglena*; the label could be removed by treatment with DNase. Leff *et al.* (1963) showed that DNA from green *Euglena* showed a satellite band in density gradient centrifugation. Edelman *et al.* (1964) extracted DNA from isolated *Euglena* chloroplasts, and found it to have a molecular weight of about 7×10^8, comparable to that found in bacterial cells. The satellite DNA has a base-pair composition of about 70% adenine + thymine, compared to about 50% for the main band (presumably nuclear in origin). The chloroplastic DNA makes up about 3% of the total cellular DNA.

Ray (1964) has also examined the satellite DNA of *Euglena*. After rebanding in a $CsCl_2$ gradient, he found at least two species of DNA in the satellite, one having a molecular weight of 3×10^6 and the other of $20\text{–}30 \times 10^6$. Bleached mutants contained the lower molecular weight DNA but not the higher. The possibility of a third satellite DNA was indicated.

The elegant experiment of Gibor and Granick (1962b) demonstrated genetic continuity of chloroplastic DNA in *Euglena*. By means of an ultraviolet microbeam technique they were able to irradiate either the cytoplasm or the nucleus of *Euglena*. In the former case, the cells were bleached; irradiation of the nucleus with the cytoplasm shielded sometimes killed the cell, but viable cells were always green.

These studies show that *Euglena* chloroplasts contain one or more species of DNA which are presumably self-replicating and, at least with respect to greening, have genetic continuity. Edelman *et al.* (1964) have cautioned that it has not yet been demonstrated that this DNA is informational for the formation of chloroplast structures; the nucleus has not been excluded from some role in the synthesis of RNA and protein associated with the development of the photosynthetic apparatus. Indeed, on the basis of genetic studies in higher plants, Gibor and Granick (1964) have proposed that the chloroplast genes are part of an inducible operon with the regulator genes under direct control of nuclear genes.

B. Synchronized Division of *Euglena*

Populations of *Euglena gracilis* are synchronized with respect to cell division by culture on a repetitive light-dark cycle at a constant temperature (Cook and James, 1960). While the procedure used in deriving the proper program of temperature and the light-dark regime is probably general for the several strains of *E. gracilis* (and perhaps for other green cells), the "best" program that can be devised varies from one strain to another. Thus, the first strain of *Euglena* to be synchronized was cultured at 20°C with a repetitive light-dark cycle of 16 to 8 hours (Cook and James, 1960). *Euglena gracilis* strain Z—used in all of the studies reported here—is best synchronized at 22°C with 14 hours of saturating light followed by 10 hours of dark.

Figure 1 shows the pattern of cell division in *E. gracilis* Z grown under these conditions. Cell divisions occur only during the 10-hour dark period. If the beginning of the light period is regarded as zero age for all cells, then the variation observed in division times during the subsequent burst is equivalent to that found in individual generation times of *Euglena* in exponential cultures (Cook and Cook, 1962). In populations of *E. gracilis* Z synchronized on this program, an approximate doubling of cell number occurs during each dark period. The program is a sufficient

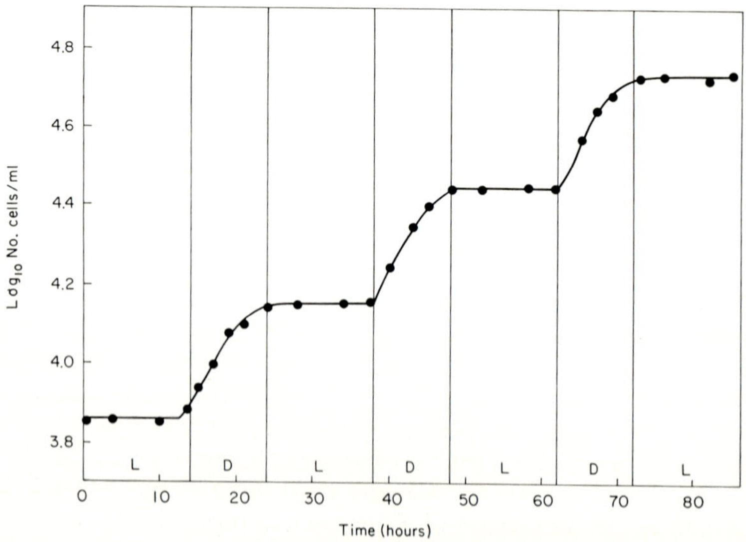

FIG. 1. Pattern of synchronous division in *E. gracilis* Z cultured on a minimal salt medium at 22°C, with a repetitive 14-hour light–10-hour dark cycle (Cook, 1965a).

mimicry of the natural ecological environment of the cell, and laboratory artefacts could hardly color studies of normal chloroplast development in synchronized *Euglena*.

II. PHYSIOLOGICAL REPRODUCTION

The capacity for photosynthesis has been examined over the life cycle of synchronized *Euglena* (Cook, 1965a), both at the light intensity of culture (1000 ft-cd) and in saturating light (2000 and 3000 ft-cd). Measurements were made with the oxygen electrode, recorded graphically, and corrected for respiration.

These results are shown in Fig. 2. At the light intensity of culture, the capacity for photosynthesis increases linearly over the whole life cycle, with perhaps a slight decrease near the end of the dark period. The photosynthetic rate in saturating light is only slightly greater than that found at 1000 ft-cd; the photosynthetic rate in synchronized *Euglena* always operates at near maximum capacity.

III. BIOCHEMICAL REPRODUCTION

The total cellular content of protein, RNA, DNA, and photosynthetic pigments has been measured in synchronized *E. gracilis* Z (Cook, 1965a). Of these, only the pigments are confined to the chloroplasts, but

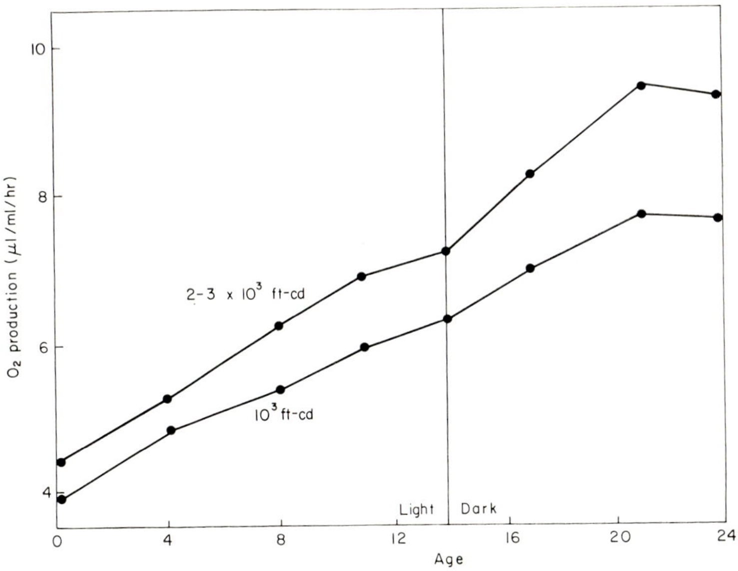

FIG. 2. The capacity for photosynthesis over one complete light-dark cycle in synchronized *Euglena*. Aliquots of cells were removed from the parent culture at regular intervals. Oxygen evolution was measured either in the light intensity of culture (1000 ft-cd) or in saturating light (2000 or 3000 ft-cd). The beginning of the light period is arbitrarily made at age 0 (Cook, 1965a).

the time course of synthesis of RNA and protein by chloroplasts can be inferred (see Section III, B) and chloroplastic DNA synthesis was examined by autoradiography.

A. PHOTOSYNTHETIC PIGMENTS

Euglena gracilis does not synthesize chlorophyll in the dark, and becomes white after extended growth in the absence of light. In synchronized cultures, chlorophyll and carotenoid synthesis is confined to the light period, being terminated instantly at the onset of the dark period (Cook, 1965a). The increase in pigment content is linear over the entire 14-hour light period. These data are summarized in Fig. 3.

B. PROTEIN AND RNA

Protein and RNA are synthesized in parallel in synchronized *Euglena*. This synthesis continues in a linear manner over the entire 24-hour light-dark cycle, when expressed per milliliter of culture (Cook, 1965a). Fig. 3 summarizes these studies.

A measure of total cellular protein and RNA gives no information of the amounts contained by the chloroplasts. Brawerman and Chargaff

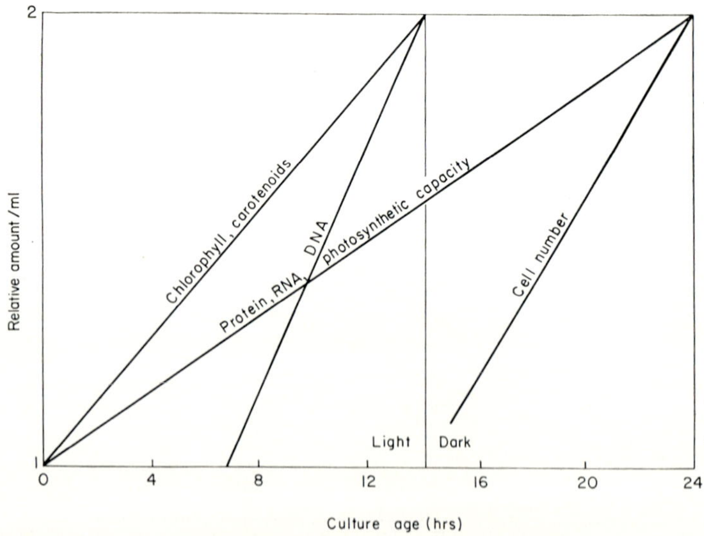

FIG. 3. Summary of the increase in cellular protein, RNA, DNA, and photosynthetic pigments over one light-dark cycle (data from Cook, 1965a).

(1959), however, have shown that *Euglena* chloroplasts contain about 40% of the total RNA and protein of the cell. If chloroplastic RNA and protein were synthesized according to temporal patterns different from that of other RNA and protein, one might expect a reflection of these differences in the rates of total synthesis. Since there is no deviation from strict linearity, it is inferred that RNA and protein of both chloroplasts and the remaining fractions of the cell are synthesized in parallel at a steady rate over the entire life cycle.

It will be noted that the capacity for photosynthesis seems to be determined more by the protein content than by the chlorophyll content. The ratio of protein to oxygen evolved is constant over the life cycle, but the ratio of chlorophyll to oxygen evolved is not. It appears likely that one or more of the enzymes of the photosynthetic pathway, rather than pigment concentration, limits the rate of photosynthesis.

C. DNA

DNA makes up about 0.2% of the dry mass of *Euglena*. This can be measured by ordinary chemical means, with an experimental error of 5–10%. Since chloroplastic DNA makes up only about 3% of the total DNA (Edelman *et al.*, 1964), replication of this fraction could hardly be detected by chemical means. Cytoplasmic incorporation of labeled DNA precursors is viewed as evidence for DNA synthesis by chloroplasts (Sagan *et al.*, 1964), however, and pulse labeling over the life cycle of synchronized *Euglena*, as well as ultraviolet inactivation studies, has yielded information on the time course of cytoplasmic DNA synthesis. This will be reviewed against the background of nuclear DNA replication, presented briefly here.

1. *Total Cellular DNA Synthesis*

The total cellular DNA in synchronized *Euglena*, as measured by chemical means, is replicated in the latter half of the 14-hour light period (Cook and Hess, 1964; Edmunds, 1964). This is shown in summary form in Fig. 3. Edmunds (1964) has shown that replication of this DNA is a necessary but insufficient event for completion of cell division in *Euglena*.

2. *Chloroplastic DNA Synthesis*

Two approaches have been followed to determine the time of chloroplastic DNA synthesis, namely autoradiography and ultraviolet inactivation of the ability to synthesize chlorophyll. Although the former procedure gives more reliable information, the ultraviolet studies are pertinent to interpretation of the tracer work.

a. Ultraviolet inactivation. Assuming that the bleaching response to low doses of ultraviolet is due to a specific inactivation of chloroplastic DNA, we have examined the bleaching response over the life cycle of synchronized *Euglena* (Cook and Hunt, 1965). Cells were removed from the parent culture at regular intervals, irradiated for varying periods of time up to several minutes, and plated. Plates were kept in the dark for about 10 days to avoid photoreactivation (Cook, 1963; Lyman *et al.*, 1959), and then placed in the light for greening in those colonies capable of doing so. The fraction of total colonies that were green was scored, yielding a regression dose-response curve for each cell age examined. Figure 4 shows the dose required to bleach 50% of the cells versus the age of the cells. *Euglena* was most sensitive to ultraviolet in terms of bleaching at the beginning of the light period, i.e., just after completion of the division burst. By assuming that DNA should be inactivated more readily when it is partially unzippered, we originally interpreted these data to mean that chloroplastic DNA is replicated shortly after the cell divides. As will be seen in the next section, it has been necessary to modify this view somewhat.

b. Cytoplasmic Incorporation of Adenine-H^3. Sagan (1965) showed that exogenous thymidine is not an effective precursor of DNA in *Euglena*, but that the purines are. Accordingly, tritiated adenine was used in pulse labeling of cytoplasmic (presumably chloroplastic) DNA in synchronized *Euglena* (Cook, 1965b). Cells were removed from the parent

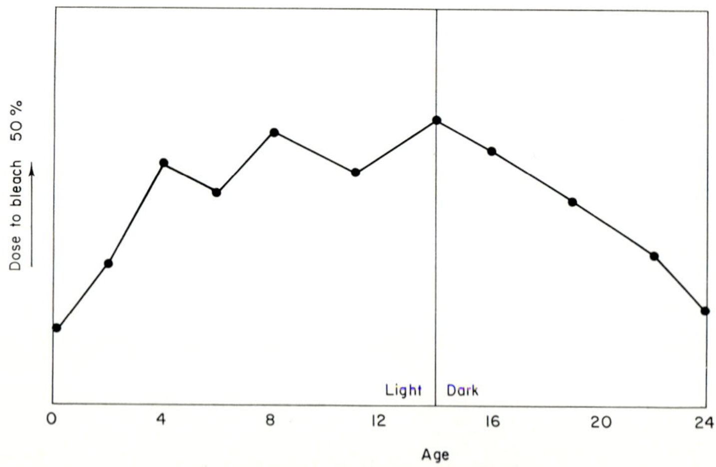

FIG. 4. The dose of ultraviolet light required to permanently bleach 50% of synchronized *E. gracilis* Z, as a function of cell age (Cook and Hunt, 1965).

culture at 2-hour intervals, concentrated by gentle centrifugation and suspended in the culture medium containing 0.2 millicurie adenine-H^3/ml. The cells were incubated in this medium for 2 hours, at the same temperature and light-dark regime as the parent culture. The cells were then packed, fixed with ethanol-acetic acid, and subsequently digested with buffered RNase. Aliquots of these cells were also digested with buffered DNase. Slides were dipped in liquid emulsion and exposed for 4 days, developed, and stained with methyl green.

Cytoplasmic grains were found over cells of all ages examined after RNase treatment. DNase removed essentially all the activity. In this particular experiment the number of cells in the parent culture increased by 106% during the dark period, so the culture was well behaved with respect to division and presumably so with respect to chloroplast development.

Figure 5 shows the average number of cytoplasmic grains found over cells of various ages. The level of incorporation is greater during the light period than in the dark. Two peaks of incorporation were found, one at the beginning and one at the end of the light period. Whether these peaks represent the *S* periods for two different species of DNA is not clear from these data, but this would be a logical conclusion if the synthesis of chloroplastic DNA is analogous to the synthesis of nuclear DNA; to my knowledge no cell has been described which has an interrupted *S* period in nuclear DNA synthesis. It is tempting to speculate that these peaks represent the temporally separated replication of the two chloroplastic satellite DNA's described by Ray (1964).

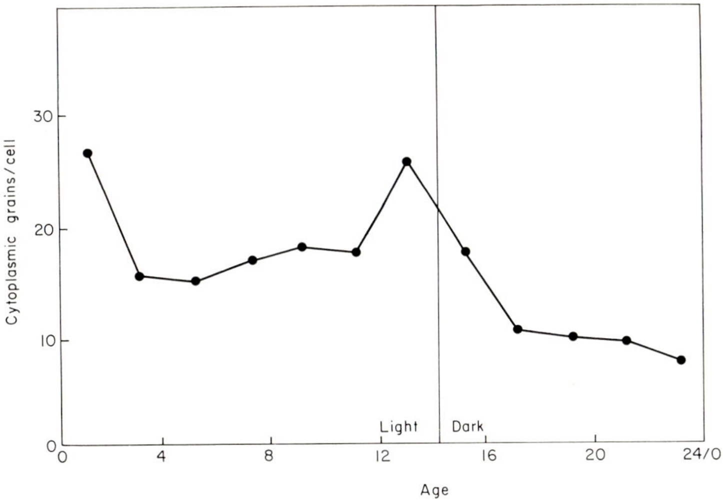

FIG. 5. Level of incorporation of tritiated adenine into cytoplasmic DNA of synchronized *Euglena*. Two peaks of incorporation are evident, one at the beginning and one at the end of the light period (Cook, 1965b).

Reference to Fig. 4 will show that *Euglena* was bleached most easily by ultraviolet at the age corresponding to the first peak of DNA synthesis. If DNA is more subject to ultraviolet inactivation during replication, then it is this DNA that is involved in the greening process. The second peak of DNA synthesis (age 12–14) corresponds to the period of maximum resistance to ultraviolet in terms of bleaching.

The level of incorporation steadily decreases to relatively low levels during the dark (division) period, and increases markedly as soon as the light period begins. There is some question as to whether this peak comes at this time because DNA synthesis is so well synchronized, or whether the light inself induces the synthesis. From the degree of synchronization of cell division, it would appear that the latter is the more likely explanation, unless chloroplast development is more highly synchronized than cell division. It wil be seen below that chloroplast division, at least, is no more highly synchronized than cell division.

IV. MORPHOLOGICAL REPRODUCTION

Because the numerous paramylum granules obscure the chloroplasts, direct counts of the plastids are not easily made, especially since the latter are connected one to another by ribbonlike structures (Epstein and Schiff, 1961). However, it was found that chloroplasts are more easily visualized after brief treatment with ether. Direct counts on the

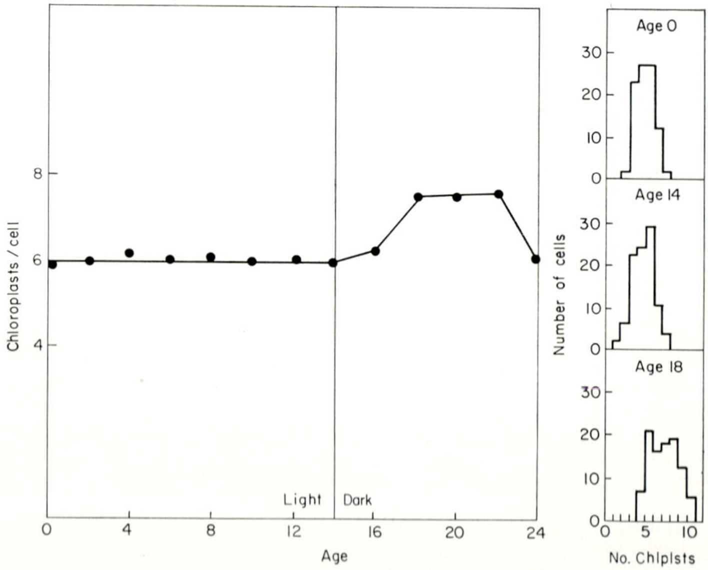

FIG. 6. Left: average number of chloroplasts per cell in synchronized *Euglena* over one complete light-dark cycle. Right: frequency histograms of the number of chloroplasts per cell at selected ages of the light-dark cycle (Cook, 1965b.)

number of chloroplasts were made on cells collected at various times during the synchronous cycle.

Figure 6 shows the average number of chloroplasts over the life cycle, and the frequency histograms from selected ages. During the light period the number of chloroplasts is quite constant at about 6 per cell, ranging for the most part between 4 and 7 per cell. The average number increases slightly but significantly to almost 8 during the dark (division) period. The frequency histograms (see e.g., age 18, Fig. 6) become broader with some indication of bimodality. These results are consistent with the view that the chloroplasts divide just prior to cytokinesis, and that this division is synchronized by the same light-dark regime which results in synchronization of cell division. A *de novo* origin, though unlikely, is not excluded by this experiment.

V. ROLE OF CHLOROPLAST DEVELOPMENT IN SYNCHRONIZING *Euglena*

The evidence reported in the preceding sections makes it clear that the light-dark regime which synchronizes cell division in *Euglena* also synchronizes chloroplast development. It may also be asked, to what extent does chloroplast development synchronize cell division? Studies with

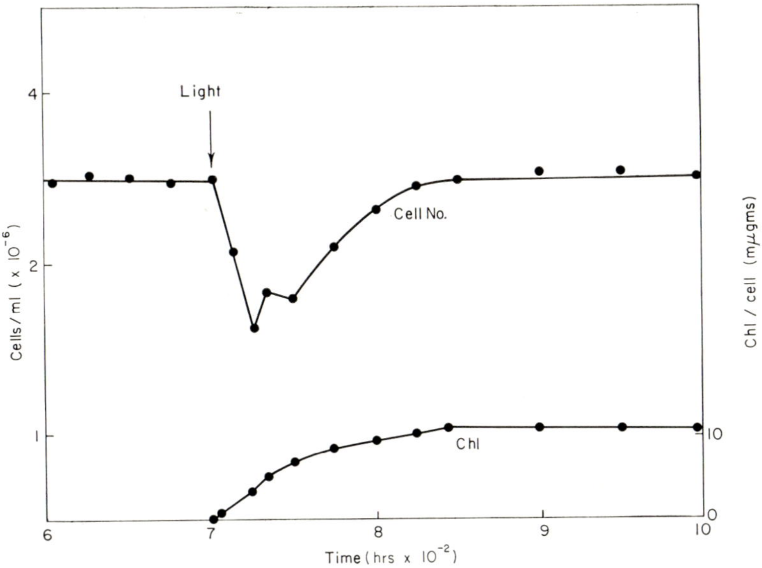

FIG. 7. Asynchronous growth of *Euglena* in the chemostat with continuous dilution, first in the dark and then in the light (500 ft-cd). Differentiation and growth of the photosynthetic apparatus, as indicated by the increasing concentrations of chlorophyll (lower curve), are initially attended by a marked inhibition of cell division, with subsequent recovery. Generation time at peak cell density about 30 hours.

asynchronous cultures indicate that development of the photosynthetic apparatus may play a significant role in this phenomenon.

Euglena was grown in the dark in a continuous culture device ("chemostat") with glucose as the limiting nutrient. Cells were maintained in exponential growth at a level of about 3.0×10^6 cells/ml (Fig. 7). Sudden exposure to light (ca. 500 ft-cd) resulted in an immediate cessation of cell division, so that the cell concentration was reduced by continued dilution with fresh media. The cell density dropped rapidly to 1.6×10^6 cells/ml before recovery of the ability to divide, after which the cells multiplied to regain the original density. This transitory inhibition of cell division was not observed with a heat-bleached *Euglena* unable to green.

The imposition of light must result in a massive reorganization of cellular activities, directed toward chloroplast development to the exclusion of processes leading to division. Chlorophyll synthesis starts immediately, and the tiny proplastids differentiate into tiny chloroplasts (Fig. 8) and finally mature chloroplasts (Fig. 9). Only the first stages of this differentiation inhibit cell division; as shown by Fig. 7, the inhibition lasts about 48 hours (under these culture conditions, about 2 generation

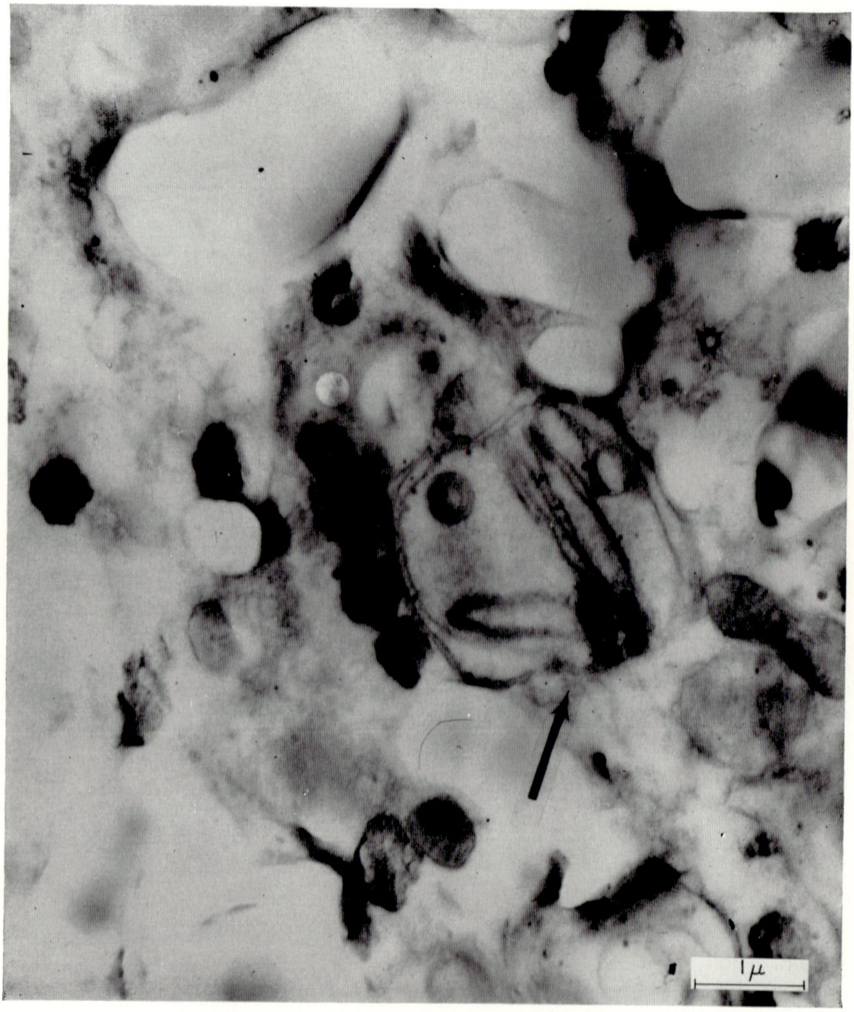

FIG. 8. Very early stage in the differentiation of a *Euglena* chloroplast, in a previously etiolated cell.

times), while 150 hours are required before chlorophyll levels become constant.

This experiment suggests that an alternating light-dark cycle may synchronize cell division in *Euglena* because it synchronizes chloroplast development. If light stimulates the synthesis of chloroplastic DNA (see Fig. 5) and in other ways initiates the growth and replication of the chloroplast—processes causing inhibition of cell division—then one would expect synchronous division on a light-dark cycle.

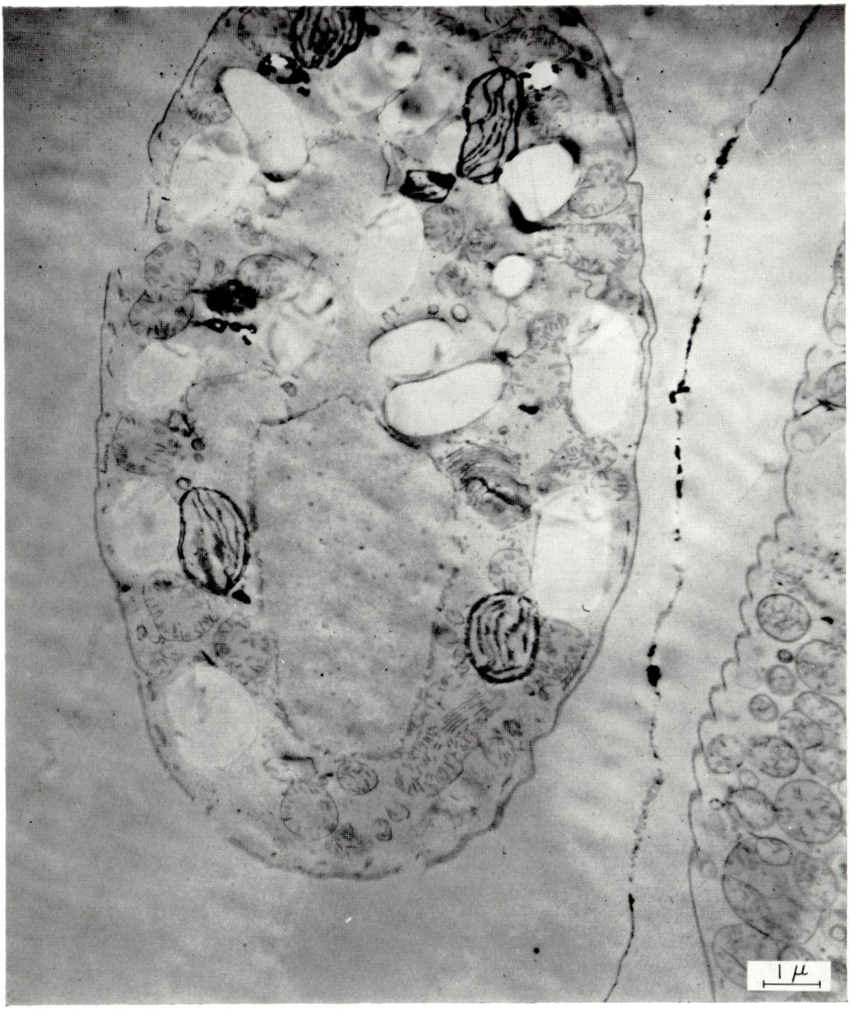

FIG. 9. Later stage in chloroplast differentiation. These chloroplasts are about ⅓ of the fully mature size and contain fewer lamellae, but otherwise appear typical.

VI. SUMMARY

Gibor and Granick (1964) have presented a model for the genetic mechanism of plastid replication and differentiation. According to these authors, the plastid contains DNA comprised of both constitutive and inducible genes. The constitutive genes are responsible for duplication of the organelle and other biochemical processes. The inducible operon determines differentiation from proplastid to chloroplast, and presumably the growth of the newly divided chloroplast to the mature size. The inducing agent is light, and the induction process may or may not involve nuclear genes. That these two processes can be separated is indicated by the fact that permanently bleached mutants of *Euglena* contain tiny proplastids capable of replication but unable to synthesize chlorophyll or enzymes of photosynthesis (Gibor and Granick, 1962a, 1964; Moriber *et al.*, 1963).

The data reviewed here tend to support this view, but further suggest that the model may be even more complex. Thus, if the DNA comprising the inducible genes is an inheritable entity, replicating either conservatively or semiconservatively, then it must replicate in dark-grown etiolated *Euglena,* where it has no immediate function. Does this DNA replicate at a specific age of the proplastid? The data reported here indicate that it may not do so in asynchronous cultures. In synchronized *Euglena,* the inducible genes (DNA) are apparently replicated early in the life of the chloroplast, since the ability to synthesize chlorophyll following ultraviolet lesion is most readily lost at this time (Fig. 4). [This DNA would correspond to the higher molecular weight satellite DNA of Ray (Section I, A), since bleached mutants lack this DNA.] In light-synchronized *Euglena,* the inducible genes (DNA) are indeed replicated at a specific age of the chloroplast. However, the degree of synchronization is so clearly greater than that of nuclear DNA synthesis, or the second peak of chloroplastic DNA synthesis, or chloroplast or cell division, that it must be assumed that light precipitates an almost immediate replication of the inducible genes (DNA), quite apart from control factors associated with the age of the cell. Chloroplasts in synchronized *Euglena* divide throughout the 10-hour dark period (Fig. 6); yet the first peak of DNA synthesis is delayed in all cells until the beginning of the light period.

The second peak of chloroplastic DNA synthesis is less well synchronized (Fig. 5), corresponding roughly to that of nuclear DNA synthesis (Fig. 3). This peak is terminated just prior to division of the chloroplast. This DNA is presumably associated with replication of the chloroplast (or proplastid), and would correspond to Ray's lower molecular weight satellite DNA, since it is found in bleached mutants which replicate proplastids but are unable to green. This DNA would then form a part of the constitutive genes of Gibor and Granick.

These considerations lead to the construction of a hypothetical model of genetic control in chloroplast growth and reproduction, modified from the original model of Gibor and Granick. It is inferred that the inducible genes replicate continuously but slowly in the dark because of a partial "leaky" repression directed by the nuclear genes or perhaps the constitutive genes of the chloroplast. The imposition of visible light derepresses replication of the "inducible" genes, resulting in immediate synthesis of DNA, but this replication can occur only once in each life cycle. Growth of *Euglena* on an appropriate light-dark cycle serves to synchronize replication of this DNA.

Replication of the constitutive genes occurs at the same time as replication of nuclear DNA. Both are therefore subject to the same control in initiation of DNA synthesis, or possibly replication of nuclear DNA induces simultaneous replication of the chloroplastic constitutive genes.

While this model of chloroplastic DNA replication is highly speculative at this time it is at least consistent with all the data. The identification of the two satellite DNA's with the time of their synthesis is tentative only, but the possibility of labeling chloroplastic DNA specifically during one of the 2 *S* periods in synchronized *Euglena* should permit a more definitive approach to the study of chloroplast inheritance, differentiation, and growth.

ACKNOWLEDGMENTS

Previously unpublished work of the author was supported in part by Public Health Service Research Grant GM-12179, from The National Institute of General Medical Sciences, and in part by a National Aeronautics and Space Administration Grant to the University of Maine.

REFERENCES

Brawerman, G. (1963). *Biochim. Biophys. Acta* **72,** 317.
Brawerman, G., and Chargaff, E. (1959). *Biochim. Biophys. Acta* **31,** 164.
Cook, J. R. (1963). *Photochem. Photobiol.* **2,** 407.
Cook, J. R. (1965a). *Plant Physiol.* (in press).
Cook, J. R. (1965b). *J. Cell Biol.* (in press).
Cook, J. R., and Cook, B. (1962). *Exptl. Cell Res.* **28,** 524.
Cook, J. R., and Hess, M. (1964). *Biochim. Biophys. Acta* **80,** 148.
Cook, J. R., and Hunt, W. (1965). *Photochem. Photobiol.* **4,** 877.
Cook, J. R., and James, T. W. (1960). *Exptl. Cell Res.* **21,** 583.
Edelman, M., Cowan, C. A., Epstein, H. T., and Schiff, J. A. (1964). *Proc. Natl. Acad. Sci. U.S.* **52,** 1214.
Edmunds, L. N. (1964). *Science* **145,** 266.
Epstein, H. T., and Schiff, J. A. (1961). *J. Protozool.* **8,** 427.
Gibor, A., and Granick, S. (1962a). *J. Protozool.* **9,** 327.
Gibor, A., and Granick, S. (1962b). *J. Cell Biol.* **15,** 599.
Gibor, A., and Granick, S. (1964). *Science* **145,** 890.
Granick, S. (1961). *In* "The Cell" (J. Brachet and A. E. Mirsky, eds.), Vol. II, pp. 489–602. Academic Press, New York.
Grenson, M. (1964). *Intern. Rev. Cytol.* **16,** 37.
Leff, J., Mandel, M., Epstein, H. T., and Schiff, J. A. (1963). *Biochem. Biophys. Res. Commun.* **13,** 126.
Lyman, H., Epstein, H. T., and Schiff, J. A. (1959). *J. Protozool.* **6,** 264.
Lyman, H., Epstein, H. T., and Schiff, J. A. (1961). *Biochim. Biophys. Acta* **50,** 301.

Moriber, L. G., Herschenov, B., Aaronson, S., and Bensky, B. (1963). *J. Protozool.* **10,** 80.

Pogo, A. O., Brawerman, G., and Chargaff, E. (1962). *Biochem.* **1,** 128.

Pringsheim, E. (1958). *Rev. Algol.* (*Paris*) [N.S.] **4,** 41.

Ray, D. S. (1964). Biophys. Lab. Rep. No. 125, from the W. W. Hansen Labs. of Physics, Stanford, California.

Sagan, L. (1965). *J. Protozool.* **12,** 105.

Sagan, L., Ben-Shaul, Y., and Schiff, J. A. (1964). *Abstr. 4th Ann. Meeting Am. Soc. Cell Biol., Cleveland* No. 167.

CHAPTER 9

Silica Shell Formation in Synchronously Dividing Diatoms

Joyce C. Lewin, Bernhard E. Reimann, William F. Busby, and Benjamin E. Volcani

I. INTRODUCTION

Diatoms are unicellular algae characterized by the presence of a silica shell encasing the cytoplasmic contents. In most forms this shell is a box-like structure constructed of 2 almost equal halves, the smaller fitting into the larger, like a petri dish. Every vegetative diatom cell has at least 4 shell parts: 2 valves which represent the top and the bottom of the dish and 2 overlapping girdles representing the sides. Each girdle may consist of one or more girdle bands. There are in some diatoms additional shell parts which are inserted between the valve and the girdle and which are called intercalary bands. The silica of the shell parts is amorphous and to some extent hydrated (Lewin, 1962).

Valve formation in most diatoms occurs immediately after cytokinesis. Geitler (1953, 1962), in discussing the sequence of events, i.e., cytoplasmic expansion, mitosis, cytokinesis, and shell formation, pointed out that it could not be decided whether and how far these individual events may have a necessary causal connection with one another. Mitosis occurs very rapidly in diatoms (von Denffer, 1949) and likewise, the time required for the formation of the new silica valve at each cell division may be short, in some species only a matter of 10–20 minutes (Reimann, 1960). Therefore, a study of the course of uptake of silicic acid, and its polymerization and consequent deposition as amorphous silica in an organized structure, could be greatly facilitated by the use of diatom cells that are dividing more or less synchronously.

The usual manner of cell division in diatoms is by binary fission, whereby a parental cell divides to form two new daughter cells. Each daughter cell inherits one of the parental silica valves, and in addition forms a new valve, which is produced while the daughter cell is still within the parental shell. Figure 1 illustrates how this takes place. The daughter cells may form their new silicified girdle bands while still within the parental girdle or after the cells have become separated from one another. The time, manner, and extent of girdle band formation (also intercalary band formation) varies considerably from species to species, whereas the formation of the valves always takes place under normal circumstances immediately following cytokinesis. The amount of silica per cell also varies from species to species, some being strongly silicified (silica—50% or more of the dry weight), while others are more weakly silicified (silica—less than 10% of the dry weight) (Lewin and Guillard, 1963). Therefore, one can expect considerable variations between species with regard to the kinetics or time course of silicic acid uptake.

II. METHODS FOR INDUCING SYNCHRONY

Two different methods for inducing synchrony in diatoms can be used. With the first method, cells are brought into the same physiological state by an extended period of darkness and then induced to divide by a light period. With the second method, cells are cultured in a medium in which silicon is the growth limiting factor. Upon depletion of the silicon, cells can then be induced to divide by addition of silicic acid to the medium.

A. Temporary Blocking of Development by a Dark Period

Probably in many diatoms growing under natural conditions, cell division is normally synchronized to some extent by the alteration of

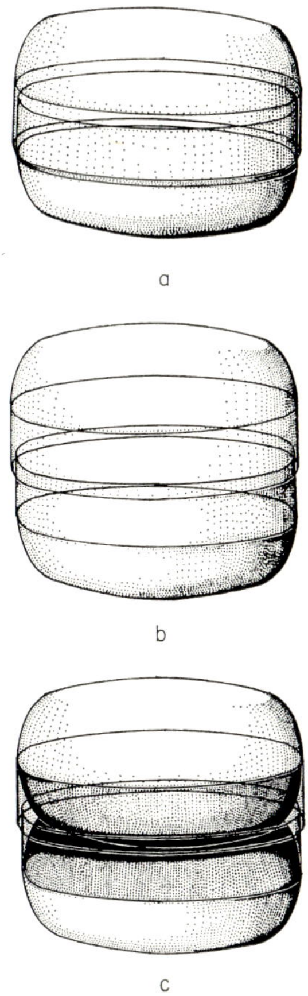

FIG. 1. Drawing to illustrate the manner of cell division in diatoms: (*a*) young cell; (*b*) growing cell; (*c*) divided cell.

daylight and darkness. This happens also in laboratory cultures kept in an artificial light-dark cycle. When Reimann (1960) studied cinematographically the timing of cell division and especially the time required to produce the silica shell in *Stephanopyxis turris* (Grev. et Arn.) Ralfs, he noted that in many filaments all of the cells divided simultaneously, probably because they were recently derived from a single cell and thus were in the same physiological stage of development. By means of an extended dark period, the cells of a culture can be brought at will into such a uniform physiological state.

Von Denffer (1949) showed that cell division in cultures of *Nitzschia palea* (Kütz.) W. Smith, could be synchronized by a light-dark cycle. Cells brought to a physiologically uniform condition by being kept in the dark for 10 days divided synchronously when reilluminated in con-

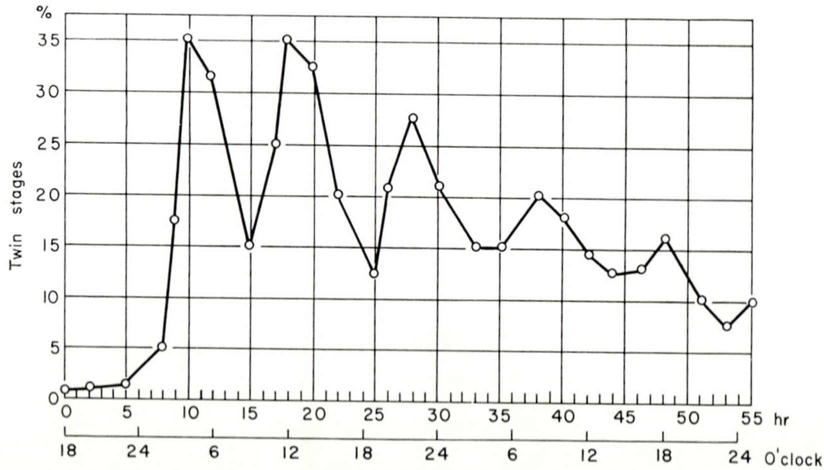

FIG. 2. Rhythmic cell division of *N. palea* as measured by number of twin stages in culture. Cells previously kept in dark for 10 days were illuminated in constant light for the period shown (from von Denffer, 1949).

tinuous light. At the end of 10 hours, and again after 20 hours, approximately 35% of the cells were in division stages (see Fig. 2). After this second division the synchronization died away, leaving about 6% of the cells in division stages at any one time. However, with a 6:6 hour light-dark rhythm, and using a light intensity of 10,000 lux, von Denffer was able to maintain repetitive synchronous cell divisions of *N. palea* showing 30% twin stages at the peak period of division (see Fig. 3). Cell divisions began in every second light period and were completed in the ensuing dark period.

Experiments with most other organisms have been made with the object of obtaining cultures undergoing repeated synchronized division. However, it has not been our aim to maintain this sort of culture. Since our research is concerned particularly with the formation of the silica shell and pertinent related biochemical and morphological phenomena, we considered it sufficient to induce cells in a rather dense culture to undergo just one rather well synchronized division. In this way enough cell material is available for investigations, e.g., on the process of silicic acid uptake, the possible occurrence of organic-silicon intermediates, the character of the membrane responsible for transport of the silicon, etc.

Since many diatoms are capable of heterotrophic growth when supplied with carbon sources such as sodium lactate or glucose (Lewin and Lewin, 1960), such substrates should not be present in the medium during the period of starvation in the dark. On the other hand, organic substrates present in the medium under photosynthetic conditions often have a

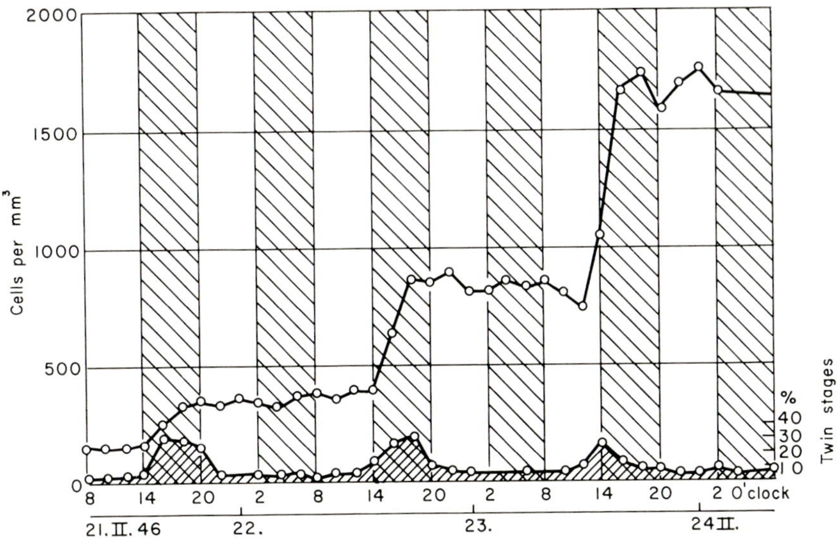

FIG. 3. Repetitive synchronous cell divisions of *N. palea* as measured by total cell numbers in culture. Synchrony induced by a 6:6 hour light-dark cycle; light intensity, 10,000 lux (from von Denffer, 1949).

favorable influence on the condition of the cells even if they do not directly affect the growth rate. Rho (1959) showed that lactic acid is one of the early products of photosynthesis in a diatom, which may help to explain the beneficial effect of lactate in the medium upon the induction of synchrony by the method described here.

Synchronously dividing cells of the marine diatom *Cylindrotheca fusiformis* Reimann and Lewin, are obtained as shown in the following tabulation:

Culture medium I[a]	Amounts/liter
Rila Marine Mix (Utility Chemical Co., Paterson, N.J.)	40 gm
$NaNO_3$	0.16 gm
$Na_2SiO_3 \cdot 9\ H_2O$	0.1 gm
2 Na·EDTA	0.012 gm
A trace element mixture to give:	
Fe	0.5 mg
Zn	0.3 mg
B, Cu, Co, Mn, Mo	0.1 mg each
Thiamine	0.5 mg
Tryptone (Difco)	1.0 gm

[a] Before autoclaving the medium is adjusted with NaOH to pH 8.0.

Culture conditions: Successful inducement of synchrony in this species depends upon the initial physiological homogeneity of the inoculum which is derived from a new cell clone isolated only a few days prior to the experiment. A 1.5-liter culture, in a flask furnished with aeration and sampling tubes as shown in Fig. 4, is suspended in a thermostatically controlled water bath at 25°C and constantly illuminated with fluorescent lamps ("cool white" and "warm white," Sylvania) at 3000 lux. When the cell number reaches approximately 6×10^4 cells/ml the light is turned off. After a 24-hour dark period, a more intense illumination is given (20,000 lux from an iodine filament lamp, "Quartzline," General Electric) and at the same time sterile solutions of L-cysteine (10^{-4} *M* final concentration), L-methionine (10^{-5} *M* final concentration), and sodium lactate (0.02 *M* final concentration) are added. The course of events is shown graphically in Fig. 5.

B. Temporary Blocking of Development by Silicon Depletion

Synchronously dividing cells of the freshwater diatom *Navicula pelliculosa* (Bréb.) Hilse can be obtained by the following procedure, using the culture medium given in the following tabulation:

Culture medium II[a]	Amounts/liter
$Ca(NO_3)_2 \cdot 4\ H_2O$	0.1 gm
K_2HPO_4	0.01 gm
$MgSO_4 \cdot 7\ H_2O$	0.025 gm
Na_2CO_3	0.02 gm
$Na_2SiO_3 \cdot 9\ H_2O$	0.08 gm
A trace element mixture to give:	
Fe	0.5 mg
Zn	0.3 mg
B, Cu, Co, Mn, Mo	0.1 mg each
Tryptone (Difco)	1.0 gm

[a] The pH is adjusted to 7 with HCl. In this medium silicon is the growth-limiting factor, permitting a final cell yield of 4–5 $\times 10^6$ cells/ml.

1. *Small-Scale Cultures*

Fifty ml of nutrient medium, sterilized by autoclaving in silica-free plastic bottles (polycarbonate, 8 ounce capacity from Nalge Co., Inc., Rochester, N.Y.) is inoculated with cells from a stock culture. It is then incubated on a reciprocating shaker with 4500 lux illumination from fluorescent lamps ("cool white" and "warm white," Sylvania) at 20°C. After the silicon in the medium has become depleted, as indicated by

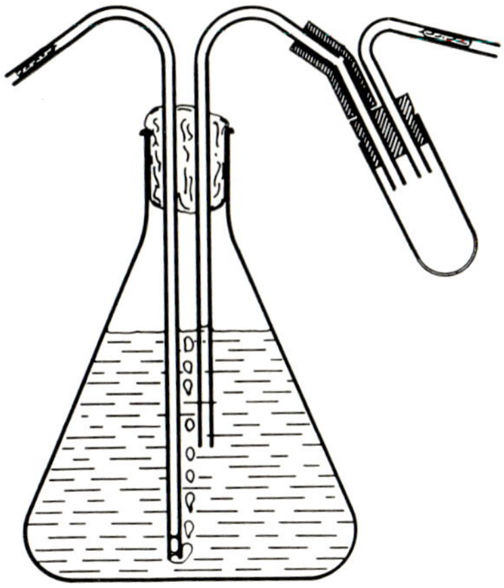

FIG. 4. Culture flask assembly used for aerating and sampling cultures of *C. fusiformis*.

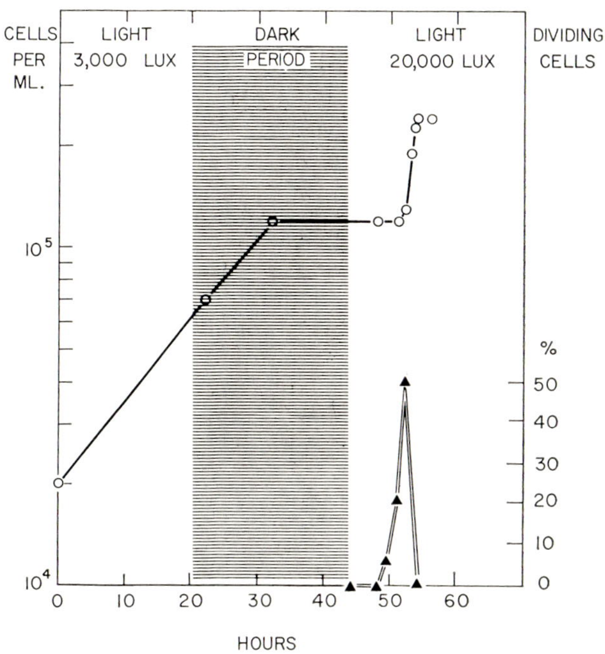

FIG. 5. Induction of a synchronized cell division in *C. fusiformis* by introduction of a 24-hour dark period. (Percent of cells in division stages included.)

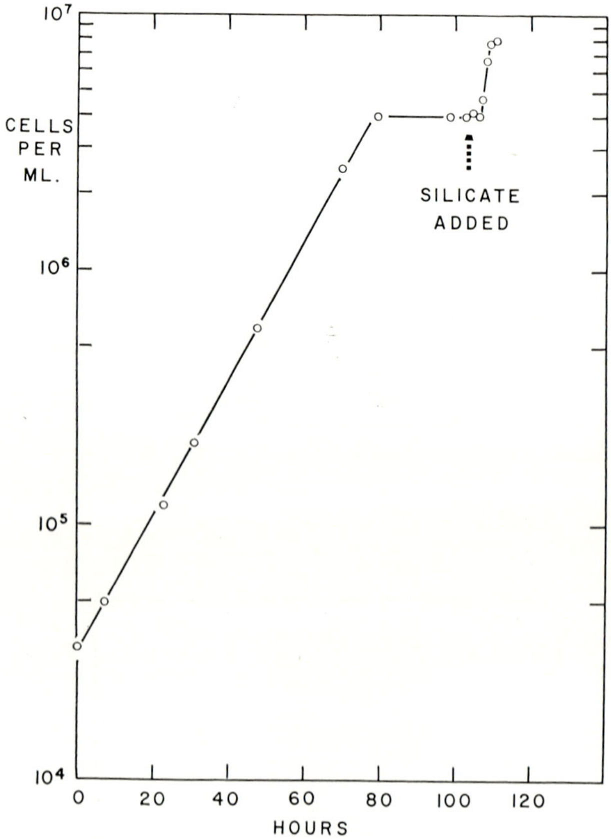

FIG. 6. Growth cessation and resumption (cell numbers) of *N. pelliculosa* upon depletion of silicate and addition of fresh silicate.

cessation of growth, the cells are kept for a further 14–24 hours in the silicon-depleted medium, and then a fresh, sterile sodium silicate solution is added to a final concentration of 20 mg silicon/liter. After a lag period of 3 hours, the cell number doubles within the ensuing 2-hour period (see Fig. 6).

2. *Large-Scale Cultures*

In order to supply sufficient material for various chemical studies, synchrony is induced in 3-liter (or occasionally 9-liter) cultures. Such large cultures pose problems not encountered on a smaller scale, particularly because of the need for increased aeration and the necessity to limit foaming of the medium and clumping and settling of the cells. Significant cell mortality (10–15%) may occur under these conditions unless a reduced sulfur compound is present.

Three liters of culture can be readily grown in a 1-gallon polycarbonate bottle (Nalge) illuminated at 17,000 lux and aerated by bubbling 10 liters of air per minute. A suitable inoculum consists of 350 ml of a late exponential-phase culture grown with 10 mg silicon/liter and containing approximately 5×10^6 cells/ml. At inoculation it has been found beneficial to add polypropylene glycol (P-2000, 5 mg/liter final concentration; Matheson, Coleman, and Bell) to reduce foaming and clumping, and sodium thiosulfate (5×10^{-4} *M*) to prevent cell mortality.

The cells are kept in a silicon-deficient condition for 14 hours. In most other respects the cultures are handled according to the procedures developed for small-scale synchronous cultures, except that it is advantageous to add cysteine and methionine (10^{-4} and 10^{-5} *M* respectively) at the same time as the silicate is supplied to induce synchronous division. The pH of the silicate solution should be adjusted to 7.5 with HCl immediately before it is added.

The course of synchronous division is similar to that induced in small-scale cultures.

III. PHYSIOLOGICAL STUDIES OF SILICON UPTAKE AND SHELL FORMATION

A. Theoretical Considerations

The most suitable biological system for a study of the mechanism of silica deposition would be one in which a large amount of silicon is incorporated within a short period of time into a population of cells at the same stage of development. Such a system would be most favorable for studying the pathway of silicic acid into the cell and then to the site where it is deposited as a new shell. If silicon at some stage should form a transitory intermediate, it might be best sought under these conditions. The choice of suitable experimental material is therefore of primary importance. The following criteria should be considered: (*a*) The species should have a short generation time; as a rule species with smaller cells reproduce more rapidly (Williams, 1964). (*b*) The cells should preferably have rather thick shells. (*c*) The cell size should be as uniform as possible; because of the manner in which cell division occurs in diatoms, many species are unsuitable since they produce cell populations composed of individuals which vary considerably in size and thus in physiological condition. (*d*) The cells should be amenable to the induction of a synchronized cell division.

Up until now all of our studies involving synchronized cultures (Lewin, 1961a) have been restricted to a single strain of *Navicula pelliculosa*, a freshwater pennate diatom. This species has certain of the desirable features mentioned above: the cells are small ($8 \times 3\ \mu$), of uniform size (cell dimensions in this species do not decrease with repeated cell division), rather heavily silicified (20% of the dry weight is silica) (Lewin, 1957), and cells from a silicon-deficient culture take up and deposit silica upon the addition of silicate. However, as discussed below (Section IV,B), this species is not so suitable for morphological studies.

B. Experimental Studies

Silicate uptake in *Navicula pelliculosa* has been shown to depend upon aerobic respiration, and thus can be inhibited by inhibitors such as cyanide, fluoride, and azide at concentrations inhibitory to respiration (Lewin, 1955). Dinitrophenol, which "uncouples" phosphorylation, also inhibits silicate uptake. Various organic substrates which stimulate respiration of silicon-deficient cells starved in the dark simultaneously stimulate silicon uptake.

Earlier experiments (Lewin, 1954) showed that silicon-deficient cells of this species, when resupplied with silicate but prevented from normal growth by being kept in the dark in the absence of a nitrogen source, each take up just sufficient silicate for the formation of two new valves. Counts of individual shells in acid-digested material, made with the light microscope before and after such an uptake, established that no additional valves were formed, although the silica taken up had apparently become associated with the insoluble shell fraction. Lewin concluded that under these conditions the silica was deposited on the previously existing valves.

Recent investigations (W. F. Busby, unpublished work) of silicon uptake and shell formation in *N. pelliculosa* have been made using cells synchronized by the silicon-deficiency technique. Uptake of silicate from the medium in such an experiment is shown in Fig. 7. In a similar experiment the formation of new shell material, represented by an increase in deposited silica, is shown in Fig. 8. Silicon uptake was determined by colorimetric analysis of the culture medium (modified method of Tůma, 1962) after removal of the cells by centrifugation, and shell formation was followed by sodium carbonate fusion of the cell residues remaining after alcohol and acid extractions of the whole cells. After a lag period of approximately 30 minutes, the rapid loss of silicate from the medium and its appearance in the shells became evident. Silicate uptake and shell formation are essentially complete before any increase in cell number is detectable.

C. Role of Sulfur in Silicon Uptake

Lewin (1954) demonstrated that sulfhydryl groups played a role in the mechanism of silicon uptake. The silicon uptake in *Navicula pelliculosa*

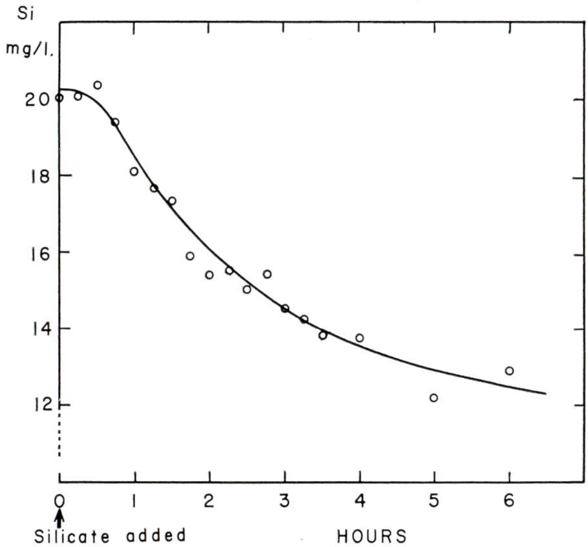

FIG. 7. Decrease of silicon in growth medium after silicate is supplied to a silicon-deficient culture of *N. pelliculosa*.

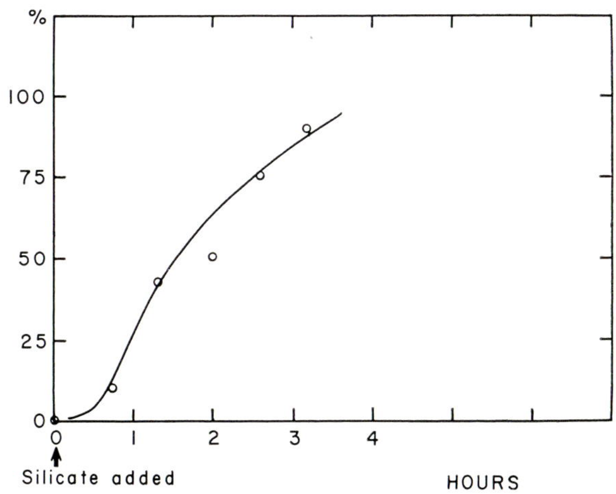

FIG. 8. Increase of silica (expressed as percent of original silicon dioxide) in the shell fraction of *N. pelliculosa* after silicate is supplied to a silicon-deficient culture.

could be completely inhibited by washing the cells with distilled water, a treatment which had no effect on cell respiration. Various reduced sulfur compounds restored the ability of washed cells to take up silicon; cells washed in dilute sodium sulfide (4×10^{-5} *M*) fully retained their ability to absorb silicon. A sulfhydryl inhibitor, cadmium chloride, inhibited silicon assimilation at a concentration (10^{-3} M) having no effect on cell respiration; this inhibitory effect was reversed by sulfhydryl compounds such as cysteine and glutathione. On the basis of these studies with *N. pelliculosa*, it was concluded that sulfhydryl groups were involved in silicon uptake (Lewin, 1954). Moreover, as mentioned previously, cysteine and methionine reduce the division time in cultures synchronized either by silicon deficiency or by darkness. In addition, the presence of thiosulfate in the medium is apparently essential to prevent cell mortality in large cultures of *N. pelliculosa* aerated under conditions of silicon depletion. All of these observations point to the importance of bivalent sulfur. Since there is no information yet available for other diatom species, we do not yet know whether this is a general phenomenon of diatom physiology.

IV. MORPHOLOGICAL DEVELOPMENT OF THE SILICA SHELL

The aim of our morphological investigations is to study the various stages which occur in the biosynthesis of the shell in order to determine which cell organelles are involved in silica deposition and how the form of the shell is controlled. Some progress has been made in this direction. In this connection we have given considerable attention to the specific properties of the cell material, since certain diatom species are obviously much more suitable for such investigations than are others.

A. Limitations Imposed by Nature of Cell Material

For comparative morphological studies of normal cells, and of cells exposed to a variety of chemical or physical treatments, it is helpful to have cultures containing cells all in exactly the same state of development. However, perfect homogeneity cannot be attained for various reasons, in addition to the innate tendency for intercellular variation. In some species a pronounced morphological polarity of each dividing cell results in inequality of the two daughter cells. In ultrathin sections of *Cylindrotheca fusiformis* the daughter cell that inherits the hypotheca of the mother cell is regularly characterized for a short period after cell division by a concentration of cytoplasmic particles higher than that in the old epitheca (see Fig. 9).

The cells of some species may exhibit a greater natural variation when kept under identical environmental conditions than do the cells of another species. In some species the time required for complete cell separation may be of different lengths in different individuals, while in other species the daughter cells may all stay together in pairs for the same length of time. In *Navicula pelliculosa*, for example, the 2 daughter cells may separate instantly after their new valves are formed or they may

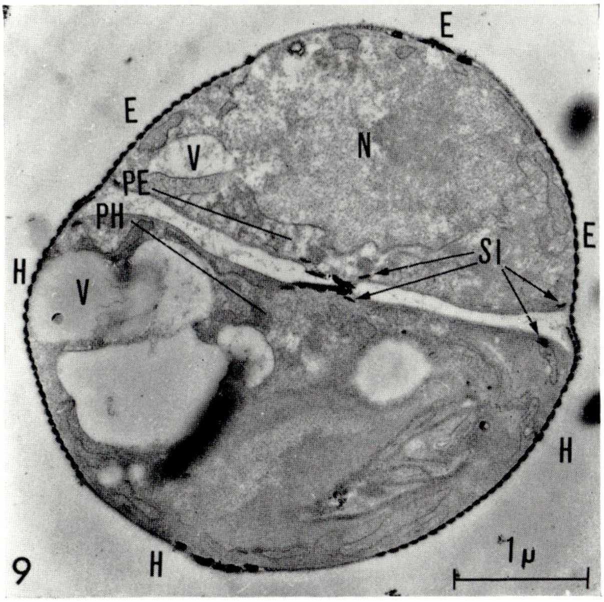

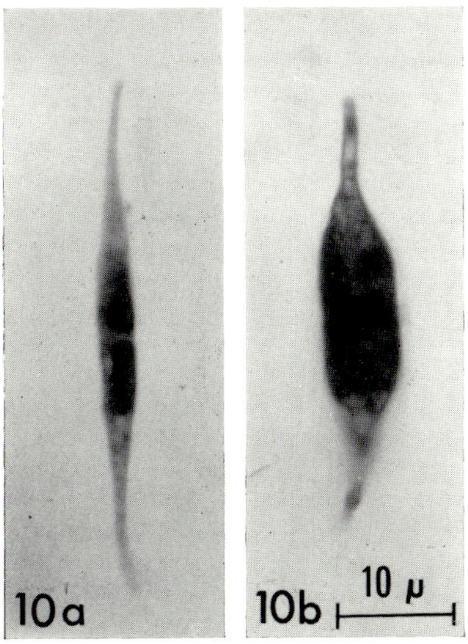

FIG. 9. Transapical section through recently formed daughter cells of *C. fusiformis*, enclosed by girdle bands of parental cell. Note that cytoplasm of hypothecal daughter cell (PH) is denser than that of epithecal cell (PE). E = epitheca; H = hypotheca; N = nucleus; SI = silica parts of new cell walls. Cells were fixed with potassium permanganate and osmic acid, post-stained with ruthenium red (× 20,000).

FIG. 10. Light micrographs of two cells of *C. fusiformis* fixed with osmic acid (× 1500). (*a*) Cell grown in presence of sufficient silicon; (*b*) cell kept for 12 days in silicon-deficient medium.

remain attached for longer periods. The same behavior was also observed in *Nitzschia linearis* (Reimann, 1960). Particularly with *N. pelliculosa* further cell divisions may occur before the cells have separated, thus 3 or 4 cells not necessarily in the same stage of development are attached to each other. Because of variations of this kind the value of such a species as a source of uniform cell material is limited.

Not every species is ideal for microscopic investigations. Obviously, for morphological studies large cells are preferable: more details can be seen by light microscopy and individual cells can be more easily oriented for sectioning for electron microscopy. Sectioning should be carried out parallel to the planes of symmetry in order to obtain useful sections showing the developing shells. It is very difficult, if not impossible, to orient diatom cells as small as those of *N. pelliculosa*, for example, and therefore with such small cells a rather extended investigation has to be made until suitable pictures through the right plane can be obtained.

B. Possible Morphological Effects of Treatment Used to Induce Synchrony

Different methods for synchronizing cell division may affect the morphology of the cells in different ways.

Cells of *Cylindrotheca fusiformis* kept for 12 days in light in silicon-deficient cultures, in which cell division had completely stopped, exhibit a morphological character entirely different from that found in actively growing cultures. The cells are shorter and wider, with a volume more than 6 times that of a normal cell (Figs. 10a, 10b, 11a, 11b). The nuclei are small and stain deeply with acetocarmine. Electron microscope studies reveal an increase in volume of the endoplasmic reticulum, which encroaches upon the nuclear area (Figs. 11b, 12). The nuclear structure is disturbed, being penetrated by a number of globular and tubular membranous elements (Figs. 11b, 12). The chromatophores are greatly extended (compare Figs. 11a and 11b) with an increase in number and length of the parallel bundles of thylacoids, though these apparently do not differ morphologically from those of normal cells. The mitochondria are considerably more numerous than in normal cells, and there is also an increase in the number of vacuoles with their associated storage products (Fig. 13). The girdle bands, under these circumstances, are very thin and quite irregularly distributed. These effects are interesting and further investigations need to be made, but they serve to illustrate how important it is to be aware of the possible adverse effects than can be caused by the treatment to be used for induction of synchrony.

The cells of *Navicula pelliculosa* kept for 14 hours in a silicon-deficient medium do not reveal indications of any specific cell damage. At

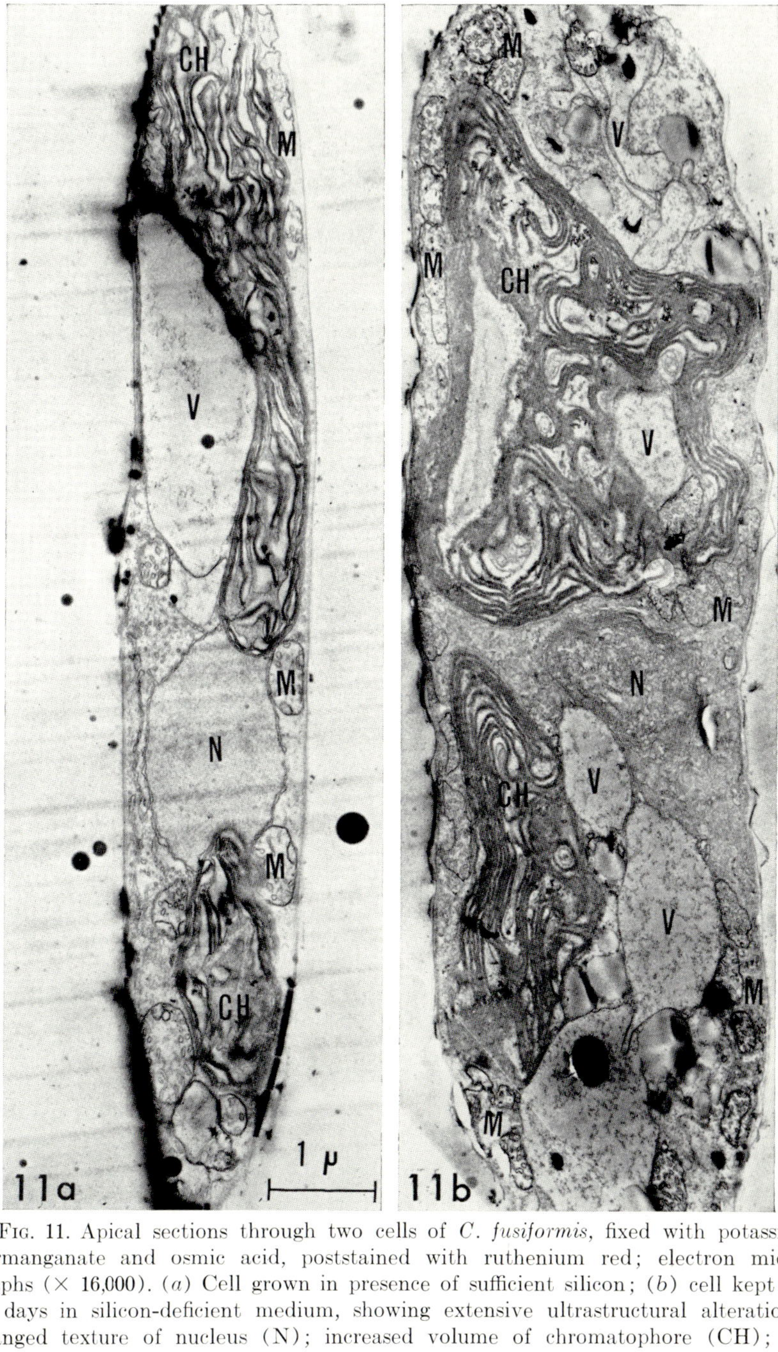

FIG. 11. Apical sections through two cells of *C. fusiformis*, fixed with potassium permanganate and osmic acid, poststained with ruthenium red; electron micrographs (× 16,000). (*a*) Cell grown in presence of sufficient silicon; (*b*) cell kept for 12 days in silicon-deficient medium, showing extensive ultrastructural alterations: changed texture of nucleus (N); increased volume of chromatophore (CH); increased numbers of mitochondria (M) and vacuoles (V).

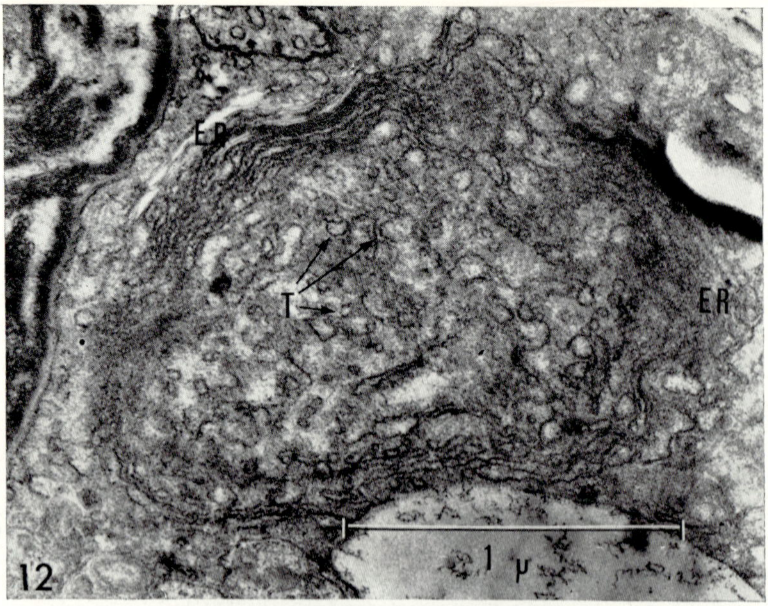

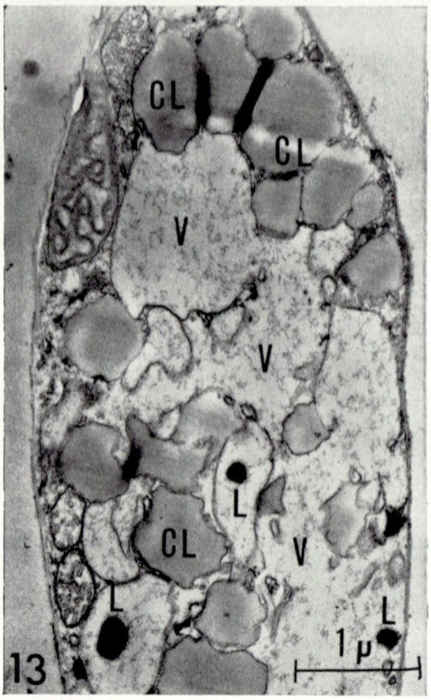

FIG. 12. Section of nucleus of cell of *Cylindrotheca fusiformis* kept for 12 days in silicon-deficient medium. Note tightly packed membranes of endoplasmic reticulum (ER) around nucleus, and globular and/or tubular membranous elements (T) inside it, penetrating the entire nucleus (as revealed by serial sections). Fixed with potassium permanganate and osmic acid; poststained with ruthenium red (× 50,000).

FIG. 13. Apical section through cell of *Cylindrotheca fusiformis* kept under silicon-deficient conditions, showing vacuolar system (V) containing lipid droplets (L), and electron-opaque bodies of polysaccharide (chrysolaminarin?) (CL). Fixed with potassium permanganate and osmic acid; poststained with ruthenium red (× 16,000).

the end of the silicon starvation period in preparations stained with acetocarmine and mounted in phenol (according to the method of von Stosch, 1952), it is possible to distinguish 5 stages of cell development. About 1% are apparently single cells. Nearly two-thirds of the cells have divided, but have not formed new shells between the daughter protoplasts. These twinned cells have the same pervalvar width (as seen in girdle view) as the single cells. An additional 20% of the cells are in compact pairs, both with complete silica shells. Ten to 15% are pairs showing some extension preparatory to a second division, and the remainder are already subdivided into quadruplets. When silicate is added to the medium, new shells are produced, and after 3½ hours all but 2% of the cell pairs have formed new shells, although 80% are still paired. About 25% of these paired cells show some pervalvar extension and 3% have already divided again, indicating the recommencement of cell division. The rest of the culture consists of single cells. This is the time at which the cell number (particle number) starts to increase. Observations on living cells do not reveal these different stages clearly enough even if an oil immersion objective is used. Although the cells of the culture do not show a uniform morphology, they certainly have a degree of physiological uniformity at the end of the starvation period and should be useful for physiological investigations.

There do not appear to be any adverse morphological effects in cells of *Cylindrotheca fusiformis* resulting from the 24-hour dark period employed to bring the cells into the same physiological state. This method for inducing synchrony is therefore perhaps to be considered preferable to the second method involving silicon-deficient cultures.

C. Formation of the Silica Shell

Investigations on the morphogenesis of the silica shell were initiated by Toman and Rozsival (1948). In an electron microscopical study of a great number of silica shells of *Nitzschia palea* from a clonal culture, they found an appreciable proportion of thinner shells which, they concluded, came from cells that had just divided. They seem to have been unaware that cells of this species regularly form "internal shells," normally thinner than the valves. Reimann (1960) extended these studies with diatom cells in which the stage of shell formation could be determined by light microscopical observations before the intact silica shell was prepared by digesting away the cell contents with pancreatin, and studied with the electron microscope. Two results were of particular interest. (*a*) Dividing cells of *Melosira varians* and *Stephanopyxis turris* are able to produce complete valves within 15–20 minutes; *Nitzschia linearis* completes its valves within 10–15 minutes. (*b*) The growth of the siliceous parts is apparently "appositional," preceding by an even peripheral extension of the polymerizing silica. Apparently the new silica shells are formed inside the cells within and under the control of the living protoplast which confers on the shell components their highly specific form.

Electron microscopic studies of the particular cell structures involved in the process of shell growth have been made with cells of *Cylindrotheca fusiformis* induced to divide simultaneously (Section II,A). Under optimum conditions up to 50% of the cells may be in various stages of cell division at one time (Fig. 5). So far, the most significant observation is that the siliceous structures are completely surrounded by a closely appressed 3-layered membrane which presumably plays a specific role in the processes of silicic acid polymerization and shell growth (Fig. 14) (Reimann, 1964). On the other hand, as soon as a cell dies, silicic acid begins to leach out of the shell (Lewin, 1961b), indicating that in the living condition the membrane may somehow prevent the outward diffusion of silica. Thus, in addition to its own extension, there are two possible activities in which the membrane may be involved: (*a*) one-way transport of silicic acid, and (*b*) control of polymerization and solidification of the silica. We are proposing that this membrane, with its comparatively limited and specific functions, be called "silicalemma."

Finally we might raise the intriguing problem: how is the specific form of the diatom shell controlled? Since the silicalemma is presumably flexible, we may assume that its shape conforms in some way to a cytoplasmic matrix, perhaps by changes in the sol-gel configuration of the cytoplasm which could serve as a semiflexible mold for the silica shell. So far this is wholly in the realms of hypothesis, but we hope that developmental studies of cells in synchronized cultures may provide clues to its eventual elucidation.

V. GENERAL CONCLUSIONS

An investigation of silicic acid uptake and its subsequent polymerization as a shell structure is facilitated by use of a simultaneously dividing cell population in which the cells are presumably passing through similar physiological and developmental stages.

Although some progress has been made in the morphological and physiological aspects of silica deposition, nothing is known of the biochemical processes involved. Currently 2 diatom species are being studied. *Cylindrotheca fusiformis,* a marine diatom in which a single synchronous division is induced by a dark period, has proved invaluable in morphological studies by virtue of its suitability for sectioning. It has been demonstrated by electron microscopy that the polymerization and solidification of the silica takes place within a membrane-limited vesicle within the cytoplasm. A study of the structural and enzymatic activity of the

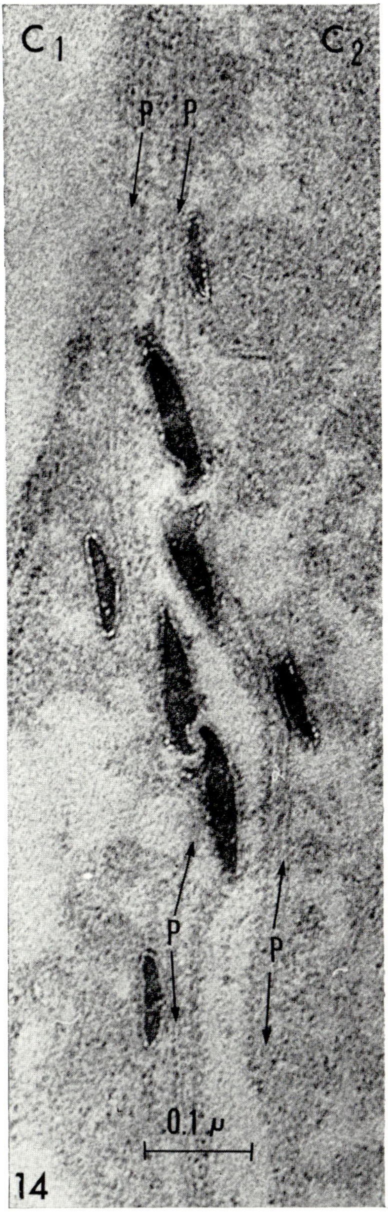

FIG. 14. Paratransapical section showing parts of two recently formed daughter cells (C_1, C_2) of *Cylindrotheca fusiformis,* showing early stages of shell formation. Each cell is delimited by its plasmalemma (P), and each siliceous structure by a "silicalemma." Silica parts that have reached their final size and shape are already apparent outside, and between, the plasmalemmae. Silica parts still in the process of growth occur within the cell, just beneath the plasmalemmae (× 160,000).

membrane system, silicalemma, involved in silica deposition would presumably reflect its role in silicic acid transport.

In *Navicula pelliculosa*, a freshwater diatom, a synchronously dividing culture can be obtained by the addition of excess silicate to the culture after a short period of silicate starvation. Although heavily silicified and not easily sectioned, it promises to be a valuable tool in the study of biochemical processes of shell formation because of the rapid rate of incorporation of relatively large amounts of silicic acid. Such a system may have important implications in the study of transport phenomena in general. The biochemical responses of this system during the period of rapid silicic acid uptake should also prove enlightening, especially in regard to components implicated in transport, e.g., phospholipids and proteins. If there are silicon intermediates, it should be possible to detect them in this system.

Acknowledgments

Some of the unpublished data presented in this chapter were obtained with support from Grant GM-08229-3 from the National Institutes of Health, United States Public Health Service, and by Grant G-15068 from the National Science Foundation. The authors are grateful to Miss Juta Kiethe for technical assistance, and to Dr. R. A. Lewin for suggestions and help in the preparation of the manuscript.

REFERENCES

von Denffer, D. (1949). *Arch. Mikrobiol.* **14,** 159.

Geitler, L. (1953). *Planta* **43,** 75.

Geitler, L. (1962). *Ber. Deut. Botan. Ges.* **75,** 393.

Lewin, J. C. (1954). *J. Gen. Physiol.* **37,** 589.

Lewin, J. C. (1955). *J. Gen. Physiol.* **39,** 1.

Lewin, J. C. (1957). *Can. J. Microbiol.* **3,** 427.

Lewin, J. C. (1961a). *Proc. 9th Intern. Botan. Congr., Montreal, 1959* Vol. 1, pp. 253–254. Univ. of Toronto Press, Toronto.

Lewin, J. C. (1961b). *Geochim. Cosmochim. Acta* **21,** 182.

Lewin, J. C. (1962). *In* "Physiology and Biochemistry of Algae" (R. A. Lewin, ed.), pp. 445–455. Academic Press, New York.

Lewin, J. C., and Guillard, R. R. L. (1963). *Ann. Rev. Microbiol.* **17,** 373.

Lewin, J. C., and Lewin, R. A. (1960). *Can. J. Microbiol.* **6,** 127.

Reimann, B. E. F. (1960). *Nova Hedwigia* **2,** 349.

Reimann, B. E. F. (1964). *Exptl. Cell Res.* **34,** 605.

Rho, J. H. (1959). *In* "International Oceanographic Congress—Preprints," (Mary Sears, ed.) p. 199 AAAS Washington, D. C.

Toman, M., and Rozsival, M. (1948). *Studia Botan. Czech.* **9,** 26.

Tůma, J. (1962). *Mikrochim. Acta* **3,** 513.

von Stosch, H. A. (1952). *Zuechter* **22,** 269.

Williams, R. B. (1964). *Ecology* **45,** 877.

CHAPTER 10

Intracellular Control of Enzyme Synthesis and Activity during Synchronous Growth of *Chlorella**

Robert R. Schmidt

I. INTRODUCTION

Synchronized cultures of microorganisms have now become a widely accepted tool for mapping the biochemical and biophysical events (and

* The following abbreviations have been used in this chapter: RNA, DNA: ribonucleic acid, deoxyribonucleic acid; AMP, ADP, ATP: adenosine monophosphate, adenosine diphosphate, adenosine triphosphate; CMP, CDP, CTP: cytidine monophosphate, etc.; GMP, GDP, GTP: guanosine monophosphate, etc.; UMP, UDP, UTP: uridine monophosphate, etc.; dTMP, dTDP, dTTP: deoxythymidine monophosphate, etc.; dAMP, dADP, dATP: deoxyadenosine monophosphate, etc.; P_i: inorganic orthophosphate; $(P_i)_x$: inorganic polyphosphate; TCA: trichloroacetic

the factors controlling these events) during the normal life history of a cell (see reviews: Burns, 1962; Campbell, 1957; Hoogenhaut, 1963; Scherbaum, 1960; Tamiya, 1964; Zeuthen, 1958).

Baker and Schmidt (1964a) have pointed out, however, that at least 2 criteria must be satisfied before the metabolic shifts observed during synchronous growth can be said to truly reflect the "normal" behavior of a cell during its development. "First, it should be demonstrated that periodism in any biochemical or biophysical event is recurrent in successive generations of synchronous growth without repeated exposure of the cells to the synchronization treatment. If the metabolism of synchronized cells is only measured in the first growth cycle following the synchronization treatment, it is possible that the observed metabolic shifts do not reflect the normal metabolic behavior of the cells, but merely a convalescence from an environmental stress or shock. Second, although culture conditions (nutrient concentration, pH, light intensity, etc.) may be changing during synchronous growth, it is essential to show that the limits, within which these change, do not at any time alter the growth rate of the cells during the course of synchronous growth. Too frequently either one or both of the above criteria are overlooked by investigators employing synchronized cells as biochemical tools. It is impossible, therefore, to speculate on the interrelationships between the metabolic events recorded for many of the organisms under investigation."

The present chapter has not been written as a critical review of the many biochemical studies resulting from the use of synchronized cultures of *Chlorella*. It is instead a discussion of the experimental approach that has been taken and the results that have been obtained, by the members of this laboratory, in the hope of elucidating the biochemical mechanisms which control the stepwise changes in the metabolic activities of a cell during its growth and division. The study of cellular control mechanisms has been approached in 3 stages. In the first stage (Section III), we have scanned different metabolic processes in search of periodism in the levels of key cellular metabolites during the synchronous growth of the unicellular green alga, *Chlorella pyrenoidosa*. After observing periodism in the level of a particular metabolite, we have progressed to the second stage of research (Section IV) which involved finding the enzyme(s) (anabolic or catabolic) whose apparent level was responsible for the fluctuations in the intracellular level of the metabolite. In the third stage (Section V), we have attempted to elucidate whether the periodism in the apparent level of the enzyme(s) was controlled at the gene (control of enzyme synthesis) or enzyme (control of enzyme activity by allosteric affectors, etc.) levels during the course

---

acid; GSH: glutathione; EDTA: ethylenediaminetetraacetic acid; Tris: tris(hydroxymethyl)aminomethane; K_m: Michaelis constant.

of cellular development. Although studies of the biochemical mechanisms which control enzyme synthesis or activity are important in themselves (Cole and Schmidt, 1964; Gorman *et al.*, 1964; Masters *et al.*, 1964), the integration of the data from these studies, with the other metabolic activities of the cell, depends largely on the aforementioned stages of research.

II. ORGANISM AND GROWTH CHARACTERISTICS

A thermophilic species of *Chlorella*, strain 7-11-05 of *Chlorella pyrenoidosa* (Sorokin, 1957; Sorokin and Myers, 1953), was selected for use in studying the growth and division cycle of a cell, for the following reasons: (*a*) The cells can be readily brought into a high degree of synchrony, using intermittent illumination, without inducing abnormal periodism in most metabolic processes (Baker and Schmidt, 1963, 1964a; Johnson and Schmidt, 1963; R. R. Schmidt and King, 1961; Spencer *et al.*, 1961). (*b*) Once synchronized, the cells maintain a high degree of synchrony for successive generations (Fig. 1) without repeated exposure to the synchronization treatment (Baker and Schmidt, 1964a; Cole and Schmidt, 1964; Cur-

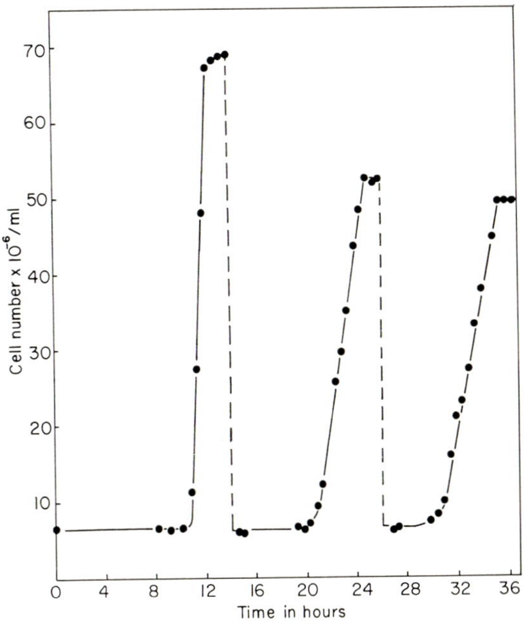

FIG. 1. Successive generations of synchronous growth of *C. pyrenoidosa* (strain 7-11-05) without repeated exposure of the cells to the synchronization treatment (intermittent illumination). The culture was diluted, while still in the light, to its original cell density after 12 and 27 hours of synchronous growth to prevent low light intensity or nutrient deficiency from limiting the growth rate of the cells. (From Johnson and Schmidt, 1963.)

nutt and Schmidt, 1964a; Herrmann and Schmidt, 1965; Johnson and Schmidt, 1963). (*c*) The organism's high growth rate (dry weight doubles approximately every 3.5 hours) but relatively long growth and division cycle (12–13 hours) provides ample cellular material for biochemical studies, as well as accurate sampling and processing of the cells during synchronous growth studies (R. R. Schmidt and Spencer, 1964). (*d*) The organism, being a photolithoautotroph (Fogg, 1953), can be cultured in an aqueous medium containing only highly oxidized salts, thereby minimizing the problem of bacterial contamination in the mass cultures needed for most biochemical studies (Baker and Schmidt, 1963). (*e*) The cells are sufficiently large in size (R. R. Schmidt and Spencer, 1964) that clearcut cytological observations can be made in parallel with biochemical measurements. (*f*) The subcellular organelles (nucleus, chloroplast, mitochondria, etc.) are similar to those in higher plants and animals (Tamiya, 1963); and (*g*) the cells divide into 8 or 16 daughter cells (depending on the cultural conditions), greatly magnifying any premitotic pools of intermediates which may accumulate to be utilized during the periods of nuclear and cellular division.

The average daughter cell of *C. pyrenoidosa*, strain 7-11-05, is approximately 3 μ in diameter and contains a single nucleus and chloroplast (R. R. Schmidt and Spencer, 1964). Cellular dry weight, phosphorus, and sulfur increase as parallel exponential functions throughout the synchronous growth cycle (Fig. 2). Total cellular nitrogen is also essentially an ex-

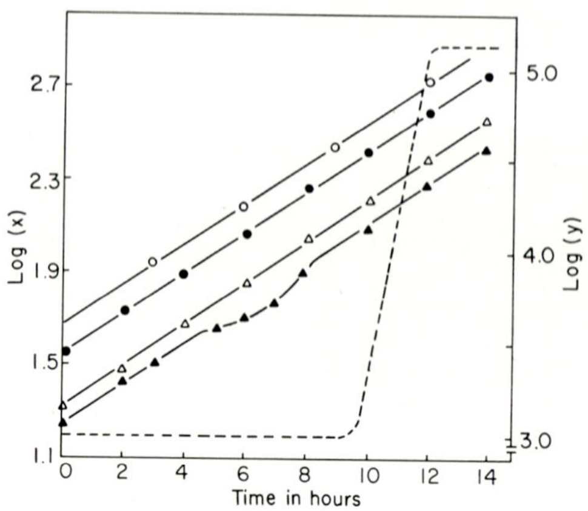

FIG. 2. Characteristics of synchronous growth of *C. pyrenoidosa* (strain 7-11-05). x = ▲——▲, μg total cellular nitrogen/ml; △——△, $10^2 \times$ mg cellular dry weight/ml; ●——●, 10 × μg total cellular phosphorus/ml. y = ○——○, counts/minute of total cellular-S^{35}/ml. The dashed line represents the typical change in cell number per milliliter during synchronous growth. (Data from Johnson and Schmidt, 1963; R. R. Schmidt and Spencer, 1964.)

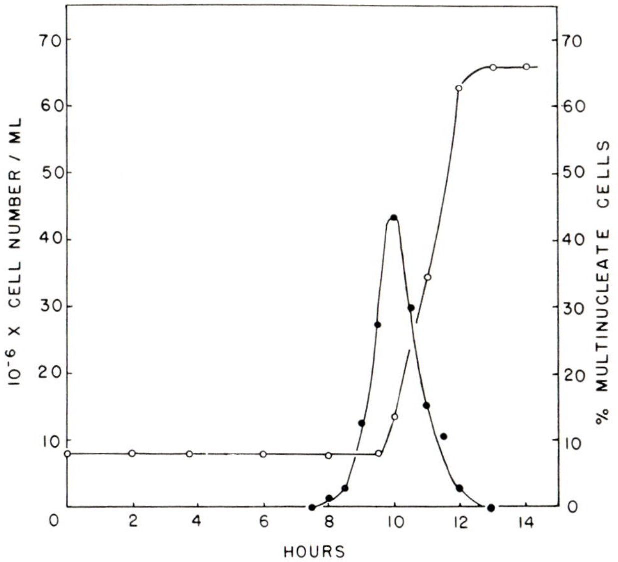

FIG. 3. Relationship of nuclear division to generation time during synchronous growth of *C. pyrenoidosa* (strain 7-11-05). ●——●, % multinucleate cells; ○——○, cell number/ml. (From Curnutt and Schmidt, 1964a.)

ponential function, but reproducible deviations from log linearity are observed between the fourth and ninth hours of cellular development. Sorokin (1957) has observed the cellular volume of this organism also to increase as an exponential function. During the period of nuclear division (initiated at approximately 7.75 hours, Fig. 3), 3 sequential divisions result in the formation of 8 daughter nuclei. (Immediately prior to this period, the single chloroplast divides into 8 daughter chloroplasts which are distributed equally between the daughter nuclei at each nuclear division.) A cell wall then is deposited around each of the daughter nuclei (each with a chloroplast), resulting in the formation of 8 daughter cells still enclosed by the original mother cell wall. At this stage of development the mother cell is approximately 6 μ in diameter (R. R. Schmidt and Spencer, 1964). Cell release (division) begins at approximately the tenth hour of development with the rupture of the mother cell wall, resulting in the release of 8 new daughter cells and the completion of the cell cycle (twelfth to thirteenth hour). The time of nuclear division and cell release has been found to vary slightly ($\pm$0.5 hours) in different synchronous growth studies; therefore, the timing of these cellular events is reported below each figure discussed in this chapter. It should also be noted that since cellular dry weight, phosphorus, and sulfur are parallel exponential functions, the concentration of any cellular component (inorganic polyphosphate, starch,

RNA, DNA, etc.) can be expressed on the basis of any of these 3 cellular parameters and still reflect the same periodism during the cell cycle.

III. METABOLIC SHIFTS DURING SYNCHRONOUS GROWTH

A. Inorganic Polyphosphate Metabolism

The dramatic cytological (chromosome condensation and migration, membrane formation, cross wall deposition, etc.) and chemical (see Sections IV and V) changes that accompany the period of nuclear division in a *Chlorella* cell would seem to require an increase in the energy generating capacity of the cell or some preformed pool (Swann, 1957) of high energy compounds which could be utilized to meet the undoubtedly increased energy demands of the cell during this event (Fig. 3). The reports that the lowest photosynthetic (Sorokin, 1957) and respiratory (Sorokin and Myers, 1957) rates, during the synchronous growth cycle of *Chlorella pyrenoidosa*, occur during the periods of nuclear and cellular division, suggested that some intracellular reserve of high energy compounds accumulated prior to these periods and was utilized during them.

Of interest to our laboratory group were the reports of deposits of a high molecular weight polymer of orthophosphate, called inorganic polyphosphate (or metaphosphate), $(P_i)_x$, which accumulated in most nonvascular plants in bodies called metachromatic or volutin granules (Wiame, 1947, 1948; G. Schmidt *et al.*, 1946, 1949; G. Schmidt, 1951; Mudd *et al.*, 1958). The nature of the anhydride bonds, linking the phosphate units together in this macromolecule, made $(P_i)_x$ a likely candidate as an intracellular energy and/or phosphate reserve. Because this macromolecule was utilized by certain bacteria for nucleic acid synthesis during phosphate starvation (Mudd *et al.*, 1958), we proposed (within our laboratory) that this compound might also be utilized during normal cellular growth when the energy and/or phosphorus demands of the cell rapidly accelerate, such as probably during the period of nuclear division. If $(P_i)_x$ was utilized as a phosphorus and/or energy source for key biosynthetic processes, it seemed probable that the level of this macromolecule could regulate the time and rate of the nuclear event.

In a preliminary study (R. R. Schmidt, 1961), the number of volutin granules per cell was estimated at certain periods during the synchronous growth of *C. pyrenoidosa*. Daughter cells were observed to contain only 1–2 granules while mother cells, immediately prior to nuclear division, contained from 20–30 per cell. Because each mother cell divided into 8 daughter cells, each containing only 1–2 volutin granules, there appeared to be a net loss of these granules during nuclear and cellular division. These data suggested that $(P_i)_x$ accumulated prior to and was utilized during these events.

The next step was to actually measure the level of $(P_i)_x$ and total nucleic acid phosphorus during a synchronous growth cycle to determine (*a*) if $(P_i)_x$ does accumulate prior to and decrease during the period of nuclear division, and (*b*) whether or not the trends for $(P_i)_x$ and total nucleic acid phosphorus reflect a precursor-product relationship. This study revealed that the $(P_i)_x$ level increased, reaching a level of approximately 30–35% of the total cellular phosphorus, for a brief period prior to nuclear division but decreased to approximately 15–18% by the end of this event (R. R. Schmidt and King, 1961; Baker and Schmidt, 1963, 1964a). Furthermore, an inverse relationship was observed between the level of $(P_i)_x$ and the levels of total nucleic acid phosphorus, acid-soluble phosphate, and phospholipid fractions (Fig. 4) of the cells (Baker and Schmidt, 1963, 1964a; Herrmann and Schmidt, 1965), suggesting that $(P_i)_x$ was utilized as a phosphorus and/or energy source for the synthesis of these compounds during normal growth and division.

The next research objective was to determine what type of biochemical mechanism was controlling the apparent periodism in the rate of synthesis and/or breakdown of $(P_i)_x$ during the cell cycle. A possible point of control appeared to be the reaction

$$x\text{ATP} + (\text{P}_\text{i})_n \rightleftharpoons x\text{ADP} + (\text{P}_\text{i})_{n+x}$$

catalyzed by polyphosphate kinase (A. Kornberg *et al.*, 1956; S. R. Kornberg, 1957; Muhammed, 1961; Harold, 1964; Iwamura and Kuwashima,

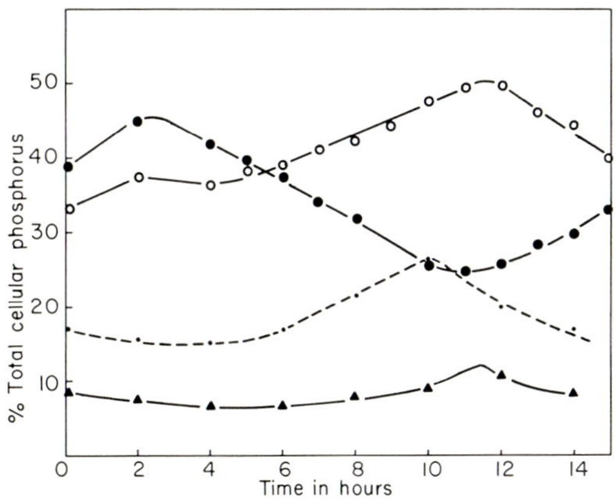

FIG. 4. Intracellular distribution of phosphorous during synchronous growth of *C. pyrenoidosa* (strain 7-11-05). ●——●, inorganic polyphosphate; ○——○, total nucleic acid phosphorus; · - - - - ·, total acid-soluble phosphorus; ▲——▲, phospholipid phosphorus. Nuclear division began at 7.75 hours and cell release at 10 hours of growth. (Data from Baker and Schmidt, 1963; Herrmann and Schmidt, 1965.)

1964). This was the only enzyme known to catalyze the synthesis of $(P_i)_x$. Because the equilibrium of this reaction was far to the left, requiring high ATP/ADP ratios for net synthesis of $(P_i)_x$, Kornberg (1957) proposed that the relative concentrations of ATP and ADP within the cell might control the net synthesis or utilization of $(P_i)_x$.

Curnutt and Schmidt (1964a) described a possible relationship of the energy-generating systems, e.g., oxidative and photosynthetic phosphorylation, to the energy-requiring systems and the synthesis and utilization of $(P_i)_x$ during the synchronous growth cycle of *Chlorella*. It was proposed that the accelerated rates of photosynthesis (Sorokin, 1957) and respiration (Fig. 5) early in cellular development elevate the ATP/ADP ratio to values conducive to the net synthesis of $(P_i)_x$. Likewise, the later decrease, in the photosynthetic and respiratory rates during the course of cellular maturation, depresses the ATP/ADP ratio slowing $(P_i)_x$ synthesis. Events such as the synthesis of nucleic acids (Herrmann and Schmidt, 1965) and proteins (Hare and Schmidt, 1965a), mitosis, cross wall deposi-

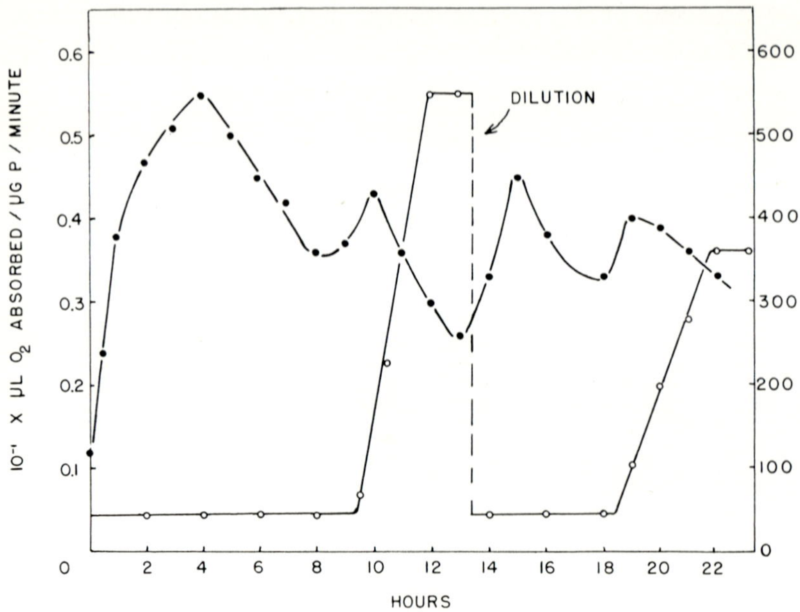

FIG. 5. Endogenous respiration rate during successive cycles of synchronous growth (under continuous illumination) of *C. pyrenoidosa* (strain 7-11-05). ●——●, endogenous respiration rate; ○——○, cell number. The culture was diluted in the light at the thirteenth hour of growth, with preequilibrated culture medium, to prevent low light intensity or nutrient deficiency from limiting the growth rate of the cells. (From Curnutt and Schmidt, 1964a.)

tion, and daughter-cell release serve to depress the ATP/ADP ratio even further, resulting in the net utilization of $(P_i)_x$.

Although this proposal seemed to be an attractive one, the intracellular ATP/ADP ratio was found (Curnutt and Schmidt, 1964b) to remain approximately 1 to 1 throughout the synchronous growth cycle of *C. pyrenoidosa* (Fig. 6). Since ATP and ADP are synthesized and metabolized in different cellular organelles (mitochondria, chloroplasts, cytoplasm, etc.), it is not surprising (in retrospect) that the total cellular levels of these nucleotides did not reflect their levels at the site of $(P_i)_x$ synthesis. This example serves to point out that if an intermediate is involved in different reactions (or the same reaction) in different cellular compartments, it is useless to measure the total cellular level of the intermediate to determine whether or not its level may be controlling a particular reaction. Thus, Kornberg's hypothesis that the relative concentrations of ATP and ADP control the synthesis and utilization of $(P_i)_x$ was not substantiated, but it was certainly not invalidated by these studies. Studies have been initiated to locate the intracellular sites of $(P_i)_x$ synthesis to determine the relationship of these sites to the energy-generating systems of the cell so that meaningful ATP/ADP ratio data can be obtained.

At approximately this stage in the $(P_i)_x$ research program, a puzzling

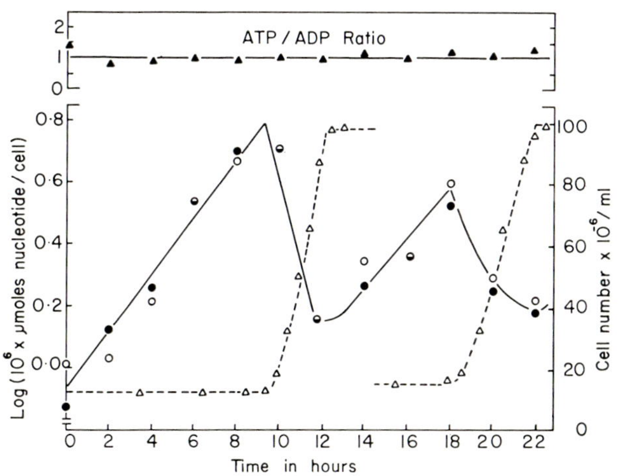

FIG. 6. The ATP/ADP ratio and the intracellular levels of these nucleotides during successive cycles of synchronous growth (under continuous illumination) of *C. pyrenoidosa* (strain 7-11-05). △——△, cell number, ▲——▲, ATP/ADP ratio; ○——○, ATP; ●——●, ADP. The culture was diluted in the light at the thirteenth hour of growth, with preequilibrated culture medium, to prevent low light intensity or nutrient deficiency from limiting the growth rate of the cells. (From Curnutt and Schmidt, 1964b.)

observation was made. Although the intracellular level of $(P_i)_x$ appeared to shift dramatically when expressed as percentage of the total cellular phosphorus, no periodism was observed when the concentration of this phosphate polymer was expressed on a per cell basis during synchronous growth of *C. pyrenoidosa* (Fig. 7). On a per cell basis, the $(P_i)_x$ level increased linearly throughout cellular development.

The apparent paradox between the 2 methods of expressing the concentration of $(P_i)_x$ results because the uptake of P_i is an exponential function while $(P_i)_x$ accumulation is a linear function during cellular development. Thus, a greater percentage of the P_i absorbed during early growth is utilized for $(P_i)_x$ synthesis than in the latter stages of cellular growth. On a percentage basis, therefore, the $(P_i)_x$ level would appear to increase during the initial phases and decrease during the latter stages of cellular development.

When the levels of RNA and DNA were also expressed on a per cell basis (Fig. 8), it appeared that $(P_i)_x$ was not an intracellular phosphorus and/or energy reserve for nucleic acid synthesis. The possibility remained, however, that the turnover rate of $(P_i)_x$ varied during the growth and division cycle and that its linear deposition was merely the net result of what had been synthesized and utilized.

To determine if the $(P_i)_x$ that accumulates in the cells prior to nuclear division is utilized during this event, the intracellular distribution of phos-

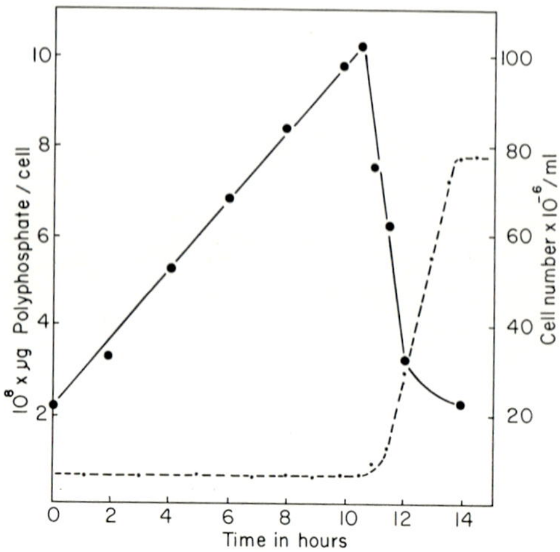

FIG. 7. Total inorganic polyphosphate per cell during synchronous growth of *C. pyrenoidosa* (strain 7-11-05). ●——●, inorganic polyphosphate; ·----· cell number. (Data from Baker and Schmidt, 1964b; R. R. Schmidt and Herrmann, 1965.)

phorus and radioactivity was measured (Baker and Schmidt, 1964b) when uniformly P^{32}-labeled cells were placed on a nonradioactive culture medium during nuclear division (Fig. 9). The changes in total radioactivity of the different phosphate fractions reflected the fate of the endogenous

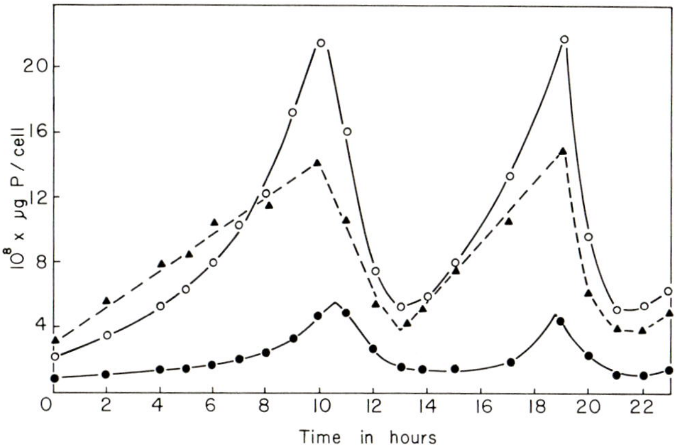

FIG. 8. The levels of inorganic polyphosphate-P, RNA-P, and DNA-P during 2 successive synchronous growth cycles (under continuous illumination) of *C. pyrenoidosa* (strain 7-11-05). ▲——▲, inorganic polyphosphate; ○——○, RNA-P; ●——●, DNA-P. (From Herrmann and Schmidt, 1965.)

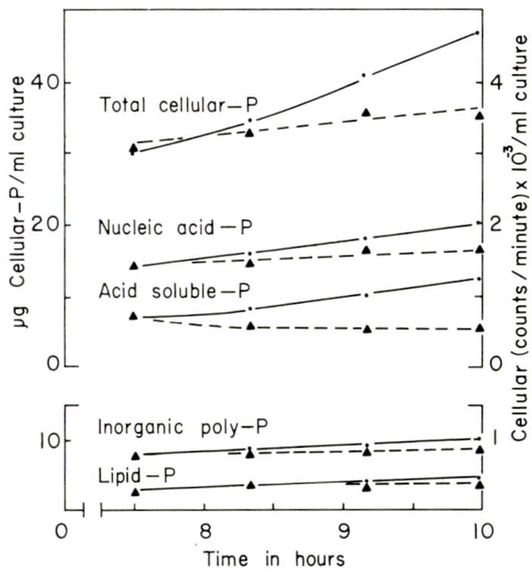

FIG. 9. The intracellular distribution of phosphorus and radioactivity when uniformly labeled (P^{32}) synchronized cells of *C. pyrenoidosa* (strain 7-11-05) were placed in a nonradioactive culture medium during the period of nuclear division. ●——●, phosphorus; ▲– – –▲, radioactivity. (From Baker and Schmidt, 1964b.)

phosphorus-containing compounds accumulated prior to nuclear division. Since there was no loss of radioactivity (but actually an increase) from the $(P_i)_x$ fraction, there appeared to be negligible turnover of this macromolecule during nuclear division. The main source of phosphorus for the synthesis of total nucleic acid appeared to be the culture medium.

To determine whether or not $(P_i)_x$ could be utilized by this organism during a period of environmental stress, synchronized cells were placed in a phosphorus deficient medium immediately prior to nuclear division and the changes in the intracellular distribution of phosphorus were measured during this event (Baker and Schmidt, 1964c). The $(P_i)_x$ was the only phosphorus fraction which immediately began to decrease in concentration under conditions of phosphate starvation (Fig. 10). The loss of phosphorus from the $(P_i)_x$ fraction was quantitatively accounted for in the increases of phosphorus in the nucleic acid and phospholipid fractions. By the end of the period of nuclear division, approximately 75% of the $(P_i)_x$ phosphorus could be accounted for as nucleic acid and phospholipid phosphorus.

$(P_i)_x$ seemed to play no role as an intracellular phosphorus and/or energy source during normal growth and division, but its utilization appeared to be inducible during certain environmental stresses (i.e., phosphate starvation), thereby increasing the cell's chances for survival. Furthermore, the accelerated respiration rate during normal nuclear division (Fig. 5),

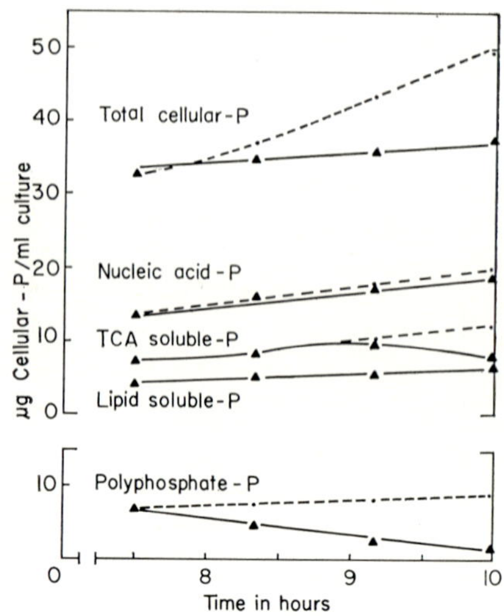

FIG. 10. Intracellular distribution of phosphorus in synchronized cells of *C. pyrenoidosa* (strain 7-11-05), when cultured in normal and phosphate-free culture medium, during the period of nuclear division. ●- - -●, normal culture medium; ▲——▲, phosphate-free culture medium. (From Baker and Schmidt, 1964c.)

not previously observed by Sorokin and Myers (1957), seemed likely to be meeting the increased energy requirements of the cell during this period of development.

This picture of the metabolic role of $(P_i)_x$, in the high temperature strain of *C. pyrenoidosa*, seemed to be complete until Miyachi *et al.* (1964) recently fractionated the total cellular $(P_i)_x$ of *C. ellipsoidea* into 4 apparently different fractions (A, B, C, and D), using differential extraction and precipitation procedures. They found these $(P_i)_x$ fractions to be metabolized differently under normal and phosphate-deficiency conditions. $(P_i)_x$ fractions A and C were used apparently for the synthesis of DNA and phosphoprotein while exogenous phosphate was the source of phosphorus for RNA synthesis under normal nutritional and photosynthetic conditions. $(P_i)_x$ fractions B and D, however, were not utilized under these conditions but were mobilized instead, along with A and C, for the synthesis of DNA, phosphoprotein, and RNA in the absence of exogenous phosphate. During phosphate sufficiency or deficiency in the dark, none of the $(P_i)_x$ fractions were appreciably utilized. Furthermore, the synthesis of these $(P_i)_x$ fractions, with the exception of fraction A, was light dependent.

Miyachi *et al.* (1964) have suggested that these 4 kinds of $(P_i)_x$ are located in different subcellular structures. Because the utilization and synthesis of $(P_i)_x$ fraction C was light dependent, it was assumed to exist in or near the chloroplast. The volutin granules were isolated (Hase *et al.*, 1963) and found to contain mainly $(P_i)_x$ fraction A. The synthesis of $(P_i)_x$ fraction A may be linked to the respiratory or glycolytic systems since the synthesis of this $(P_i)_x$ occurred in the dark.

Certain of the observations made by Miyachi and coworkers (Miyachi, 1961; Miyachi and Miyachi, 1961; Miyachi and Tamiya, 1961; Miyachi *et al.*, 1964) do not agree well with the current concepts for nucleic acid synthesis. Current evidence (see Section IV, Fig. 19) indicates that the deoxyribonucleoside diphosphates are synthesized by reduction of the corresponding ribonucleoside diphosphates. Therefore, the phosphate which is incorporated into RNA and DNA has a common origin, namely, ribose 5′-phosphate (Warren, 1961). Since phosphate from $(P_i)_x$ would have to enter DNA or RNA via ribose 5′-phosphate, it does not seem reasonable that phosphate from $(P_i)_x$ fractions A and C can be incorporated into DNA, while phosphate from the culture medium is simultaneously incorporated into RNA. One explanation is that $(P_i)_x$ fractions A and C are located in a cellular organelle (mitochondrion, chloroplast, etc.) where RNA synthesis is not occurring simultaneously with the synthesis of DNA in that organelle. Therefore, if RNA (and even DNA) synthesis was occurring, at the same time, in other cellular organelles at the expense of exogenous phosphate, the results described by these workers could have been obtained.

The present author believes, however, that the most probable explanation for the apparent transfer of phosphate from $(P_i)_x$ fraction C to DNA, stems from the inadequate procedure (Miyachi and Miyachi, 1961) used to estimate DNA phosphorus (and P^{32}). Since DNA, phosphoprotein, and $(P_i)_x$ fraction C were precipitated together following extraction, and then the phosphate content of each component determined by difference (after nonspecific hydrolysis steps), the probability of the contamination of the DNA by phosphoprotein or $(P_i)_x$ fraction C seems very high. Before it can be unequivocally stated that any $(P_i)_x$ fraction is utilized for DNA and phosphoprotein synthesis during normal growth of *Chlorella*, better methods for estimating the phosphate content of these macromolecules must be utilized.

In a recent study (R. R. Schmidt and Herrmann, 1965), the total cellular $(P_i)_x$ of the high temperature strain of *C. pyrenoidosa* was fractionated into $(P_i)_x$ A, B, C, and D using the procedure described for *C. ellipsoidea* by Miyachi and Miyachi (1961; Miyachi *et al.*, 1964). The major $(P_i)_x$ fractions in *C. pyrenoidosa* were found to be D and C (Fig. 11). These fractions represented approximately 30 and 70% of the total $(P_i)_x$, respectively. Although $(P_i)_x$ fraction A was apparently a major component of the volutin granules in *C. ellipsoidea*, it amounted to less than 1% of the total $(P_i)_x$. $(P_i)_x$ fraction B was virtually nonexistent in *C. pyrenoidosa*.

Although it is premature to speculate on the factors controlling the syn-

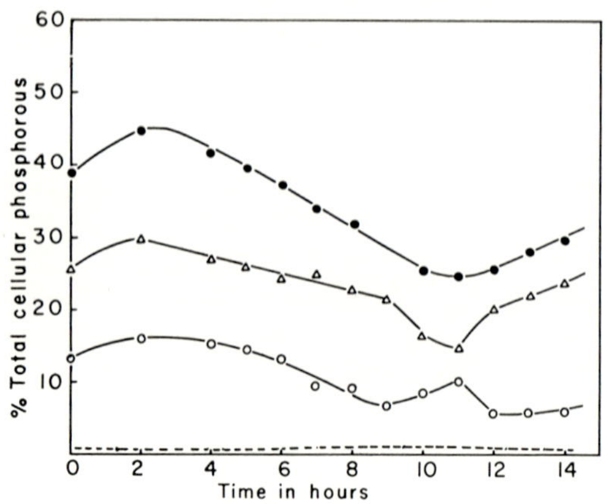

FIG. 11. Total inorganic polyphosphate and inorganic polyphosphate fractions A (acid-soluble), C (acid-insoluble, soluble in 2 *N* KOH, coprecipitates with potassium perchlorate), and D (same solubility characteristics as C but not precipitable with potassium perchlorate) during synchronous growth of *C. pyrenoidosa* (strain 7-11-05). ●——●, total polyphosphate; ·-----·, polyphosphate A; △——△, polyphosphate C; ○——○, polyphosphate D. Nuclear division began at 7.75 hours and cell release at 10 hours. (Data from R. R. Schmidt and Herrmann, 1965.)

thesis (and perhaps utilization) of $(P_i)_x$ fractions C and D in *C. pyrenoidosa*, it is interesting to note that $(P_i)_x$ fraction C follows a trend similar to the photosynthetic rate while $(P_i)_x$ fraction D exhibits periodism similar to the endogenous respiration rate during synchronous growth of this organism.

The differences in the pool sizes of the $(P_i)_x$ fractions in *C. pyrenoidosa* and *C. ellipsoidea*, the suggestion of differences in the ability of these 2 organisms to utilize $(P_i)_x$ under normal conditions, and the findings by Harold (1964) that certain mutants of *Aerobacter aerogenes* lack the enzymes for either synthesis or utilization of $(P_i)_x$ yet still grow normally, serve to illustrate the need for a comparative biochemical study of $(P_i)_x$ metabolism in the genus *Chlorella* before generalizations are made concerning $(P_i)_x$ metabolism in this group of organisms.

B. Carbohydrate Metabolism

The accelerated rate of respiration (Curnutt and Schmidt, 1964a), synthesis of RNA and DNA (R. R. Schmidt and Herrmann, 1965), and deposition of cell wall, etc., during nuclear division in the high temperature strain of *Chlorella pyrenoidosa*, when the rate of photosynthesis was at a minimum (Sorokin, 1957), suggested that a previously accumulated carbohydrate reserve was mobilized during this event to supply the carbon precursors (and energy) for organic synthesis. This suggestion indeed appeared to be true when the total starch levels were measured during 2 consecutive synchronous growth cycles of this organism (Fig. 12). The

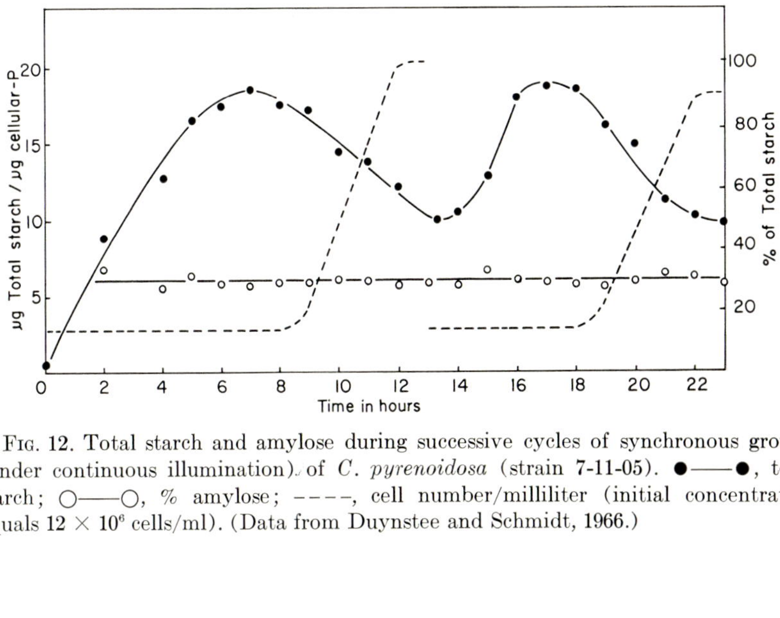

FIG. 12. Total starch and amylose during successive cycles of synchronous growth (under continuous illumination) of *C. pyrenoidosa* (strain 7-11-05). ●——●, total starch; ○——○, % amylose; ----, cell number/milliliter (initial concentration equals 12 × 10^6 cells/ml). (Data from Duynstee and Schmidt, 1966.)

total starch level increased rapidly, with the accelerated photosynthetic rate early in cellular development, reaching its maximum level approximately ½ hour prior to the initiation of the period of nuclear division. Its level then decreased throughout the periods of nuclear and cellular division, approaching a minimum by the end of these events. The amylose level was approximately 30% of the total starch, with the remainder (70%) as amylopectin, throughout the synchronous growth cycles. It should be noted that the total cellular phosphorus and cellular dry weight increase as parallel functions during the cell cycle; therefore, the total starch level will reflect the same periodism when expressed on the basis of either parameter.

Although the apparent decrease in the total starch level suggests its net utilization during the periods of nuclear and cellular division, the observed periodism may be reflecting only changes in the rate of starch synthesis rather than its utilization. This question can be answered from studies which measure the turnover rate of the starch, using C^{14} as a tracer, throughout the cell cycle.

C. Sulfur and Nitrogen Metabolism

Hase *et al.* (1958) have studied the deficiency of the macronutrients (nitrogen, phosphorus, magnesium, potassium, and sulfur) on the synchronous growth of *Chlorella ellipsoidea*. It was revealed that the effect of sulfur deficiency was unique in causing complete cessation of cellular division. The nucleus of these sulfur-deficient cells divided into 2, instead of the usual 8. Growth halted at this stage with the cells being unable to undergo further nuclear or cellular division.

This preliminary finding by Hase and coworkers concerning the possible involvement of sulfur in nuclear or cellular division, coupled with the finding of Rapkine (1931), who observed marked accumulations of acid-soluble compounds rich in sulfhydryl groups prior to cellular division in synchronized sea urchin eggs, along with the study (Mazia, 1954) in which the mitotic apparatus of sea urchin eggs was shown to be rich in sulfur-containing proteins possibly bound together by disulfide linkages, prompted our laboratory to initiate a research program to elucidate the relationship of sulfur metabolism to nuclear and cellular division in the high temperature strain of *C. pyrenoidosa*.

In the initial study (Johnson and Schmidt, 1963), the intracellular distribution of sulfur-S^{35} was measured during synchronous growth, after nonradioactive synchronized daughter cells were placed in a culture medium containing sulfate-S^{35}. The total cellular S^{35} (% of cellular dry weight) remained essentially constant throughout cellular development. However, the percentage of S^{35} (% total cellular S^{35}) in the total cellular protein fraction, cold TCA extract, and sulfolipid fractions of the cells showed definite periodism. Between the third and fifth hours of cellular development, the cold TCA-soluble fraction showed a marked increase in percentage of S^{35} while the protein fraction showed a decrease of equal magnitude (Fig. 13). The decrease in protein-S^{35} was traced to a decrease in the percentage of protein cystine-cysteine-S^{35} (Fig. 14). The percentage of protein methionine-S^{35} increased during this period but not enough to compensate for the decrease in protein cystine-cysteine-S^{35}. Thus, the percentage of total protein-S^{35} decreased. Between the fifth and seventh hours of cellular maturation, the percentage of S^{35} in the protein and TCA-soluble fractions returned to their original (daughter cell) levels. The protein sulfur-amino acids likewise returned to their initial levels. Approximately ½ hour after these shifts in the TCA and protein fractions of the cells had occurred, mitosis was initiated. A close relationship between these shifts and the initiation of nuclear division seemed possible.

Because this initial experiment measured the incorporation (and intracellular distribution) of sulfate-S^{35} into initially nonradioactive cells, the periodism in percentage of S^{35} of the protein sulfur-amino acids could have reflected changes in the concentration and/or fluctuations in the turnover rates of these amino acids in the protein.

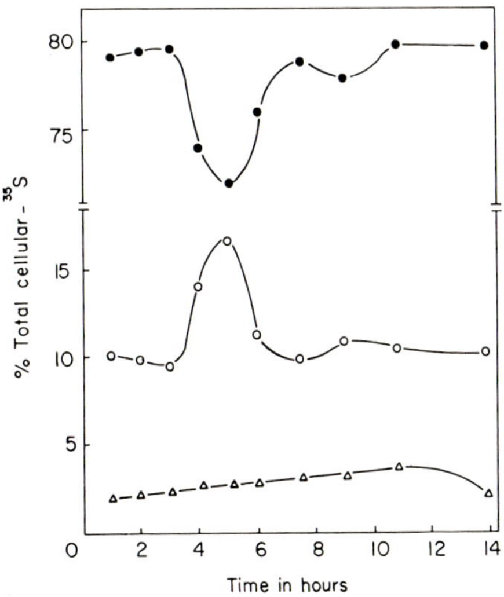

FIG. 13. Intracellular distribution of S^{35} during synchronous growth of *C. pyrenoidosa* (strain 7-11-05). ●——●, protein-S^{35}; ○——○, acid-soluble-S^{35}; △——△, sulfolipid-S^{35}. Nuclear division began at 7.50 hours and cell release at 9.75 hours of growth. (From Johnson and Schmidt, 1963.)

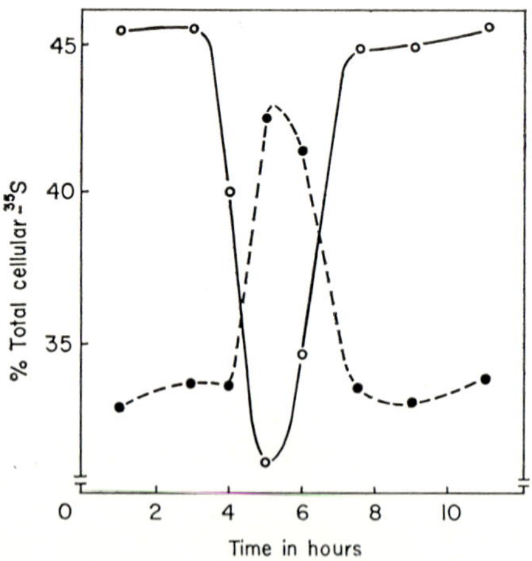

FIG. 14. Distribution of S^{35} in the total cellular protein during synchronous growth of *C. pyrenoidosa* (strain 7-11-05). ○——○, cysteine-plus cystine-S^{35}; ●——●, methionine-S^{35}. Nuclear division began at 7.5 hours and cell release at 9.75 hours of growth. (From Johnson and Schmidt, 1963.)

When the actual level of each protein amino acid was measured, the protein sulfur-amino acids were observed to exhibit negligible periodism in concentration (Fig. 15A–C). Therefore, the dramatic premitotic periodism in the percentage of S^{35} of the protein sulfur-amino acids represented changes in their turnover rate within the protein rather than shifts in their concentration therein. In view of these results, fluctuations in the turnover rates of the other protein amino acids, which remained at essentially constant levels (leucine, valine, isoleucine, phenylalanine, etc.), can be also anticipated during the course of cellular development.

Simultaneous with the studies on the distribution of the sulfur-amino acids in the protein fraction of the cells, experiments were initiated to identify the S^{35}-containing compound(s) of the TCA extracts (Johnson and Schmidt, 1963; Hare and Schmidt, 1965b). The TCA could be removed from the extracts with ethyl ether without loss of radioactivity from the original extract. Paper chromatography of the aqueous extracts (after removal of TCA) in solvent systems which moved all amino acids revealed that essentially all (greater than 98%) of the S^{35} was associated with a single, ninhydrin-positive, ultraviolet light-absorbing spot which remained at the origin of the chromatograms (only a trace of S^{35} was associated with the free sulfur-amino acids).

Acid hydrolysis (6 *N* HCl, 120°C, 24 hours, under nitrogen in a sealed tube) of the compound(s) remaining at the origins of these chromatograms yielded 6–9 different amino acids (glutamic acid, glycine, aspartic

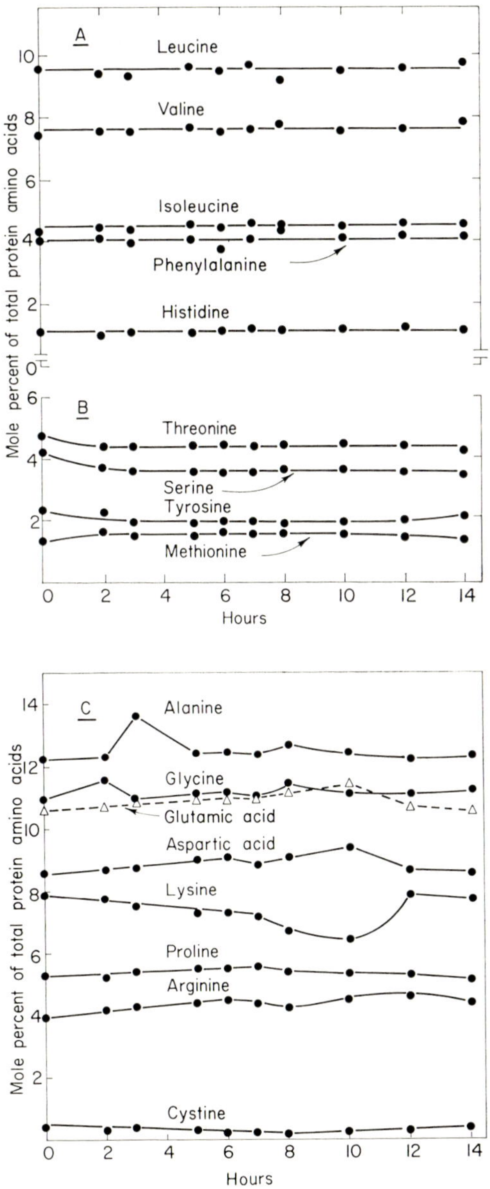

FIG. 15. A–C. Levels of protein amino acids during synchronous growth of *C. pyrenoidosa* (strain 7-11-05). Nuclear division began at 7.75 hours and cell release at 10 hours of growth. (From Hare and Schmidt, 1965a.)

$$\text{mole percent} = \frac{(\mu\text{mole amino acid}}{\mu\text{mole total protein amino acids)}} \times 100$$

acid, alanine, serine, leucine, isoleucine, threonine, and cysteic acid), and several ultraviolet light-absorbing compounds. All of the radioactivity was associated with cysteic acid which conceivably could have arisen from cysteine through oxidation by peroxides in the ethyl ether used to remove the TCA from the original extract.

Hase *et al.* (1959a), in a similar study with synchronous *C. ellipsoidea*, reported essentially no periodism in the S^{35} level of the protein fraction of the cells. Periodism was observed, however, in the S^{35} content of the TCA and ethanol extracts of the cells during cellular development. The major S^{35}-containing compounds in these extracts were tentatively identified as nucleotide-peptides or polynucleotide-peptides (Hase *et al.*, 1959b,c,d, 1960a,b).

The preliminary results obtained with the TCA extracts of the high temperature strain of *C. pyrenoidosa* suggested that these extracts might also contain peptides or nucleotide-peptides. Isolation, purification, and identification of these peptides or nucleotide-peptide derivatives began with TCA extraction (at 0°C) of approximately a 100 ml pellet of S^{35}-labeled cells harvested at the fifth hour of cellular development when the percentage of S^{35} was highest in the extract. The TCA was removed from the extract with benzene to avoid any oxidation of cysteine by peroxides. The extract was always maintained below pH 4.0 to render maximum stability (Raacke, 1961) to compounds containing mixed acid anhydride bonds, such as amino acid adenylates or similar nucleotide derivatives of peptides. The extract was concentrated by lyophilization and subjected to anion-exchange column chromatography using gradient elution (Fig. 16). Approximately 95% of the total radioactivity of the extract was eluted in the first 400 fractions. Upon acid hydrolysis, all major radioactive peaks released amino acids, suggesting the peptide nature of the radioactive compounds. Some of these radioactive peptides appeared to have an ultraviolet light-absorbing component associated with them (for example, fractions 45–65 and 280–300) while others seemed to exist as free peptides (see asymmetry between absorbancy at 260 mμ and radioactivity for fractions 155–185 and 355–385).

When fractions 280–300 were combined and further purification was attempted by cation-exchange column chromatography using gradient elution, the radioactive peptide (peptide A) and the ultraviolet light-absorbing component were eluted together (note symmetry of radioactivity and absorbancy at 260 mμ), suggesting a covalent linkage between these 2 components (Fig. 17). Similar column chromatography of the radioactive peptide in fractions (peptide B) 155–185 revealed that it probably exists as a free peptide (Fig. 18).

Although structural characterization of these 2 apparently different

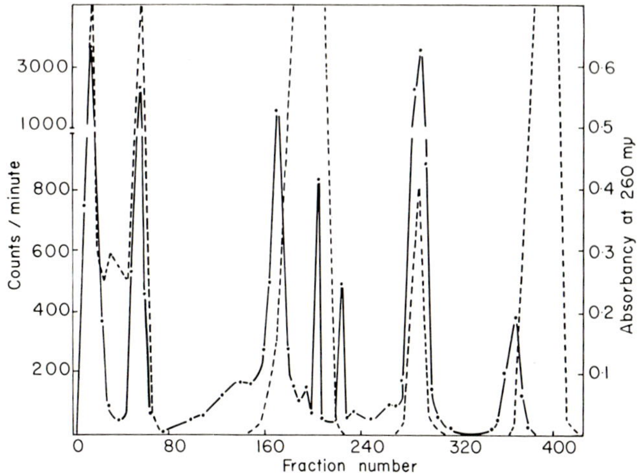

FIG. 16. Separation of the S^{35}-containing compounds, in a TCA extract of *C. pyrenoidosa* (strain 7-11-05), using anion-exchange column chromatography. ●——●, radioactivity; ----, absorbancy. Resin: Dowex-1-x4, formate form (200–400 mesh), sized by floatation to a range of 30–50 μ. Column size: 2.5 × 35 cm. Linear gradient elution: 0.005 *N* → 0.3 *N* formic acid in 400 fractions. Fraction size: 60 drops (approximately 4–5 ml). (Data from Hare and Schmidt, 1965b.)

types of peptides (peptides A and B) has not been completed, certain important findings can be reported. Peptide A contains three amino acids and a maximum absorbancy at 272 mμ. Furthermore, this peptide derivative reacts readily with hydroxylamine at pH 7.0, a property also exhibited by carboxyl-activated amino acids (i.e., amino acid adenylates; Raacke, 1961). In fact, because hydroxylamine only reacts rapidly with acid anhydride linkages under the conditions employed (Lippmann and Tuttle, 1945), this reaction has often been accepted as good supporting evidence for the existence of this type of bond in a molecule. These preliminary data indicate that peptide A may be a derivative of a nucleotide. In addition, the reaction of this peptide derivative with hydroxylamic suggests that the peptide is linked to the nucleotide by a mixed acid anhydride bond involving the C-terminal carboxyl group of the peptide and the phosphate of the nucleotide as reported for nucleotide-peptides in yeast (Schuur and Koningsberger, 1960; Harris and Wisman, 1962a,b,c).

A very surprising observation was that all radioactive peptides, obtained by anion-exchange column chromatography (Fig. 16) reacted readily with hydroxylamine (the column eluant itself was unreactive). Thus, 2 classes of "activated" peptides may have been discovered in *C. pyrenoidosa:* (*a*) a class in which the peptides are "activated" by nucleotides, and (*b*) a class in which the peptides are "activated" possibly by esterification with orthophosphate (transferred from GTP, etc.) as proposed by Noll *et al.* (1963).

Nucleotide-peptides have been observed in the acid extracts of microorganisms (Brown, 1958; Comb, 1962; Comb *et al.*, 1961; Cooper *et al.*, 1963; Gilbert and Yemm, 1958; Habermann, 1959; Harris and Davies, 1959; Harris and Wisman, 1962a,b,c; Hase *et al.*, 1959b,c,d, 1960a,b; Ito and Strominger, 1960a,b; Koningsberger *et al.*, 1957; Park, 1952a,b,c; Potter and Dounce, 1956; Strominger and Threnn, 1959) and mammalian tissues (Ondarza and Aubanel, 1960; Steinberg *et al.*, 1960; Strominger, 1959; Szafranski and Bagdasarian, 1961; Wilken and Hansen, 1961), however, the biochemical function of these compounds, in most cases, is not clear. There is evidence that the nucleotide-peptides, isolated from penicillin-treated *Staphylococcus aureus*, are involved in the synthesis of the bacterial cell wall (Park and Strominger, 1957; Strominger, 1959; Strominger and Threnn, 1959).

Although nucleotide-peptides may function in cell wall synthesis of certain bacteria, it has been postulated (Koningsberger *et al.*, 1957; Steinberg *et al.*, 1960; Szafranski and Bagdasarian, 1961) that they may also play an important role in the biosynthesis of other cellular proteins.

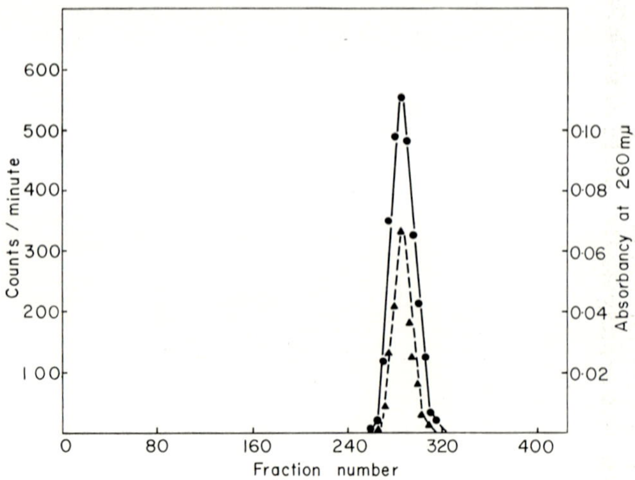

FIG. 17. Cation-exchange column chromatography of a S^{35}-containing peptide-derivative isolated from the TCA extract of *C. pyrenoidosa* (strain 7-11-05). ●——●, radioactivity; ▲– – –▲, absorbancy. (Peptide derivative obtained from fractions 280–300 of a Dowex-1-x4 column, see Fig. 16.) Resin: Dowex-50-x8 (200–400 mesh), sized by floatation to 40–60 μ range. Column size: 1 × 23 cm. Linear gradient elution: 0.005 $N \rightarrow 1.5\ N$ HCl in 400 fractions. Fraction size: 60 drops (approximately 4–5 ml). (Data from Hare and Schmidt, 1965b).

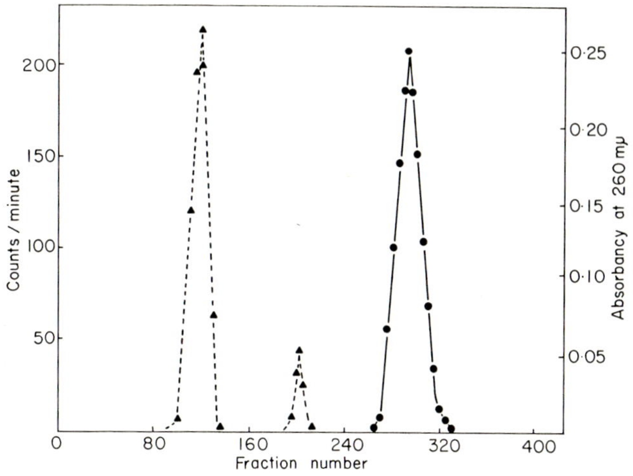

FIG. 18. Purification of a S^{35}-containing peptide by cation-exchange column chromatography after isolation from the TCA extract of *C. pyrenoidosa* (strain 7-11-05) by anion-exchange column chromatography. ●——●, radioactivity; ▲——▲, absorbancy. (Peptide obtained from fractions 155–185 of a Dowex-1-x4 column, see Fig. 16.) Resin: Dowex-50-x8 (200–400 mesh) sized by floatation to 40–60 μ range. Column size: 1 × 23 cm. Linear gradient elution: 0.005 $N \rightarrow$ 1.5 N HCl in 400 fractions. Fraction size: 60 drops (approximately 4–5 ml). (Data from Hare and Schmidt, 1965b.)

It has been difficult, however, to see how these nucleotide-peptides could fit into the scheme of protein biosynthesis as it is currently pictured (Watson, 1963).

We picture 2 major pathways for the synthesis of new functional cellular proteins (R. R. Schmidt and Hare, 1965): (*a*) synthesis from amino acids by the currently accepted pathway (amino acid activation, transfer RNA, mRNA, etc.; Watson, 1963), and (*b*) synthesis from peptides originating from pre-existing cellular proteins by nucleotide dependent transfer reactions analogous to those of glucose in starch biosynthesis involving uridine diphosphate glucose.

The relatively brief intervals during which conspicuous cytological changes occur within cells (nuclear division, cross-wall deposition, etc.) do predict that certain metabolic processes must be periodic or discontinuous during normal cellular development. There are undoubtedly enzymes and structural proteins that only function during these events. It would seem illogical, from the standpoint of the conservation of energy, that these proteins would remain as nonfunctional entities or would be completely disassembled to amino acids once they have functioned in these periodic metabolic events.

It is our hypothesis that many enzymes and structural proteins with dissimilar functional roles have a similar peptide subunit structure and that by rearrangement and/or by subtraction or addition of peptides from within or between these proteins that their specific biochemical function could be altered with a minimum change in free energy to the system (as compared to protein synthesis from amino acids). These peptide rearrangements within or between proteins are pictured to be, in many cases, nucleotide dependent, explaining the existence of nucleotide-peptides in most, if not all, organisms. By this mechanism, proteins having functioned at one stage of cellular development, could be utilized readily to function in new metabolic events of the next phase of cellular maturation. It is also visualized that as a cell grows there must be net protein synthesis. This protein synthesis, of course, would have to proceed from the amino acid level.

The above hypothesis is supported in part by (*a*) the reports concerning the subunit structure of certain enzymes (Monod *et al.*, 1963), (*b*) the observations that substrate specificity can change by changing the ratio of the contributing peptide subunits in certain enzymes (Markret *et al.*, 1963; Yielding and Tomkins, 1963), (*c*) the finding that cellular protein amino acids are in rapid turnover with the peptide and/or nucleotide-peptide fraction of the cells during certain stages of the cell cycle (Johnson and Schmidt, 1963; Hare and Schmidt, 1965a), and (*d*) the observation that synchronized cells of *C. pyrenoidosa*, placed in a nitrogen deficient medium at the middle of their growth cycle, continue growth at the same rate and divide at the same time as nitrogen sufficient cells (indicating a highly efficient and rapid use of preexisting protein sources because soluble nonprotein nitrogen reserves are very low in this organism; R. R. Schmidt, 1965; Hare and Schmidt, 1965a).

It can be anticipated, therefore, that if C^{14} is used as a tracer, instead of S^{35}, that a multitude of different nucleotide-peptides will be revealed to transiently accumulate and disappear during the synchronous growth cycle of *C. pyrenoidosa*.

IV. INTRACELLULAR CONTROL OF RNA AND DNA SYNTHESIS

Metabolic (Rose and Schweigert, 1953; Roll and Weliky, 1955; Roll *et al.*, 1956; Reichard, 1957; Amos and Magasanik, 1957; Edmonds, 1958; McNutt, 1958a,b; Reichard, 1958, 1959; Thomson *et al.*, 1962) and enzymatic (Abrams *et al.*, 1960; Moore and Hurlbert, 1960, 1962; Reichard *et al.*, 1961; Reichard, 1961, 1962; Moore and Reichard, 1963, 1964; Bertani *et al.*, 1963; Larsson, 1963; Laurent *et al.*, 1964; Moore *et al.*, 1964; Abrams and Duraiswami, 1965) evidence, from a number of differ-

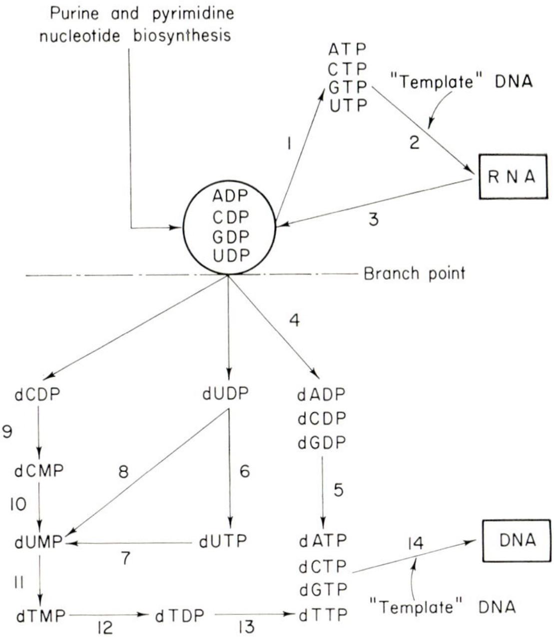

FIG. 19. Scheme for the biosynthesis of RNA and DNA. 1, ribonucleoside diphosphate kinases; 2, RNA polymerase; 3, polynucleotide phosphorylase; 4, ADP, CDP, etc., reductases; 5, deoxyribonucleoside diphosphate kinases; 6, dUDP kinase; 7, dUTP pyrophosphatase; 8, dUDP phosphatase; 9, dCDP phosphatase; 10, dCMP deaminase (aminohydrolase); 11, dTMP synthetase; 12, dTMP kinase; 13, dTDP kinase; 14, DNA polymerase. References supporting each enzymatic step are reported in the text.

ent organisms, strongly indicates that the nucleotide precursors for the biosynthesis of both RNA and DNA have a common biosynthetic pathway which branches at the ribonucleoside diphosphate level (Fig. 19). At this branch point, the ribonucleoside diphosphates can be phosphorylated by appropriate kinases (Berg and Joklik, 1953; Krebs and Hems, 1953), and then polymerized by RNA polymerase(s) to form RNA (Weiss and Gladstone, 1959; Weiss, 1960; Hurwitz *et al.*, 1960; Huang *et al.*, 1960; Edmonds and Abrams, 1960; Burma *et al.*, 1961; Weiss and Nakamoto, 1961; Chamberlin and Berg, 1962), or they can be reduced to their corresponding deoxy derivatives (Larsson, 1963; Laurent *et al.*, 1964). dATP, dCTP, and dGTP can be formed through phosphorylation (Canellakis and Mantsavinos, 1958; Bollum, 1958; Bertani *et al.*, 1963; Kielley, 1963) of their corresponding deoxyribonucleoside diphosphates, while dTTP can be formed from dUMP by a sequence of reactions which includes methylation of dUMP to form dTMP (Friedkin and Kornberg, 1957; Wahba and Friedkin, 1961; Blakley and McDougall, 1962; Friedkin *et al.*, 1962; Greenberg *et al.*, 1962) and phosphorylation of dTMP in apparently 2 steps to form dTTP (Friedkin, 1959; Kielley, 1963). dUMP is formed from deoxyuridine and/or deoxycytidine diphosphate by a series of reactions (Bertani *et al.*, 1961, 1963; G. F. Maley and Maley, 1959; Scarano, 1960). The 4 deoxyribonucleoside triphosphates can then be polymerized by DNA polymerase to form DNA (Lehman *et al.*, 1958; Bessman *et al.*, 1958).

It would appear that fluctuations in the level of one or more of the ribonucleoside diphosphates, within any subcellular organelle engaged in both RNA and DNA synthesis, would affect the rates of synthesis of both nucleic acids. Thus, the rate of DNA synthesis could not be limited by the level of the nucleotides in the nucleoside diphosphate pool without causing a corresponding limitation in the rate of RNA synthesis, unless the substrate affinities of one or more of the ribonucleoside diphosphate reductases were much lower (larger K_m) than those of the ribonucleoside diphosphate kinases.

Of interest from a metabolic control standpoint are the differential rates of accumulation of RNA and DNA in the high temperature strain of *Chlorella pyrenoidosa*. The level of DNA (% of total cellular phosphorus or dry weight) was observed (Herrmann and Schmidt, 1965) to decrease steadily while the RNA level increased continuously during premitotic growth in synchronous cultures of this organism (Fig. 20). Thus,

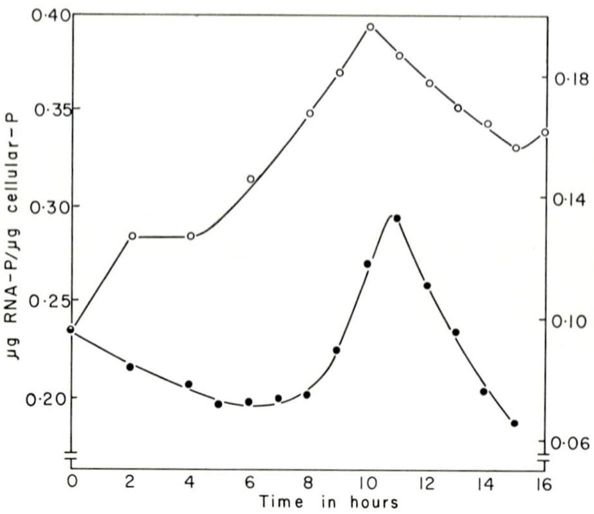

FIG. 20. The intracellular levels of RNA and DNA during synchronous growth of *C. pyrenoidosa* (strain 7-11-05). ○——○, RNA; ●——●, DNA. Nuclear division began at 7.75 hours and cell release at 10 hours of growth. (Data from Herrmann and Schmidt, 1965.)

if the ribonucleoside diphosphate pool is the branch point for the biosynthetic pathways for RNA and DNA in *C. pyrenoidosa,* the rate-limiting step in DNA synthesis, during premitotic development would likely be (*a*) one or more of the enzymatic steps converting the ribonucleoside diphosphates to the deoxyribonucleoside triphosphates, (*b*) the activity of the DNA polymerase, (*c*) the availability of the DNA "template", or (*d*) combinations of these possibilities.

To determine whether the rate-limiting step in DNA synthesis is one or more of the enzymatic steps converting the ribonucleoside diphosphates to the deoxyribonucleoside triphosphates, research was initiated to measure the apparent levels of the enzymes (prior to the branch point) involved in the biosynthesis of the ribonucleoside diphosphates as well as the enzymes (after the branch point) involved in the conversion of these nucleotides, to the deoxyribonucleoside triphosphates during the cell cycle.

Because of the rapid turnover of the bulk of the cellular RNA that is synthesized (Watson, 1963), it was anticipated that the apparent levels of the enzymes, involved in the synthesis of the ribonucleoside diphosphates, would increase at a rate equal to or greater than the rate of RNA accumulation (net result of synthesis and degradation) during cellular development. However, the apparent levels of the enzymes involved in the conversion of the ribonucleoside diphosphates to the deoxyribonucleoside triphosphates were anticipated to increase at a rate greater than the rate of accumulation of DNA (which exhibits negligible turnover compared to RNA) if they were not limiting DNA synthesis, while they would be expected to increase at a rate equal to the rate of DNA accumulation if they were limiting the synthesis of this macromolecule.

Because of its early position in the proposed biosynthetic pathway for RNA and DNA synthesis, the apparent level of aspartate transcarbamylase (the first enzyme in pyrimidine nucleotide biosynthesis) was measured throughout cellular development (Cole, 1964; Cole and Schmidt, 1964). The periodism in the apparent level of this enzyme was almost identical to the trend obtained for the accumulation of RNA (Fig. 21). It can be concluded, therefore, that there is essentially no rate-limiting step on the pyrimidine biosynthetic pathway between aspartate transcarbamylase and RNA, and that the apparent level of this enzyme may be controlling the rate of RNA synthesis. Also, the apparent levels of the enzymes, involved in the biosynthesis of the purine nucleotide precursors must increase at a rate equal to or greater than the rate of increase in the apparent level of aspartate transcarbamylase; otherwise, the apparent level of aspartate transcarbamylase and RNA accumulation could not be essentially identical functions during the cell cycle. Using

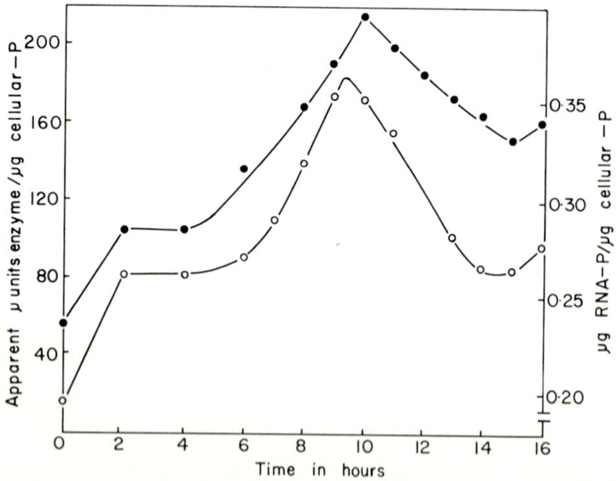

FIG. 21. Relationship between the level of cellular RNA phosphorus and the apparent level of aspartate transcarbamylase during synchronous growth of *C. pyrenoidosa* (strain 7-11-05). ●——●, RNA phosphorus; ○——○, apparent units of aspartate transcarbamylase. One enzyme unit equals 1 μmole product formed per minute. The RNA phosphorus level was measured in a synchronous growth cycle almost one hour longer (nuclear division began at 7.75 hours and cell release at 10 hours of growth) than the cycle during which the apparent level of aspartate transcarbamylase was measured. (Data from Herrmann and Schmidt, 1965; Cole and Schmidt, 1964.)

the same line of reasoning, the turnover rate of RNA must be nearly constant in order for the rate of RNA accumulation to reflect the activity of aspartate transcarbamylase throughout cellular development. Furthermore, since the apparent levels of the enzymes involved in the synthesis of both the purine and pyrimidine ribonucleoside diphosphates appear to increase at a rate equal to or greater than the rate of increase in the apparent level of aspartate transcarbamylase (which increases at a rate greater than the rate of DNA accumulation during premitotic development), it can be concluded that the rate of DNA accumulation is limited somewhere beyond the ribonucleoside diphosphate branch point which the present evidence now strongly supports.

A clue as to a possible rate-limiting step in the conversion of the ribonucleoside diphosphates to the deoxyribonucleoside triphosphates was revealed in studies on the apparent levels of the enzymes associated with deoxyribonucleotide metabolism in normal and regenerating liver. The kinases that phosphorylate the monophosphates of deoxycytidine, deoxyadenosine, and deoxyguanosine to the corresponding triphosphates were observed to be at approximately the same levels in both normal and regenerating rat liver (Mantsavinos and Canellakis, 1959; Canellakis *et al.*, 1959; Gray *et al.*, 1960). However, the levels of the enzymes (dCMP deaminase, dTMP synthetase, and dTMP kinase), involved in the synthesis of dTTP, increased markedly with the onset and/or the continuation of DNA synthesis in regenerating liver (Canellakis and Mantsavinos, 1958; Bollum and Potter, 1959; Hiatt and Bojarski, 1960; Weissman *et al.*, 1960; F. Maley and Maley, 1960; Myers, *et al.*, 1961; Myers, 1962; Beltz, 1962). Furthermore, it was demonstrated that all of the deoxyribonucleoside monophosphates except dTMP were actively phosphorylated in normal mouse liver; whereas, in mouse ascites hepatoma, a very actively regenerating tissue, dTMP was extensively phosphorylated (Kielley, 1963). Thus, it appeared that the regulation of the rate of dTTP synthesis might play an important role in controlling the rate of DNA synthesis, inasmuch as DNA polymerase requires the presence of all 4 deoxyribonucleoside triphosphates in order to synthesize DNA at a measurable rate (Bessman *et al.*, 1958).

The possibility that the levels or activities, of the enzymes on the dTTP biosynthetic pathway, may be limiting the rate of DNA synthesis during the cell cycle of *C. pyrenoidosa* is strongly suggested by the observation that periods of increased or decreased rates of DNA accumulation are preceded by periods during which the apparent levels of dCMP deaminase (Shen, 1965; Shen and Schmidt, 1966) and dTMP kinase (Johnson, 1964; Johnson and Schmidt, 1966) exhibit corresponding rates of increase or decrease, respectively (Fig. 22). However, before it can be unequivocally stated that the availability of dTTP is the only deoxyribonucleoside triphosphate limiting DNA synthesis, the apparent levels of the CDP, GDP, and ADP reductases and the kinases for dCDP, dGDP, dADP must be measured throughout cellular development. The measurement of the apparent levels of these enzymes along with the remaining enzymes in the dTTP biosynthetic pathway (CDP and UDP reductases; dCDP, dUDP, and dUTP phosphatases; dTMP synthetase; dTDP kinase) are currently in progress in this laboratory.

Although the apparent levels of the enzymes, involved in the biosynthesis of one or more of the deoxyribonucleoside triphosphates, may appear in themselves to be limiting DNA synthesis during a particular period of cellular development, other factors (the activity of DNA polymerase; availability of the DNA template) might be simultaneously limiting. Therefore, all potential levels of control of DNA synthesis should be evaluated before final conclusions are drawn concerning the rate-limiting step(s) in DNA synthesis.

A comparison, as mentioned earlier, of the apparent level of dTMP kinase with the apparent levels of the other deoxyribonucleoside mono-

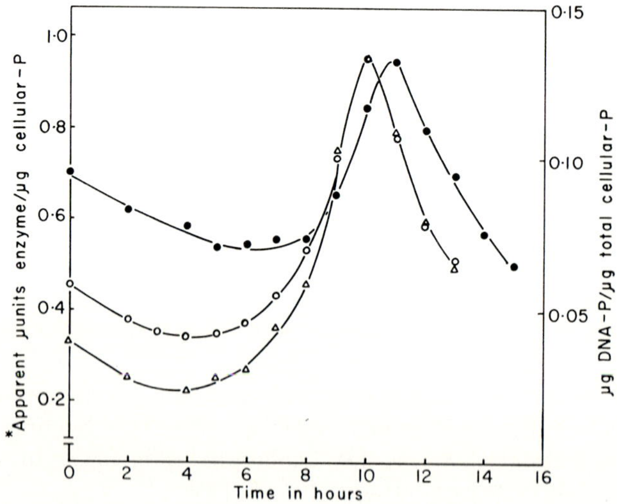

FIG. 22. Relationship between the level of cellular DNA phosphorus and the apparent levels of dCMP deaminase and dTMP kinase during synchronous growth of *C. pyrenoidosa* (strain 7-11-05). ●——●, DNA phosphorus; ○——○, dCMP deaminase; △——△, dTMP kinase. One enzyme unit equals 1 μmole product formed/minute. Nuclear division began at 7.75 hours and cell release at 10 hours of growth. (Data from Herrmann and Schmidt, 1965; Shen and Schmidt, 1966; Johnson and Schmidt, 1966.) Asterisk: The apparent level of dCMP deaminase was multiplied by a factor of 10^4 (i.e., $10^4 \times$ apparent μ units dCMP deaminase/μg cellular-P) in order to graph it with the apparent level of dTMP kinase.

phosphate kinases in normal and regenerating rat liver, may have little or no value for interpreting which enzymatic steps may be limiting DNA biosynthesis *in vivo*. Since the deoxyribonucleoside monophosphate kinases other than dTMP kinase do not appear to be on the direct pathway in the *de novo* synthesis of DNA (Fig. 19), these enzymes may fluctuate (or show no periodism) in apparent level in a manner quite independent of the enzymes on the direct pathway. Perhaps, it is not surprising that the activities of the deoxyribonucleoside monophosphate kinases, other than dTMP kinase, remain at approximately the same apparent level in normal and regenerating rat liver. A more valid comparison of enzyme activities, in lieu of our current knowledge, would be the apparent levels of the deoxyribonucleoside diphosphate kinases and dTMP kinase.

A related problem is the use of deoxythymidine (or free purine or pyrimidine bases) incorporation into DNA to estimate the time, rate, and duration of DNA synthesis during the cell cycle (Hornsey and Howard, 1956; Firket and Verly, 1958; Stanners and Till, 1960; Prescott, 1960; Sisken and Kinosita, 1961). Since deoxythymidine does not appear to be a normal intermediate in the *de novo* synthesis of DNA (Fig. 19), the activity or level of deoxythymidine kinase (which converts deoxythymidine to dTMP) may have no relationship to the levels or activities of the enzymes on the direct pathway, and may, therefore, be a rate-limiting step in the incorporation of deoxythymidine into DNA even during periods of active synthesis of this polymer. Thus, deoxythymidine incorporation in itself should not be used to elucidate the DNA cycle of a cell.

At present, the apparent levels of the deoxyribonucleoside monophosphate kinases and the deoxyribonucleoside kinases are also being measured, in this laboratory during the synchronous growth cycle of *C. pyrenoidosa,* to determine whether or not the problems outlined above are demonstrable.

The measurement of the apparent levels of enzymes *in vitro,* after rupturing the cellular integrity appears to have at least 2 serious shortcomings.

First, current evidence indicates that mitochondria, chloroplasts, and nuclei contain DNA which is undoubtedly genetically distinct (Chun *et al.*, 1963; Gibor and Izawa, 1963; Gibor and Granick, 1964; Kirk, 1964; Luck and Reich, 1964; Pollard, 1964; Sinclair *et al.*, 1965). In addition to DNA, each of these cellular organelles is endowed with ribosomes undoubtedly involved in the synthesis of specific organelle proteins using information transcribed from the DNA by a mRNA system (Lyttleton, 1962; App and Jagendorf, 1963; Brawerman, 1963a,b; Brawerman and Eisenstadt, 1964; Eisenstadt and Brawerman, 1964; Muramatsu *et al.*, 1964; Clark, 1965). Since each of these organelle types probably reproduces at a different time during the cell cycle (i.e., the chloroplast divides immediately prior to each nuclear division in *Chlorella*), the DNA replication and the RNA biosynthetic cycles (as well as the apparent levels of the enzymes associated with these processes) of these different organelle types are also undoubtedly asynchronous during cellular development. Thus, when the cellular RNA and DNA and the apparent levels of the enzymes associated with the synthesis of these macromolecules are measured in cell homogenates (or enzymes isolated from the same), a summation of the chemical nature and metabolic activities of all the subcellular organelles is obtained. However, since the bulk of the cellular DNA (and therefore RNA biosynthetic system) is nuclear in origin, *in vitro* studies with cell homogenates probably reflect the nature or activity of the nuclear system, except for those periods during the cell cycle when the rate of chloroplastic or mitochondrial DNA and RNA synthesis is proceeding maximally. If the detailed and subtle mechanisms and interrelationships which control DNA and RNA synthesis and degradation are to be realized, techniques must be developed to distinguish between the activity of the same metabolic process occurring within the same and different subcellular organelle types, during the cell cycle.

Second, different enzymes, even within the same metabolic pathway, by virtue of their inherent chemical and physical properties, suffer different degrees of denaturation during identical cell homogenization procedures. It is difficult, therefore, without extensive enzyme stability studies, to relate the apparent levels of different enzymes assayed *in vitro* to their actual related level *in vivo*. This problem was particularly evident when the apparent levels of dCMP deaminase and dTMP kinase were compared *in vitro* (Fig. 22). The ten thousandfold difference in the apparent levels of these 2 enzymes appears to be due, in large part, to the lability (G. F. Maley and Maley, 1959) of dCMP deaminase during the homogenization (sonication) procedure used to rupture this particular strain of *Chlorella*. Thus, generalizations should not be made directly from *in vitro* enzyme assays, concerning whether or not the absolute level of an enzyme may be limiting in a particular pathway, but rather generalizations should be made concerning only the rate of change in the apparent levels of enzymes within a pathway relative to the rate of change in the level of the final product of the pathway (i.e., apparent levels of enzymes in dTTP pathway versus dTTP or DNA level).

V. INTRACELLULAR CONTROL OF ENZYME SYNTHESIS AND ACTIVITY

From the excellent discussions of control phenomena by Jacob and Monod (1961), Monod *et al.* (1963), Pitot and Heidelberger (1963), and Umbarger (1964), it is evident that fluctuations in the enzymatic capacity of a cell can result from changes in the intracellular amount (actual level) of enzyme by control of enzyme synthesis at the gene level (i.e., induction or repression of enzyme synthesis), or by changes in the activity of a constant amount of enzyme by (*a*) activators or inhibitors acting at the enzyme level (i.e., end-product inhibition, etc.), or (*b*) factors imposed by the location of the enzyme within the cell and/or its organelles (pH, availability of substrates or cofactors, etc.).

Because one or more of the above control parameters may be operative at the same or different stages of cellular development, enzyme assays *in vivo* or *in vitro* reflect only an apparent enzyme level which may or may not be equal to the actual level of the enzyme. Therefore, after observing the dramatic periodism in the apparent levels of aspartate transcarbamylase (Fig. 21), and dCMP deaminase and dTMP kinase (Fig. 22), research was initiated to determine whether the observed periodism was due to changes in the rate of enzyme synthesis (or degradation), or due to changes in enzyme activity. Cellular compartmentation would not be, of course, a factor controlling enzyme activity in the *in vitro* system.

The approaches and results of this research endeavor are presented in the following sections.

A. Aspartate Transcarbamylase

Although the apparent level of aspartate transcarbamylase could be observed to exhibit dramatic periodism when expressed as a percentage of the total cellular phosphorus (Fig. 21), equally revealing information was obtained when the apparent level of this enzyme was expressed in its logarithmic form (Fig. 23). The apparent enzyme level, when measured during 2 consecutive synchronous growth cycles, was observed to increase at alternating exponential rates in each growth cycle. The trends in apparent enzyme level observed during the first synchronous growth cycle (with the exception of the first 2 hours of growth; were recurrent in the second growth cycle, without repeated exposure of the cells to the synchronization treatment, suggesting (Baker and Schmidt, 1964a) that these trends are associated with the normal development of a cell. Therefore, to facilitate discussion, only the trends in the first growth cycle will be considered in detail.

The rate of increase in the apparent level of aspartate transcarbamylase during early cellular development changed (increased) at the fifth hour. The period between the fifth and ninth hours of cellular development is interpreted as representing a period of accelerated synthesis of

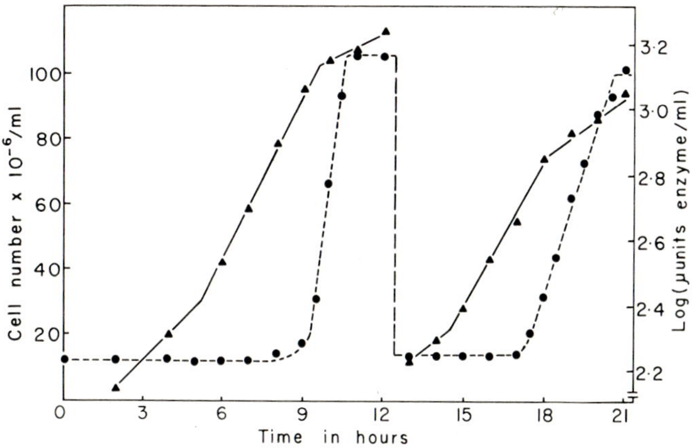

FIG. 23. The apparent level of aspartate transcarbamylase during successive synchronous growth cycles (under continuous illumination) of *C. pyrenoidosa* (strain 7-11-05). ▲——▲, apparent units of aspartate transcarbamylase; ●——●, cell number. One enzyme unit equals 1 μmole product formed/minute. (Data from Cole and Schmidt, 1964.)

this enzyme. This conclusion is partially supported by the fact that neither inhibition nor activation effects could be demonstrated during this period (Table I). If small molecule (heat stable) inhibitor(s) were present at either the fifth, sixth, or seventh hours, one would expect a decrease in enzyme activity by increasing the concentration of these factor(s) (by further addition of heat-treated supernatants from these points). Conversely, if (heat stable) activator(s) were present in increasing concentration, one would expect a progressive increase in activation by addition of the heat-treated supernatants during this same time interval. The possibility that an activator is synthesized in saturating amounts is not ruled out; however, in such a case, it would not be controlling the rate of increase in the apparent level of the enzyme.

At the ninth hour of cellular growth, there was a marked depression in the rate of increase in the apparent level of the enzyme. This depression suggested either a depressed rate of enzyme synthesis, the inception of an inhibition phenomenon, removal of activator(s), or any combinations thereof. Since end-product inhibition has been implicated by several workers (Yates and Pardee, 1957; Bresnick, 1962; Neumann and Jones, 1962) as a factor regulating the activity of this enzyme, the peak in acid-soluble nucleotide phosphorus (Baker and Schmidt, 1963) observed just prior to the depression in the apparent level of aspartate transcarbamylase, suggested the possibility of end-product inhibition beginning at the ninth hour of cellular development. To test this possibility, the studies reported in Tables I, II, and III were performed. The data presented in Table I clearly demonstrate the functioning of inhibitor(s) during the period when the rate of increase in apparent enzyme level was depressed (ninth, tenth, and twelfth hours). This observation is confirmed in the second growth cycle (sixteenth versus twentieth hours). Evidence that the inhibition is unique to certain periods of cellular development is obtained from the observation that heat-treated supernatant from the twelfth hour inhibits enzyme activity at the sixth hour. This observation is also confirmed in the second growth cycle (sixteenth versus twentieth hours). Furthermore, it should be noted that addition of heat-treated supernatant from the sixteenth hour had no effect upon twentieth hour activity.

The data in Table II further suggest end-product inhibition in *Chlorella pyrenoidosa* and are in good agreement with the data of Neumann and Jones (1962) for lettuce seedlings. It should be noted that the principal inhibitors of the enzyme from these 2 plant sources are pyrimidine nucleoside monophosphates which would be stable to the heat treatment used in these studies. (The cell homogenates from *C. pyrenoidosa*, possessing inhibitory properties, were found to retain 100% of their inhibi-

TABLE I

INHIBITION OF ASPARTATE TRANSCARBAMYLASE ACTIVITY AT CERTAIN PERIODS DURING CELL DEVELOPMENT OF *C. pyrenoidosa* (STRAIN 7-11-05)

Untreated crude enzyme[a]	Heat-treated crude enzyme[b]		Inhibition[c] (%)
Hour of growth	Hour of growth	Amount added to untreated crude enzyme (ml)	
5	5	0.2	0
		0.4	0
6	6	0.2	0
		0.4	<1
7	7	0.2	0
		0.4	<1
9	9	0.2	5.4
		0.4	8.4
10	10	0.2	6.1
		0.4	10.8
12	12	0.2	5.5
		0.4	10.2
16	16	0.4	<1
20	20	0.4	7.9
6	12	0.4	13.5
16	20	0.4	10.7
20	16	0.4	0

[a] For each study 0.2 ml of crude enzyme (2 hour, 100,000 *g* supernatant of a cell homogenate was used in the following incubation mixture: aspartic acid-2-C^{14}, 15.1 μM; carbamyl phosphate, 20 μM; 0.2 *M* Tris-HCl buffer (pH 9.2) containing 1 m*M* EDTA and GSH; final volume adjusted to 0.5 ml. The incubation time was 15 minutes at 37.5°C.

[b] 2.0 ml of crude enzyme was exposed to 100°C for 2 minutes, cooled in ice, centrifuged, and the specified volumes of the supernatant were used to test for inhibition. The supernatants of the heat-treated crude enzyme were in all cases shown to have an activity not greater than that of controls having no enzyme.

[c] Calculated as follows

$$100 - (\text{activity with addition} \times 100)/(\text{activity without addition})$$

(Data from Cole and Schmidt, 1964.)

tory activity during 4 minute exposures to 100°C). It is also interesting to note that UMP is the principal inhibitor of aspartate transcarbamylase activity in *C. pyrenoidosa* and lettuce seedlings, whereas CTP is the principal inhibitor in *Escherichia coli* (Gerhart and Pardee, 1963).

If the inhibitory compound(s) in *C. pyrenoidosa* are indeed nucleotide end products, it would be expected that they could be removed from the heat-treated supernatants by Norit-A adsorption. The data presented in Table III demonstrate that the inhibition observed in the studies pre-

TABLE II

END PRODUCT INHIBITION OF ASPARTATE TRANSCARBAMYLASE FROM *C. pyrenoidosa* (STRAIN 7-11-05)

Compound[a]	Final concentration (*M*)	Percentage of inhibition[b]
UMP	0.01	88.9
UDP	0.01	26.0
UTP	0.01	10.9
CMP	0.01	22.0
CTP	0.01	12.5
GMP	0.01	4.8
TMP	0.01	11.7
AMP	0.01	7.4
P_i	0.01	1.9

[a] In these experiments 0.2 ml of enzyme was assayed in the presence of the added compound as described in Table I.

[b] Calculated as in Table I. (Data from Cole and Schmidt, 1966.)

TABLE III

REMOVAL OF INHIBITOR(S) OF ASPARTATE TRANSCARBAMYLASE ACTIVITY FROM HEAT-TREATED CELLULAR EXTRACTS OF *C. pyrenoidosa* (STRAIN 7-11-05) WITH NORIT-A

Untreated crude enzyme[a]: Hour of growth	Heat-treated crude enzyme[a]: Hour of growth	Norit-A addition[b]	Inhibition (%)
20	20	−	8.1
20	20	+	0
16	16	−	0
16	16	+	0
16	20	−	10.6
16	20	+	0

[a] As in Table I.

[b] 0.2 ml of a Norit-A suspension (10 mg/ml) was added to 1.0 ml of the heat-treated enzyme supernatants (See Table I). After centrifugation, 0.4 ml of clear supernatant was used to test for inhibition. Studies without Norit-A addition were performed using 0.4 ml of heat-treated crude enzyme as in Table I. (Data from Cole and Schmidt, 1966.)

sented in Table I can be removed by addition of small quantities of Norit-A. The limited quantities of cellular material available in these studies (Cole, 1964; Cole and Schmidt, 1964, 1966) would not allow a more definitive identification of these inhibitors; however, characterization studies are currently in progress using extracts from large quantities of cells harvested at the eleventh hour of cellular development (Shen and Schmidt, 1966).

Though the inhibition phenomenon alone (by extrapolation) apparently cannot explain the magnitude of the reduced rate of increase in apparent level of aspartate transcarbamylase during the period following the ninth hour, it should be borne in mind that the concentration of both enzyme and inhibitor(s) are greatly altered during preparation of the enzyme (disruption of intracellular organelles, fifteenfold dilution of enzyme and inhibitor during sonication procedure, etc.), and the effect of such treatment cannot be estimated in this study. Therefore, concurrent with the control of aspartate transcarbamylase activity by Norit-A adsorbable compound(s), there might also exist a depressed rate of enzyme synthesis during this period (ninth through twelfth or seventeenth through twentieth hours). Since no stimulation of activity was observed at the twentieth hour by addition of supernatant from the sixteenth hour, coupled with the observation that no activation phenomena could be demonstrated at any point by further addition of their respective heat-treated supernatants, the removal of an activator seems least likely as a control factor in these studies.

Although some insight concerning the time and nature of the control of aspartate transcarbamylase activity has been obtained, the real biochemical challenge is to determine the nature of the factors controlling the synthesis of this enzyme during the cell cycle.

B. Deoxythymidylate Kinase

When the apparent level of dTMP is expressed on a per milliliter of culture basis (Fig. 24) or as percentage of the total cellular phosphorus (Fig. 22), the apparent level of this enzyme decreases during the period of reduced DNA accumulation (Johnson, 1964; Johnson and Schmidt, 1966). Although aspartate transcarbamylase shows a lower rate of increase during the same period, it never decreases in apparent level when expressed on a per milliliter basis (Fig. 23) as does the apparent level of dTMP kinase. The decrease in the apparent level of dTMP kinase suggests than an actual loss of enzyme occurred. However, the presence of an inhibitor(s) or the loss of an activator(s) could also conceivably cause the decrease in apparent enzyme level during this period of cellular development.

To determine whether or not small molecule inhibition or activation is operative in controlling dTMP kinase activity, boiled supernatant preparations from different periods of cellular development were added to enzyme assay mixtures (5 hour boiled preparation to fifth hour assay, 8 hour to fifth hour, 7 hour to seventh hour, 12 hour to seventh hour, 11 hour to eleventh hour, 13 hour to thirteenth hour, 11 hour to thirteenth hour) using a similar experimental design as outlined in the aspartate

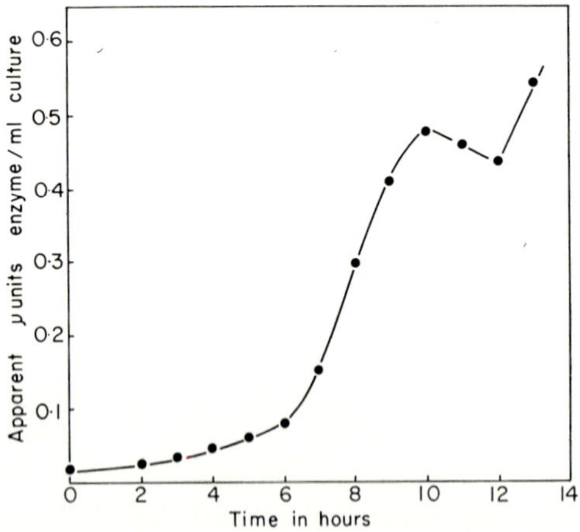

FIG. 24. The apparent level of dTMP kinase per milliliter of culture during synchronous growth of *C. pyrenoidosa* (strain 7-11-05). One unit equals 1 μmole product formed per minute. Nuclear division began at 7.75 hours and cell release at 10 hours of growth. (Data from Johnson and Schmidt, 1966.)

transcarbamylase studies (Cole and Schmidt, 1964; Table I). Surprisingly enough, no statistically significant alterations of enzyme activity were obtained by these additions. Furthermore, the failure of these boiled preparations to inhibit dTMP kinase indicates that the inhibition obtained with aspartate transcarbamylase is selective and is not a general inhibition phenomenon. Thus, it appears that very little, if any, small molecule (heat stable) inhibition or activation is operative in controlling dTMP kinase activity during the cell cycle.

The apparent level of dTMP kinase in homogenates of normal mouse liver was increased markedly by freezing and thawing these preparations in the presence of dTMP (Kielley, 1963). Kielley (1963) suggested that a reversible binding of this enzyme to subcellular particles may be controlling its activity. However, attempts to increase the apparent level of this enzyme by freezing and thawing homogenates of *C. pyrenoidosa*, in the presence of dTMP, always resulted in losses of enzyme activity. Thus, the fluctuations in apparent level of dTMP kinase during the cell cycle of *C. pyrenoidosa* are probably not a reflection of reversible particulate binding.

It seems most likely that the dramatic shifts in the apparent level of dTMP kinase in *C. pyrenoidosa* are a reflection of variations in the rate of enzyme synthesis. However, the decrease in enzyme level, between the tenth and twelfth hours, suggests that enzyme breakdown may also play a role in regulating the intracellular level of this enzyme. Although the loss of dTMP kinase may represent simply enzyme denaturation during a period of reduced enzyme synthesis or substrate availability (Hiatt and Bojarski, 1961), the possibility that there is an enzyme (under gene control) which is involved in the inactivation or hydrolysis of dTMP kinase should not be overlooked.

C. Deoxycytidylate Deaminase (Aminohydrolase)

Because the apparent levels of dCMP deaminase and dTMP kinase fluctuate similarly during cellular development (Fig. 22), it seemed likely that the rate of enzyme synthesis (and breakdown), not small molecule inhibition or activation, was also controlling the periodism in the apparent level of dCMP deaminase. However, because certain deoxyribonucleoside triphosphates, added *in vitro,* have been reported (G. F. Maley and Maley, 1962, 1963) to inhibit or activate dCMP deaminase, the possibility of small molecule regulation of this enzyme during the cell cycle could not be disregarded without further experimentation.

A new experimental approach was designed to measure the relative degree of inhibition or activation of enzyme activity by the intracellular small molecule fraction throughout the cell cycle. This approach consisted of dialyzing a 100,000 *g* supernatant of a cell homogenate to obtain a preparation of dCMP deaminase free of small molecules. Using aliquots of constant volume from the dialyzed enzyme preparation plus the components of the normal incubation mixture, the percent inhibition or activation of enzyme activity, by the addition of appropriate size aliquots of boiled cell homogenates (undialyzed) from different hours of the cell cycle, could be obtained.

Because the boiled cell homogenates were prepared from cell pellets which differed slightly in size (total cellular phosphorus), adjustments were made in the volumes of the boiled homogenate additions so that they represented an equivalent amount of cellular phosphorus throughout the synchronous growth cycle. By this technique, it was possible to relate the effect of the small molecule fraction on enzyme activity to the increase in cellular phosphorus (which has the same rate of increase as the cellular dry weight) throughout the cell cycle.

Upon dialysis of a 100,000 *g* supernatant of a cell homogenate, in preparation for the experiment outlined above, approximately a 60% increase in dCMP deaminase activity was observed. This initial observation suggested that the activity of this enzyme might be under small molecule control during cellular development. Because aliquots of the small molecule fraction, representing the same amount of cellular phosphorus, produced the same percent inhibition of enzyme activity (Table IV), it can be concluded that the level of inhibitor(s) increased at the same expo-

TABLE IV

INHIBITION OF dCMP DEAMINASE ACTIVITY BY HEAT-TREATED CELL HOMOGENATES FROM DIFFERENT PERIODS OF CELLULAR DEVELOPMENT OF *C. pyrenoidosa* (STRAIN 7-11-05)[a]

Hour of growth	Heat-treated cell homogenates[b] (ml)	Inhibition[c] (%)
0	0.14	66.7
2	0.12	66.1
3	0.16	66.7
4	0.15	66.9
5	0.18	66.1
6	0.20	69.3
7	0.18	67.8
8	0.20	66.5
9	0.17	65.6
10	0.18	67.9
11	0.19	67.5
12	0.16	65.3
13	0.15	66.9
14	0.17	66.1

[a] Incubation mixture: 0.2 ml enzyme (obtained by dialyzing an 100,000 g supernatant of a cell homogenate for 24 hours in Tris-buffer), 6 μM dCMP-2-C^{14}, 100 μM Tris-buffer, pH 7.7. The final volume was adjusted to 0.6 ml with H_2O and incubated at 38.5° C for 35 minutes.

[b] Treated the same as the crude enzyme in Table I. Each aliquot represents the same amount of cellular phosphorus prior to heat treatment.

[c] Calculated as in Table I. (Data from Shen and Schmidt, 1966.)

nential rate as the cellular phosphorus during the synchronous growth cycle. Furthermore, the percent inhibition of enzyme activity by the small molecule fraction remained constant (Table IV), while the apparent level of dCMP deaminase was periodic (Fig. 22), when both are expressed on a phosphorus basis. These observations strongly suggest that the inhibitor(s) level in the homogenates was not responsible for the fluctuations in dCMP deaminase activity during cellular development.

It seemed possible, however, that the inhibition of dCMP deaminase was of the incomplete type similar to that observed by Gerhart and Pardee (1963) with aspatate transcarbamylase and CTP. The percent inhibition of aspartate transcarbamylase approached a finite value less than 100% regardless of increases in the concentration of CTP over a certain level. If this type of inhibition was operative in the present study, the percent inhibition of enzyme activity may not have reflected the actual level of inhibitor during the cell cycle. The levels of inhibitor, in the different boiled preparations may have all saturated the amount of dialyzed enzyme used in these assays, resulting in the constant percent inhibition of enzyme activity observed in Table IV. However, this possibility was ruled out because increases or decreases, in the amount of boiled cell homogenate, added to the assay mixtures produced corresponding increases or decreases in the percent inhibition of enzyme activity (Fig. 25). Although the data in Table IV does indeed reflect the level of inhibitor(s) throughout the cell cycle, the inhibition does appear to be of the incomplete type because the percent inhibition of enzyme activity plateaus at approximately 74% as the concentration of inhibitor is increased (Fig. 25). This observation suggests that the inhibition may be allosteric in nature.

Although the activity of dCMP deaminase appears to be suppressed approximately 60% during the cell cycle, the possibility remains that this inhibition observed *in vitro* is not operative *in vivo* but only results from the exposure of the enzyme, after cell disruption, to an inhibitor which has a different intracellular location than the enzyme. However, if the inhibition is operative *in vivo*, it may mean that the formation of dUMP from dCMP is suppressed favoring the formation of dUMP from dUDP or dUTP (Fig. 19).

Because dCMP deaminase and dTMP kinase are separated by one enzymatic step in the dTTP biosynthetic pathway and the apparent levels of both enzymes still exhibit almost identical periodism during the cell cycle, it seems unlikely that the synthesis of each enzyme in this pathway is sequentially induced at the availability of the product of the previous enzymatic step in the pathway. It is more probable that the structural genes for the enzymes on this pathway lie in the same operon. Thus,

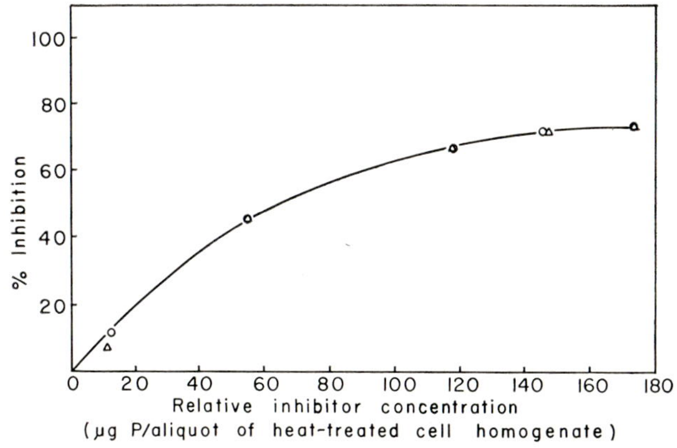

FIG. 25. Inhibition of dCMP deaminase activity by different concentrations of intracellular inhibitor(s) from the ninth and eleventh hours of cellular development of *C. pyrenoidosa* (strain 7-11-05). △——△, ninth hour; ○——○, eleventh hour. (Data from Shen and Schmidt, 1966.)

the rate of transcription of the genetic information in this operon could almost simultaneously control the rate of synthesis of all enzymes whose structural genes are located therein.

Masters *et al.* (1964) have suggested that the periods of the cell cycle during which certain enzymes double in concentration (in bacteria) may be closely correlated with the replication of the corresponding structural genes of these enzymes. Furthermore, they indicate that it may be possible, if the timing of enzyme synthesis parallels the genetic map, to map certain genes whose mutations are usually lethal and which therefore cannot be mapped by the usual procedures.

Although their proposal is an attractive one and may have merit for certain gene-enzyme relationships, the enzymatic studies with *Chlorella pyrenoidosa* described in the present chapter, as well as the *in vivo* studies of the factors controlling inorganic polyphosphate synthesis and breakdown (Herrmann and Schmidt, 1965), serve to illustrate that control mechanisms at the enzyme level (allosteric affectors, substrate availability, etc.) may obscure the actual period of enzyme synthesis during the cell cycle. Furthermore, periodism in the activity of a key enzyme, such as aspartate transcarbamylase, which can regulate the overall rate of RNA synthesis may limit the rate of transcription (mRNA synthesis) of the structural gene(s) even though the cell's potential for enzyme synthesis has been increased by gene replication. Because of these and other interrelated factors regulating the cell's enzymatic capacity, it seems doubtful that there generally will be a close relationship between the time of gene replication and the rate of enzymatic expression during the cell cycle.

Acknowledgments

The author wishes to express his deepest thanks to his former (Dr. A. L. Baker, Dr. F. E. Cole, Dr. R. A. Johnson, Mrs. S. C. Furman, and Mr. E. C. Herrmann) and present (Mr. T. A. Hare, Miss S. R. C. Shen, Mr. E. E. Duynstee, Mr. F. S. Baechtel, and Mr. T. H. Liu) graduate students who have assisted in making this chapter a reality. Thanks are also due to Miss M. K. Greene and Mrs. G. H. Rayfield for their expert technical assistance during the course of these studies.

The author also wishes to express his gratitude to Dr. R. W. Engel, Department Head, for creating and maintaining an academic climate which has been conducive to maximum intellectual and professional development.

The research discussed in this chapter has been supported in large part by grants from the National Institutes of Health (PHS GM 12042-01 and 12042-02) and the National Science Foundation (NSF G-15837 and GB-1960).

REFERENCES

Abrams, R., and Duraiswami, S. (1965). *Biochem. Biophys. Res. Commun.* **18,** 409.

Abrams, R., Libenson, L., and Edmonds, M. (1960). *Biochem. Biophys. Res. Commun.* **3,** 272.

Amos, H., and Magasanik, B. (1957). *J. Biol. Chem.* **229,** 653.

App, A. A., and Jagendorf, A. T. (1963). *Biochim. Biophys. Acta* **76,** 286.

Baker, A. L., and Schmidt, R. R. (1963). *Biochim. Biophys. Acta* **74,** 75.

Baker, A. L., and Schmidt, R. R. (1964a). *Biochim. Biophys. Acta* **82,** 336.

Baker, A. L., and Schmidt, R. R. (1964b). *Biochim. Biophys. Acta* **82,** 624.

Baker, A. L., and Schmidt, R. R. (1964c). *Biochim. Biophys. Acta* **93,** 180.

Beltz, R. E. (1962). *Arch. Biochem. Biophys.* **99,** 304.

Berg, P., and Joklik, W. K. (1953). *Nature* **172,** 1008.

Bertani, L. E., Häggmark, A., and Reichard, P. (1961). *J. Biol. Chem.* **236,** PC67.

Bertani, L. E., Häggmark, A., and Reichard, P. (1963). *J. Biol. Chem.* **238,** 3407.

Bessman, M. J., Lehman, I. R., Simms, E. S., and Kornberg, A. (1958). *J. Biol. Chem.* **233,** 171.

Blakley, R. L., and McDougall, B. M. (1962). *J. Biol. Chem.* **237,** 812.

Bollum, F. J. (1958). *J. Am. Chem. Soc.* **80,** 1766.

Bollum, F. J., and Potter, V. R. (1959). *Cancer Res.* **19,** 561.

Brawerman, G. (1963a). *Biochim. Biophys. Acta* **72,** 317.

Brawerman, G. (1963b). *Biochim. Biophys. Acta* **76,** 322.

Brawerman, G., and Eisenstadt, J. M. (1964). *J. Mol. Biol.* **10,** 403.

Bresnick, E. (1962). *Biochim. Biophys. Acta* **61,** 598.

Brown, A. D. (1958). *Biochim. Biophys. Acta* **30,** 447.

Burma, D. P., Kröger, H., Ochoa, S., Warner, R. C., and Weill, J. D. (1961). *Proc. Natl. Acad. Sci. U.S.* **47,** 749.

Burns, V. W. (1962). *Progr. Biophys. Biophys. Chem.* **12,** 1.

Campbell, A. (1957). *Bacteriol. Rev.* **21,** 263.

Canellakis, E. S., and Mantsavinos, R. (1958). *Biochim. Biophys. Acta* **27,** 643.

Canellakis, E. S., Jaffe, J. J., Mantsavinos, R., and Krakow, J. S. (1959). *J. Biol. Chem.* **234,** 2096.

Chamberlin, M., and Berg, P. (1962). *Proc. Natl. Acad. Sci. U.S.* **48,** 81.

Chun, E. H. L., Vaughan, M. H., and Rich, A. (1963). *J. Mol. Biol.* **7,** 130.

Clark, M. F. (1965). *Biochim. Biophys. Acta* **91,** 671.

Cole, F. E. (1964). Dissertation, Virginia Polytechnic Institute (Blacksburg).

Cole, F. E., and Schmidt, R. R. (1964). *Biochim. Biophys. Acta* **90,** 616.

Cole, F. E., and Schmidt, R. R. (1966). *Arch. Biochem. Biophys.* in preparation.

Comb, D. G. (1962). *J. Biol. Chem.* **237,** 1601.

Comb, D. G., Chin, W., and Roseman, S. (1961). *Biochim. Biophys. Acta* **46,** 394.

Cooper, A. H., Harris, G., Neal, G. E., and Wisman, A. (1963). *Biochim. Biophys. Acta* **68,** 68.

Curnutt, S. G., and Schmidt, R. R. (1964a). *Exptl. Cell Res.* **36,** 102.

Curnutt, S. G., and Schmidt, R. R. (1964b). *Biochim. Biophys. Acta* **86,** 201.

Duynstee, E. E., and Schmidt, R. R. (1966). *Plant Physiol.* (in press).

Edmonds, M., and Abrams, R. J. (1960). *J. Biol. Chem.* **235,** 1142.

Edmonds, M. (1958). *Federation Proc.* **17,** 215.

Eisenstadt, J. M., and Brawerman, G. (1964). *J. Mol. Biol.* **10,** 392.

Firket, H., and Verly, W. G. (1958). *Nature* **181,** 274.

Fogg, G. E. (1953). "The Metabolism of Algae." Methuen, London (Wiley, New York).

Friedkin, M. (1959). *In* "The Kinetics of Cellular Proliferation" (F. Stohlman, ed.) p. 97. Grune & Stratton, New York.

Friedkin, M., and Kornberg, A. (1957). *In* "The Chemical Basis of Heredity" (W. D. McElroy and B. Glass, eds.), pp. 609–613. Johns Hopkins Press, Baltimore, Maryland.

Friedkin, M., Crawford, E. J., Donovan, E., and Pastore, E. J. (1962). *J. Biol. Chem.* **237,** 3811.
Gerhart, J. C., and Pardee, A. B. (1963). *Cold Spring Harbor Symp. Quant. Biol.* **28,** 491.
Gibor, A., and Granick, S. (1964). *Science* **145,** 890.
Gibor, A., and Izawa, M. (1963). *Proc. Natl. Acad. Sci. U.S.* **50,** 1164.
Gilbert, D. A., and Yemm, E. W. (1958). *Nature* **182,** 1745.
Gorman, J., Taruo, P., LaBerge, M., and Halvorson, H. (1964). *Biochem. Biophys. Res. Commun.* **15,** 43.
Gray, E. D., Weissman, S. M., Richards, J., Bell, D., Keir, H. M., Smellie, R. M. S., and Davidson, J. N. (1960). *Biochim. Biophys. Acta* **45,** 111.
Greenberg, G. R., Somerville, R. L., and DeWolf, S. (1962). *Proc. Natl. Acad. Sci. U.S.* **48,** 242.
Habermann, V. (1959). *Biochim. Biophys. Acta* **32,** 297.
Hare, T. A., and Schmidt, R. R. (1965a). *J. Cellular Comp. Physiol.* **65,** 63.
Hare, T. A., and Schmidt, R. R. (1965b). Unpublished data.
Harold, F. M. (1964). *J. Gen. Microbiol.* **35,** 81.
Harris, G., and Davies, J. W. (1959). *Nature* **184,** 788.
Harris, G., and Wisman, A. (1962a). *Biochim. Biophys. Acta* **55,** 374.
Harris, G., and Wisman, A. (1962b). *Biochim. Biophys. Acta* **61,** 395.
Harris, G., and Wisman, A. (1962c). *Biochim. Biophys. Acta* **61,** 478.
Hase, E., Morimura, Y., Mihara, S., and Tamiya, H. (1958). *Arch. Mikrobiol.* **32,** 87.
Hase, E., Otsuka, H., Mihara, S., and Tamiya, H. (1959a). *Biochim. Biophys. Acta* **35,** 180.
Hase, E., Mihara, S., Otsuka, H., and Tamiya, H. (1959b). *Biochim. Biophys. Acta* **32,** 298.
Hase, E., Mihara, S., Otsuka, H., and Tamiya, H. (1959c). *Arch. Biochem. Biophys.* **83,** 170.
Hase, E., Mihara, S., and Otsuka, H. (1959d). *J. Gen. Appl. Microbiol.* **5,** 43.
Hase, E., Mihara, S., and Tamiya, H. (1960a). *J. Gen. Appl. Microbiol.* **6,** 61.
Hase, E., Mihara, S., and Tamiya, H. (1960b). *Plant Cell Physiol.* (*Tokyo*) **1,** 131.
Hase, E., Miyachi, S., and Mihara, S. (1963). *Plant Cell Physiol.* (*Tokyo*) **4,** 619.
Herrmann, E. C., and Schmidt, R. R. (1965). *Biochim. Biophys. Acta* **95,** 63.
Hiatt, H. H., and Bojarski, T. B. (1961). *Cold Spring Harbor Symp. Quant. Biol.* **26,** 367.
Hoogenhaut, H. (1963). *Phycologia* **2,** 135.
Hornsey, S., and Howard, A. (1956). *Ann. N.Y. Acad. Sci.* **63,** 915.
Huang, R. C., Maheshwari, N., and Bonner, J. (1960). *Biochem. Biophys. Res. Commun.* **3,** 689.
Hurwitz, J., Bresler, A., and Diringer, R. (1960). *Biochem. Biophys. Res. Commun.* **3,** 15.
Ito, E., and Strominger, J. L. (1960a). *J. Biol. Chem.* **235,** 553.
Ito, E., and Strominger, J. L. (1960b). *J. Biol. Chem.* **235,** 555.
Iwamura, T., and Kuwashima, S. (1964). *J. Gen. Appl. Microbiol.* **10,** 83.
Jacob, F., and Monod, J. (1961). *Cold Spring Harbor Symp. Quant. Biol.* **26,** 193.
Johnson, R. A. (1964). Dissertation, Virginia Polytechnic Institute (Blacksburg).
Johnson, R. A., and Schmidt, R. R. (1963). *Biochim. Biophys. Acta* **74,** 428.
Johnson, R. A., and Schmidt, R. R. (1966). *Biochim. Biophys. Acta* (in press).
Kielley, R. K. (1963). *Cancer Res.* **23,** 801.
Kirk, J. T. O. (1964). *Biochem. Biophys. Res. Commun.* **16,** 233.
Koningsberger, V. V., von der Grinten, C. O., and Overbeek, J. T. G. (1957). *Biochim. Biophys. Acta* **26,** 483.

Kornberg, A., Kornberg, S. R., and Simms, E. S. (1956). *Biochim. Biophys. Acta* **20,** 215.
Kornberg, S. R. (1957). *Biochim. Biophys. Acta* **26,** 294.
Krebs, H. A., and Hems, R. (1953). *Biochim. Biophys. Acta* **12,** 172.
Larsson, A. (1963). *J. Biol. Chem.* **238,** 3414.
Laurent, T. C., Moore, E. C., and Reichard, P. (1964). *J. Biol. Chem.* **239,** 3436.
Lehman, I. R., Bessman, M. J., Simms, E. S., and Kornberg, A. (1958). *J. Biol. Chem.* **233,** 163.
Lippmann, F., and Tuttle, L. C. (1945). *J. Biol. Chem.* **159,** 21.
Luck, D. J. L., and Reich, E. (1964). *Proc. Natl. Acad. Sci. U.S.* **52,** 931.
Lyttleton, J. W. (1962). *Exptl. Cell Res.* **26,** 312.
McNutt, W. S. (1958a). *J. Biol. Chem.* **233,** 189.
McNutt, W. S. (1958b). *J. Biol. Chem.* **233,** 193.
Maley, F., and Maley, G. F. (1960). *J. Biol. Chem.* **235,** 2968.
Maley, G. F., and Maley, F. (1959). *J. Biol. Chem.* **234,** 2975.
Maley, G. F., and Maley, F. (1962). *J. Biol. Chem.* **237,** PC3311.
Maley, G. F., and Maley, F. (1963). *Science* **141,** 1278.
Mantsavinos, R., and Canellakis, E. S. (1959). *J. Biol. Chem.* **234,** 628.
Markret, C. L., Tomkins, G., Kaplan, N. O. (1963). In "Symposium on Protein Structure and Function during Differentiation", Federation Meetings, Atlantic City, New Jersey.
Masters, M., Kuempel, P. L., and Pardee, A. B. (1964). *Biochem. Biophys. Res. Commun.* **15,** 38.
Mazia, D. (1954). *In* "Glutathione" (S. P. Colowick *et al.*, eds.), p. 209. Academic Press, New York.
Miyachi, S. (1962). *Plant Cell Physiol.* (*Tokyo*) **3,** 1.
Miyachi, S., and Miyachi, S. (1961). *Plant Cell Physiol.* (*Tokyo*) **2,** 415.
Miyachi, S., and Tamiya, H. (1961). *Plant Cell Physiol.* (*Tokyo*) **2,** 405.
Miyachi, S., Kanai, R., Mihara, S., Miyachi, S., and Aoki, S. (1964). *Biochim. Biophys. Acta* **93,** 625.
Monod, J., Changeux, J., and Jacob, F. (1963). *J. Mol. Biol.* **6,** 306.
Moore, E. C., and Hurlbert, R. B. (1960). *Biochim. Biophys. Acta* **40,** 371.
Moore, E. C., and Hurlbert, R. B. (1962). *Biochim. Biophys. Acta* **55,** 651.
Moore, E. C., and Reichard, P. (1963). *J. Biol. Chem.* **238,** PC2244.
Moore, E. C., and Reichard, P. (1964). *J. Biol. Chem.* **239,** 3453.
Moore, E. C., Reichard, P., and Thelander, L. (1964). *J. Biol. Chem.* **239,** 3445.
Mudd, S., Yoshida, A., and Koike, M. (1958). *J. Bacteriol.* **75,** 224.
Muhammed, A. (1961). *Biochim. Biophys. Acta* **54,** 121.
Muramatsu, M., Hodnett, J. L., and Busch, H. (1964). *Biochim. Biophys. Acta* **91,** 592.
Myers, D. K. (1962). *Can. J. Biochem. Physiol.* **40,** 619.
Myers, D. K., Hemphill, C. H., and Townsend, C. M. (1961). *Can. J. Biochem. Physiol.* **39,** 1043.
Neumann, J., and Jones, M. E. (1962). *Nature* **195,** 709.
Noll, H., Staehelin, T., and Wettstein, F. O. (1963). *Nature* **198,** 632.
Ondarza, R. N., and Aubanel, M. (1960). *Biochim. Biophys. Acta* **44,** 381.
Park, J. T. (1952a). *J. Biol. Chem.* **194,** 877.
Park, J. T. (1952b). *J. Biol. Chem.* **194,** 885.
Park, J. T. (1952c). *J. Biol. Chem.* **194,** 897.
Park, J. T., and Strominger, J. L. (1957). *Science* **125,** 99.
Pitot, H. C., and Heidelberger, C. (1963). *Cancer Res.* **23,** 1694.
Pollard, C. J. (1964). *Arch. Biochem. Biophys.* **105,** 114.

Potter, J. L., and Dounce, A. L. (1956). *J. Am. Chem. Soc.* **78,** 3078.

Prescott, D. M. (1960). *Exptl. Cell Res.* **19,** 228.

Raacke, I. D. (1961). *In* "Metabolic Pathways" (D. M. Greenberg, ed.), 2nd ed., Vol. II, p. 263. Academic Press, New York.

Rapkine, L. (1931). *Ann. Physiol. Physicochim. Biol.* **9,** 383.

Reichard, P. (1957). *Acta Chem. Scand.* **11,** 11.

Reichard, P. (1958). *Biochim. Biophys. Acta* **27,** 434.

Reichard, P. (1959). *J. Biol. Chem.* **234,** 1244.

Reichard, P. (1961). *J. Biol. Chem.* **236,** 2511.

Reichard, P. (1962). *J. Biol. Chem.* **237,** 3513.

Reichard, P., Baldesten, A., and Rutberg, L. (1961). *J. Biol. Chem.* **236,** 1150.

Roll, P. M., and Weliky, I. (1955). *J. Biol. Chem.* **213,** 509.

Roll, P. M., Weinfeld, H., and Carroll, E. (1956). *J. Biol. Chem.* **220,** 455.

Rose, I. A., and Schweigert, B. S. (1953). *J. Biol. Chem.* **202,** 635.

Scarano, E. (1960). *J. Biol. Chem.* **235,** 706.

Scherbaum, O. H. (1960). *Ann. Rev. Microbiol.* **14,** 283.

Schmidt, G. (1951). *In* "Phosphorus Metabolism" (W. D. McElroy and B. Glass, eds.), Vol. I, p. 443. Johns Hopkins Press, Baltimore, Maryland.

Schmidt, G., Hecht, L., and Thannhauser, S. J. (1946). *J. Biol. Chem.* **166,** 775.

Schmidt, G., Hecht, L., and Thannhauser, S. J. (1949). *J. Biol. Chem.* **178,** 733.

Schmidt, R. R. (1961). *J. Exptl. Cell Res.* **23,** 209.

Schmidt, R. R. (1965). Unpublished data.

Schmidt, R. R., and Hare, T. A. (1965). Unpublished data.

Schmidt, R. R., and Herrmann, E. C. (1965). Unpublished data.

Schmidt, R. R., and King, K. W. (1961). *Biochim. Biophys. Acta* **47,** 391.

Schmidt, R. R., and Spencer, H. T. (1964). *J. Cellular Comp. Physiol.* **64,** 249.

Schuur, A. H. W. M., and Koningsberger, V. V. (1960). *Biochim. Biophys. Acta* **44,** 167.

Shen, S. R. C. (1965). M.S. Thesis, Virginia Polytechnic Institute (Blacksburg).

Shen, S. R. C., and Schmidt, R. R. (1966). *Arch. Biochem. Biophys.* (in press).

Sinclair, J., Rabinowitz, M., DeSalle, L., Haselkorn, R. R., and Swift, H. H. (1965). *Federation Proc.* **24,** 539 (Abstract).

Sisken, J. E., and Kinosita, R. (1961). *J. Biophys. Biochem. Cytol.* **9,** 509.

Sorokin, C. (1957). *Physiol. Plantarum* **10,** 659.

Sorokin, C., and Myers, J. (1953). *Science* **117,** 330.

Sorokin, C., and Myers, J. (1957). *J. Gen. Physiol.* **40,** 579.

Spencer, H. T., Schmidt, R. R., Kramer, C. Y., Moore, W. E. C., and King, K. W. (1961). *Exptl. Cell Res.* **25,** 485.

Stanners, C. P., and Till, J. E. (1960). *Biochim. Biophys. Acta* **37,** 406.

Steinberg, D., Vaughan, M., Sherman, F. G., and O'Dell, B. L. (1960). *Biochim. Biophys. Acta* **40,** 225.

Strominger, J. L. (1959). *Compt. Rend. Trav. Lab. Carlsberg* **31,** 181.

Strominger, J. L., and Threnn, R. H. (1959). *Biochim. Biophys, Acta* **33,** 280.

Swann, M. M. (1957). *Cancer Res.* **17,** 727.

Szafranski, P., and Bagdasarian, M. (1961). *Nature* **190,** 719.

Tamiya, H. (1963). *J. Cellular Comp. Physiol.* **62,** Suppl. 1, Part II, 157.

Tamiya, H. (1964). *In* "Synchrony in Cell Division and Growth" (E. Zeuthen, ed.), p. 247. Wiley, New York.

Thomson, R. Y., Scotto, G. T., and Brown, G. B. (1962). *J. Biol. Chem.* **237,** 3510.

Umbarger, H. E. (1964). *Science* **145,** 674.

Wahba, A. J., and Friedkin, M. (1961). *J. Biol. Chem.* **236,** PC11.

Warren, L. (1961). *In* "Metabolic Pathways" (D. M. Greenberg, ed.), 2nd ed., Vol. II, p. 459. Academic Press, New York.

Watson, J. D. (1963). *Science* **140,** 17.

Weiss, S. B. (1960). *Proc. Natl. Acad. Sci. U.S.* **46,** 1020.

Weiss, S. B., and Gladstone, L. (1959). *J. Am. Chem. Soc.* **81,** 4118.

Weiss, S. B., and Nakamoto, T. (1961). *J. Biol. Chem.* **236,** PC18.

Weissman, S. M., Smellie, R. M. S., and Paul, J. (1960). *Biochim. Biophys. Acta* **45,** 101.

Wiame, J. M. (1947). *J. Am. Chem. Soc.* **69,** 3146.

Wiame, J. M. (1948). *J. Biol. Chem.* **178,** 919.

Wilken, D. R., and Hansen, R. G. (1961). *J. Biol. Chem.* **236,** 1051.

Yates, R. A., and Pardee, A. B. (1957). *J. Biol. Chem.* **227,** 677.

Yielding, K. L., and Tomkins, G. M. (1963). *Federation Proc.* **22,** 289 (Abstract).

Zeuthen, E. (1958). *Advan. Biol. Med. Phys.* **6,** 37.

CHAPTER 11

Energetics and the Synchronized Cell Cycle

B. W. Wilson and T. W. James*

I. INTRODUCTION

Knowledge of cell metabolism has increased greatly in the last few decades: The enzymes within cells have been catalogued; the pathways of metabolism have been mapped out; substrate and enzyme mediated feedback control systems have been investigated. However, many metabolic activities which depend upon the presence of intact cells and tissues for their existence are still not well understood. Metazoan systems are excellent experimental subjects for the study of many such patterns of metabolism. For example, some somatic cells accumulate lipids when cultured *in vitro* (MacKenzie *et al.*, 1964); embryonic tissues show progressive changes in their enzyme subunits (Kaplan and Cahn, 1962), and free amino acid pools (Wilson *et al.*, 1965), and several cell types exhibit pronounced Pasteur and Crabtree effects (Wu and Racker, 1963). In contrast, free-living cell populations are convenient tools for the study

* Supported by N.S.F. Grant No. 1643.

of patterns of metabolism associated with logarithmic growth, and synchronized cell populations are useful in the study of metabolic states related to the cell cycle.

In this chapter work done on 2 free-living flagellate cells, *Astasia longa* and a colorless strain of *Euglena gracilis* var. *bacillaris*, involving the relationships between growth, respiration, division, and the ordering in time of the rates of energy utilization and synthesis over a synchronous cell cycle is discussed.

II. RESPIRATION AND GROWTH OF SYNCHRONIZED *Astasia*

The development of synchronized populations of the cell *Astasia longa* which replicate their number and mass with each cycle of division (Blum and Padilla, 1962; Padilla and James, 1960) in continuous culture opened the way to studies of the growth and oxidative metabolism of these cells throughout a cell cycle. The first study of the oxidative metabolism of synchronized *Astasia* was designed to measure the changes in mass and respiration of the cells. Synchronized *Astasia* were grown in a continuous culture device described elsewhere (Wilson and James, 1963) using a time-temperature cycle of 6.5 hours at 28.5°C and 17.5 hours at 14.8°C. An average of 91% of the cells divided during the warm period; most of the cells divided in 2–3 hours.

The respirations of the cells were measured polarographically by removing samples from the culture apparatus and placing the unconcentrated, uncentrifuged cell suspensions in a magnetically stirred water-jacketed flask containing a Clark polarographic "oxygen" electrode connected to an amplifier and a continuous recorder. The oxygen consumptions of cells sampled at hourly intervals through the warm period of a single cycle are shown in Fig. 1. The lower section of the figure shows the increase in cell number during the time in which oxygen consumption was measured. The cells increased 85%, from approximately 32,000 cells/ml to 59,000 cells/ml. The middle section of the figure illustrates the oxygen consumptions expressed on a per cell basis. The respiration rate remained relatively constant during the cold period, averaging approximately 32 μl O_2/hr/10^6 cells at 14.8°C and increased to approximately 55 μl O_2/hr/10^6 cells in the warm period at 28.5°C. The respiration rate of the cells stayed at this level during the warm period and then returned to the respiration rate found in the previous cold period. The results were plotted on a per culture basis to reveal the changes in the respiration of the culture during the warm period. This is shown in the upper section of Fig. 1. The oxygen consumption of the culture increased 89% from approximately 1.80 μl O_2/hr/ml to 3.40 μl O_2/hr/ml during the warm period. Thus, the "respira-

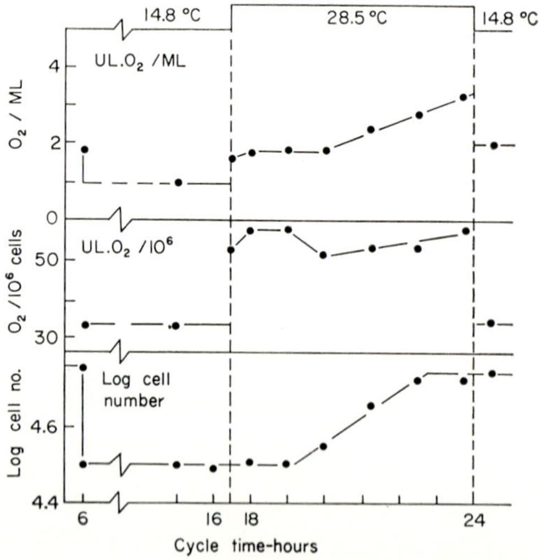

FIG. 1. The respiration of *Astasia* during a single synchronous cycle. The oxygen consumption of the cells was measured polarographically on samples incubated at the cycle temperatures. Results are expressed in μl O_2/hr/10^6 cells and μl O_2/hr/ml culture.

tory capacity" of the culture was replicated during the warm period of the cell cycle. In this experiment, the oxygen consumption of the culture did not begin to increase until 1 hour after the cells began to increase in number. Since the respiration device was not sterile, respiration measurements could not be continuously obtained from a single sample during the entire warm period. For this reason it was not possible to examine the relationship between the increase in respiration and the division of the cells in great detail.

A composite picture of the respiration rates of a number of samples taken over 5 synchronous cycles is shown in Fig. 2. The cell concentrations ranged from 30,000 cells/ml to 120,000 cells/ml. The results were expressed in terms of the initial cell numbers in the cold period. The data showed that the respiration rate was constant during the cold period, averaging 28 μl O_2/hr/10^6 cells. The average respiration rate per unit volume of culture increased 94% and the average cell number increased 96% during the warm period of the cycle. The average respiration during the warm period remained relatively unchanged during the time when the cells divided, averaging 66 μl O_2/hr/10^6 cells.

The increase in dry weight in mg/10^6 cells and in milligrams per unit volume of culture is shown in Fig. 3. The average dry weight per cell increased during much of the cold period, leveled off at the start of the warm period and then decreased per cell, but increased per culture as the cells

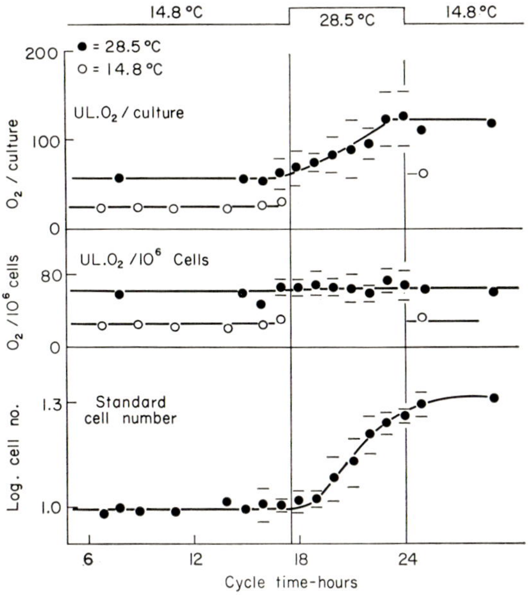

FIG. 2. The respiration of synchronized *Astasia*. A composite of results from 5 synchronous cycles. Black dots are averages of the oxygen consumptions of cells removed from the culture and tested at 28.5%; open circles are the oxygen consumptions of cells tested at 14.8°C. Results are expressed in $\mu l\ O_2/hr/10^6$ cells and $\mu l\ O_2/hr/ml$ of standard culture. (From Wilson and James, 1963.)

divided. During the cold period the dry weight of the "standard" culture increased from 1.19 mg to 1.54 mg, a synthesis of 35 mg per 10^6 cells, and then increased another 0.71 mg to 2.25 mg during the warm period. Thus, the cells synthesized one third of their dry weight in the cold period and two thirds of their dry weight in the warm period of the synchronous cell cycle. It was concluded that the doubling of the respiration rates in the warm period indicated that synchrony in *Astasia*. . . "involved an ordering in time of more than the division processes of the individual cells in the populations" (Wilson and James, 1963). The authors did not feel that the temporal correlation between cell respiration and cell division reflected a simple casual relationship between these parameters or in any way indicated which was the dependent and independent variable. Further evidence on this point will be discussed later.

The data were also used to compare the growth and energy metabolism of synchronized and exponentially grown cells. The respiration rates of synchronous and exponentially grown *Astasia* are shown in Table I. The respiration rates of synchronized cells on acetate were almost 30% higher than the respiration rates of cells grown exponentially at the cycle tem-

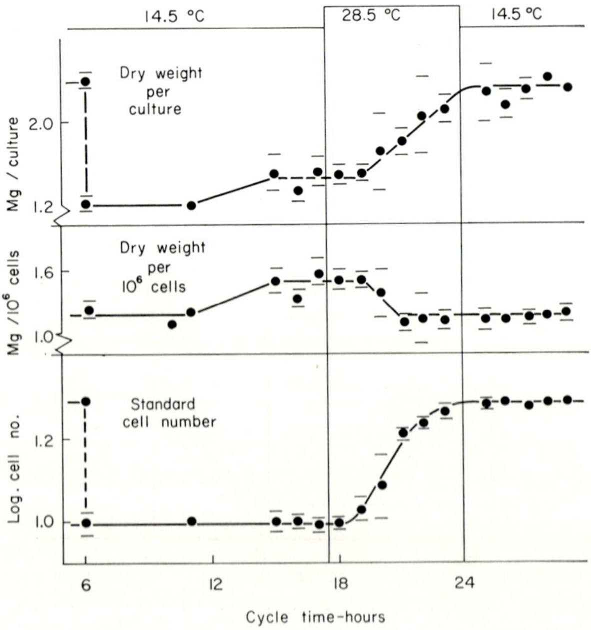

FIG. 3. The dry weight changes of synchronized *Astasia*. Data are a composite of results from 6 synchronous cycles. (From Wilson and James, 1963.)

TABLE I

A COMPARISON OF THE RESPIRATION AND GROWTH OF SYNCHRONIZED EXPONENTIALLY GROWN *Astasia*[a]

Temperature (°C)	Oxygen consumption ($\mu l\ O_2$/hr/10^6 cells) 10 mm Acetate Syn.	10 mm Acetate Exp.	Endogenous Syn.	Endogenous Exp.	Dry weight production (mg/hr/10^6 cells) Syn.	Exp.
28.5	62.3 ±5.3	49.7 ±2.1	14.7 ±3.7	8.4 ±0.5	6.7×10^{-2}	6.9×10^{-2}
14.8	26.3 ±4.3	19.1 ±1.9	4.6 ±1.4	6.5 ±0.7	2.0×10^{-2}	1.9×10^{-2}

peratures. The endogenous respirations did not behave in the same way; the endogenous respiration rate of synchronized cells at 28.5°C was almost double that of cells grown exponentially at that temperature, while the endogenous respiration rate of synchronized cells at 14.8°C was, if anything, slightly lower than that of cells grown at the same temperature. Thus, the respiration rates of synchronous and exponentially grown cells were quantitatively different on a per cell basis and the endogenous respiration rates did not exhibit the same responses to synchrony and temperature as did the respiration rates of the cells in the presence of an exogenous oxidizable substrate.

Synchronized *Astasia* were larger than exponentially grown *Astasia*, although as shown in Table I the rate of dry weight production of synchronous and exponentially grown cells were the same. Thus, unlike the cell respirations, the rates of dry weight production were not affected by the synchronization process. Considering the 30% increase in the rate of respiration of synchronized as compared to exponentially grown cells and the similarity in their rates of dry weight production Wilson and James (1963) stated, "One could almost say that *Astasia* purchased synchrony at the expense of an increased oxygen consumption." When the data were analyzed in terms of the oxygen consumption per dry weight a picture of how the cells consumated this purchase was obtained. The cells responded to the temperature cycle in such a way as to alter their respiration rate per milligram mass to correspond to the rate per milligram mass of cells grown exponentially at the cycle temperatures. During the cold period cell division ceased, the respiration rate remained constant and the dry weight of the cells increased so that the respiration per mass fell from 23.5 $\mu l\ O_2/mg/hr$ at the start of the cold period to 18.2 $\mu l\ O_2/mg/hr$ for cells grown exponentially at 15°C. During the warm period the cells divided and increased in respiration and mass so that the $\mu l\ O_2/mg/hr$ rose from 42.7 at the start of the warm period to 55.7 at its close, essentially the same oxygen/mass ratio shown by cells grown exponentially at 28.5°C that of 59.5.

III. OXYGEN TENSION AND THE RESPIRATION AND DIVISION OF SYNCHRONIZED *Astasia*

The above study revealed the gross changes in the growth and respiration of synchronized *Astasia* and provided useful information on the stoichiometry of the processes involved. However, the transient changes in the respiration of the cells were not critically examined. These were examined in the experiments of James (1965) where respiration was measured within the culture vessel itself. By so doing the relationships between temperature, oxygen tension, cell density, rate of respiration, and extent of cell synchrony were obtained.

The experimental method involved inserting a sterilized "oxygen" electrode into a large culture flask which was open to the atmosphere and equipped with a cell number monitoring device and a temperature recorder. The theory underlying the oxygen measurements is fully described elsewhere (James, 1965); it involved determining the rate of diffusion of

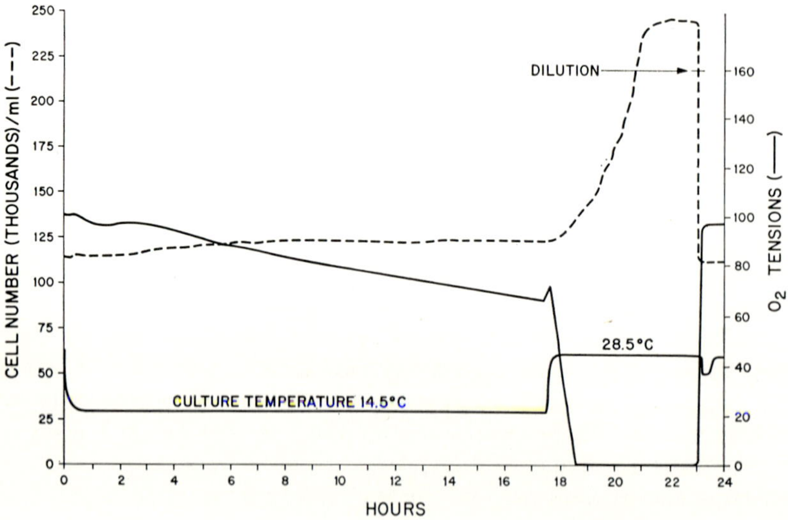

FIG. 4. Tracing of a strip chart recording of the changes in cell number, temperature, and oxygen tension of a synchronized culture of *Astasia* recorded *in situ*. (From James, 1965.)

oxygen into the medium, alone, at the warm and cold period temperatures. Since the oxygen tension in the culture is the resultant of the rate of oxygen consumption of the cells and the rate of diffusion of oxygen into the medium, knowledge of the rate of diffusion of oxygen and the oxygen tension of the culture enables one to calculate the oxygen consumption of the cells in the culture flask.

Figure 4 illustrates a strip chart recording of the cell number, oxygen tension, and temperature of a culture of initially 120,000 cells/ml determined through one synchronous cycle. The figure shows that the oxygen tension in the medium decreased gradually in the cold period and then fell to almost zero at the onset of the warm period. Zero oxygen tension in the open system forced the cells to respire at a rate limited by the maximum diffusion rate of oxygen divided by the cell number present. In this way the oxygen consumption of the cells could be controlled by varying the population density of cells within the culture.

Figure 5 contains an analysis of the data gained from experiments conducted at different cell concentrations. The cell number and the respiration rates per cell are illustrated. Culture I represents a culture in the process of changing from logarithmic to synchronized growth since it was the first in a series to be exposed to the temperature regime. Cultures II, III, and IV were selected from 15 such recordings all of which showed the same general patterns. Initial population levels ranged from 70,000

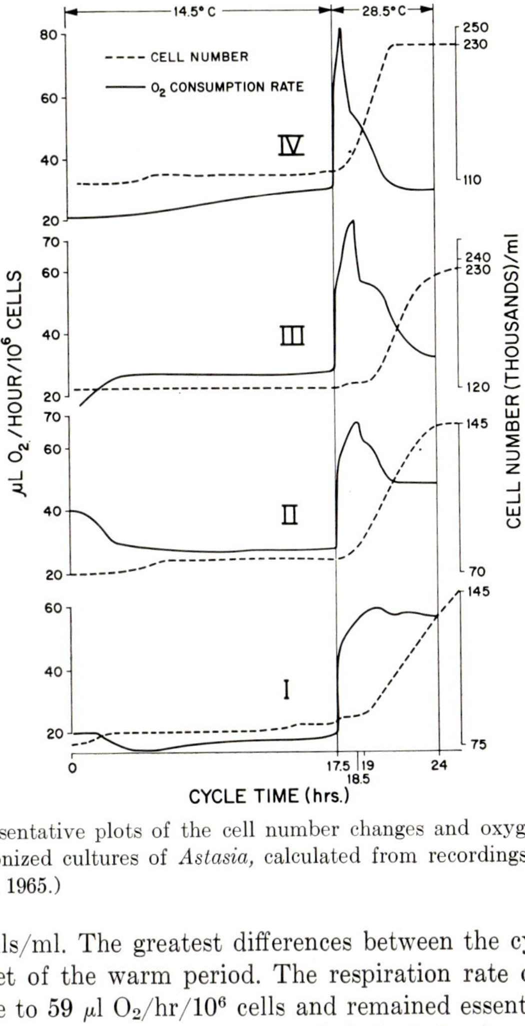

FIG. 5. Representative plots of the cell number changes and oxygen consumption rates of synchronized cultures of *Astasia,* calculated from recordings similar to Fig. 4. (From James, 1965.)

to 120,000 cells/ml. The greatest differences between the cycles occurred after the onset of the warm period. The respiration rate of the cells in Culture I rose to 59 µl O_2/hr/10^6 cells and remained essentially constant during the cell divisions in the warm period; its behavior was similar to that described in the previous study of Wilson and James (1963). The maximum rate of division in the warm period was equivalent to a doubling time of 4.7 hours. The respiration rate of Culture II rose initially to 68 µl O_2/hr/10^6 cells and then decreased midway through the division period to a value of 48 µl O_2/hr/10^6 cells. The minimum doubling time of the cells in this culture was 4.3 hours. At no time was the diffusion rate limiting to respiration rate in Cultures I and II. Cultures III and IV were set at population densities sufficient to lower the oxygen tension to zero by the nineteenth and 18.5th hour, respectively, so that the oxygen consumptions of the cells were then limited by the rate of diffusion of oxygen into the medium. Both showed the type of response obtained in Culture II, the respiration in the warm period rose initially to a very high level and then fell off drastically even with *ad libitum* oxygen available. However, the decline of the respiration rate in Cultures III and IV beyond the nineteenth and 18.5th hour was due, in part, to the production of zero oxygen tensions in the medium so that the cells respired at rates less than they would have if an excess of oxygen had been present. Furthermore, the division of the cells which increased the number at this time caused a further reduction of the respiration rate per cell. The minimum doubling time achieved in Cultures III and IV was 2.9 hours.

James considered the correlation between the respiration and division rates in the warm period for the cells examined in the study of Wilson and James (1963) and in Culture I to be somewhat fortuitous since other data (Cultures II, III, IV) indicated that this situation was not always obtained. He pointed out that it could be due to a "rather curious qualitative accident,"... "while some cells are dividing, others are synthesizing new respiratory machinery at a rate that just matches the doubling rate." This interpretation of the data is plausible when it is considered that the pattern of respiration characterized by a constant respiration per average cell in the warm period occurred in cultures which did not show a sharp burst of cell division. Marked changes in the respiration of the individual cells in this population which might have been associated with the division process would have been hidden by the lack of exact alignment of the individual cell cycles in the population and thus, the average respiration rate would fluctuate less through time than the true respiration rates of the individual cells.

Some of the early changes in respiration in the cold period shown by cultures such as I, II, and III could be attributed to the addition of fresh nutrients, particularly, thioglycolate, at the start of the cold period when the continuous cultures were diluted with fresh medium. However, in populations such as Culture IV, there was a steady increase in the respiration rate of the cells throughout the cold period, in this case it increased from 20 μl O_2/hr/10^6 cells to 30 μl O_2/hr/10^6 cells, indicating that respiratory machinery was constructed during the cold period. The fact that populations such as Culture IV showed the highest rate of respiration in the cold period and the best synchrony of cell division suggested that "cases of good synchrony can be predicted from knowledge of the level and course the respiratory rate takes in the cold period" (James, 1965).

In addition to the above, those cultures with the best synchrony and the highest rate of respiration in the cold period showed a rapid rise of respiration per cell at the onset of the warm period, and a subsequent decline in respiration rate prior to the time that zero oxygen tension was approached in the cultures. Cultures which showed the shortest duration of the division phase showed the greatest decrease in respiration rate following the onset of the warm period. The time of this decrease in respiration rate corresponded to the time at which the maximum number of prophase cells were present. This observation was similar to the results of Zeuthen (1960), and Stern and Kirk (1948) who observed a reduction in respiration just prior to the initiation of division.

The results reviewed here show that the respiration changes of synchronized *Astasia* are ordered in time over the cell cycle, and that their exact course may be altered by environmental conditions such as oxygen tension. Similar findings have been reported by Scopes and Williamson (1964) for synchronously budding populations of *Sacchromyces.* The oxygen consumption of the yeast cells increased in a stepwise fashion during the budding cycle, while the mass of the cells increased continuously. The respiration increase did not start at precisely the same time in each budding cycle; the respiration of the culture did not double with each doubling of the cell number and tended to occur earlier in successive generations.

These investigations of synchronous cell populations, and those conducted with single cells (Zeuthen, 1960) suggest that the respiratory machinery of a cell is not always produced continuously throughout the growth cycle. However, the experiments do not permit us to distinguish synthesis from activation. For example, in the study of Wilson and James (1963), some, or even all of the respiratory machinery could have been synthesized in an inactive form in the cold period and activated in the next warm period. Such an activation could be brought about in many ways such as the synthesis of a single rate limiting component in the warm period, or the removal of an inhibitory metabolic product. Scopes and Williamson considered that the discontinuous synthesis of a particular rate limiting compound was unlikely on "... kinetic grounds" (1964). However, there is no experimental evidence available to exclude such a mechanism, at least with respect to *Astasia.*

The possibility that the abrupt respiration changes are associated with the formation of new mitochondria was discussed in the original reports (James, 1965; Scopes and Williamson, 1964; Wilson and James, 1963). Unfortunately, the time course and the mechanisms of mitochondrial production are not yet known. Recent reports suggesting that mitochondria contain DNA and may replicate autonomously (Gibor and Granick, 1964; Nass and Nass, 1963) hint at exciting opportunities for future research on synchronized cells.

Several of the differences between the pattern of respiration of synchronized *Astasia* found by Wilson and James (1963), and James (1965) might be explained as being due to variation in synthetic patterns i.e., an out of phase nuclear and mitochondria cycle of duplications. Or perhaps, it is more plausible to assume that the data reported by Wilson and James (1963) reflected the average respiration of the cells *in situ* and that it represented an average of several patterns of respiration exhibited by synchronized *Astasia*. However, some of the differences between the results are also due to improvements in the cell number and oxygen determinations and the flask design and temperature control introduced by James (1965). For example, the decrease in the rate of respiration of synchronized *Astasia* just prior to the onset of cell division could not have been detected without continuously recording the temperature, cell number, and oxygen tension of the culture and achieving sharp bursts of cell division.

In addition, it was pointed out that the production of steady-state conditions of zero oxygen tension during the warm period improved the synchronization of the cells (James, 1965). One possibility is that oxygen tensions limiting to the respiration of the cells shifted the time of formation of the respiratory machinery into the cold period and, in this way, improved the synchrony of division of the cells in the next warm period. Two questions are raised by this hypothesis. How could low oxygen tensions shift the production of cell constituents from one part of the cell cycle to another? Second, how could such a shift lead to an improvement in the synchronization of the cells? To deal with these questions the relationships between the respiration, growth, and acetate utilization of *Astasia* must be considered. These relationships have been studied by short-term experiments on the oxidative assimilation of acetate by *Astasia* and *Euglena*.

IV. RESPIRATION AND SUBSTRATE UTILIZATION

The oxygen consumptions of cells utilizing oxidizable substrates are usually less than would occur if the substrates were complete oxidized; the same carbon source simultaneously donates electrons for cell reductions and molecules for cell syntheses; part of the substrate molecules are oxidized, part are assimilated. The ratio of the oxygen consumed to the substrate utilized is usually studied by short-term experiments in which a small amount of substrate is added to a cell suspension and the oxygen consumption is measured until the substrate is exhausted from the medium. The measured oxygen consumption is then compared to that which would have occurred if all the substrate had been oxidized. A major problem in the interpretation of such experiments is that of the true value of the exogenous respiration rates and whether its value should be corrected for the oxygen consumed due to the oxidation of endogenous carbon. Studies on this problem with C^{14}-labeled substrates yielded data which seemed at odds with the results gained from manometric experiments (Blumenthal *et al.*, 1957; Clifton, 1963; Danforth and Wilson, 1961). However, evidence has been obtained which strongly indicates that the carbon tracer studies provide information concerning the behavior of the substrate molecules and the short-term manometric studies yielded data on the kinetics of the substrate oxidation pathways (Danforth, 1961; Danforth and Wilson, 1961; Marzullo and Danforth, 1964; Wilson, 1963). Thus, the results gained from the 2 methods complement rather than contradict each other. The topic has been discussed elsewhere (Danforth and Wilson, 1961; Wilson, 1963).

The respiration and C^{14} experiments performed on *Euglena* and *Astasia* have resulted in the following picture of the oxidative metabolism of these cells when acetate is their sole source of carbon.

The oxidation of cellular reserve materials continues during the utilization of acetate by *Astasia* (Danforth and Wilson, 1961; Wilson, 1963; Wilson and Danforth, 1958). Therefore, the respiration due to the oxidation of intracellular reserves should be subtracted from the total amount of oxygen consumed during the utilization of a known amount of acetate to obtain the extra oxygen consumed due to the oxidation of the acetate.

During the period of acetate utilization the intracellular reserves are increasingly derived from the carbon molecules of the exogenous substrate. Those molecules which are last assimilated are the first to be oxidized (Danforth and Wilson, 1961).

The polysaccharide paramylum, a polymer of glucose, is the major, but perhaps not the sole component of the intracellular reserves (Danforth and Wilson, 1961; Marzullo and Danforth, 1964).

The carbon tracer and gas exchange data are consistent with the view that *Astasia* and *Euglena* possess 2 major respiratory systems referred to as (*a*) the reserve oxidation system and (*b*) the substrate dependent oxidation system (Danforth and Wilson, 1961; Wilson, 1963). The reserve oxidation system oxidizes reserve materials at a rate approximately equal to the endogenous metabolism of the reserves. The substrate dependent oxidation system oxidizes a constant proportion of the exogenous substrate at a rate which is dependent upon the concentration of substrate at low concentrations.

The equation for acetate utilization of nongrowing *Euglena* and thus probably for *Astasia* as well is described by the empirical equation

$$CH_3COOH + 0.84\ O_2 = 0.84\ CO_2 + 0.84\ H_2O + 1.16\ CH_2O \qquad (1)$$

In other words the extra oxygen consumed via the substrate dependent pathway is 42% of the acetate available, and carbohydrate is the major assimilatory product (Danforth, 1961; Wilson and Danforth, 1958). The empirical equation for ethanol oxidation by *Euglena* is not consistent with carbohydrate being the major assimilatory product (Eshleman and Danforth, 1964).

The kinetics of the respiration of *Astasia* and *Euglena* have been described in terms of a mathematical model (Wilson, 1963). The total amount of oxygen consumed by the cells during the utilization of a small amount of acetate is

$$O_t = \frac{Q_e}{k_1} \ln \frac{A_1}{A_t} + \frac{A_0 - A_1}{k} Q_e + k_0(A_0 - A_k) \qquad (2)$$

Where A_1 is the amount of acetate limiting to the rate of substrate utilization, A_t is the lowest amount of acetate yielding a rate of respiration measureably above the endogenous respiration rate, k is a rate constant in μmoles/hr, k_1 is a rate constant in hr^{-1}, k_0 is a constant expressing the oxygen; substrate ratio of the substrate dependent oxidation system and Q_e is the respiration rate of the cells due to the reserve oxidation system in μmole O_2/hr/10^6 cells. The results of an experiment measuring the oxygen/acetate ratios of synchronized *Astasia* sampled from the cold period and tested at 14.8°C are shown in Fig. 6. The respirations were measured polarographically in the presence and absence of acetate, the exogenous carbon source. The open circles represent the total μmoles of oxygen consumed by the cells during the utilization of 2–15 μmoles of acetate; the solid circles represent the "extra" oxygen consumptions, that is, the oxygen consumptions obtained by correcting for the activity of the reserve oxidation system. The theoretical curves which resulted when Eq. (1) was applied to the respiration of synchronized *Astasia* are shown in Fig. 7. Q_e and k_0 were given values similar to those determined for the cells and k, k_1, A_1 and A_t were estimated from the results of the experiments. The suitability of the equation as a first approximation is suggested by the fact that similar correspondences between theoretical curves and experimental results were obtained when published data on the oxidative assimilation of carbon by *Euglena* (Wilson and Danforth, 1958), *Penicillium* (Blumenthal *et al.*, 1957), *Prototheca* (Barker, 1936), and *Saccharomyces* (Winzler, 1940) were recalculated.

The results of several studies on *Euglena* and *Astasia* (Danforth, 1961; Wilson, 1963; Wilson and Danforth, 1958) indicated that k_0, the proportion of the acetate utilized which is oxidized by the substrate dependent pathway, denoted hereafter as the corrected oxygen to acetate ratio, is remarkably constant; it is unaffected by the type of growth media, the incubation pH, adaptation of the cells to acetate, and the

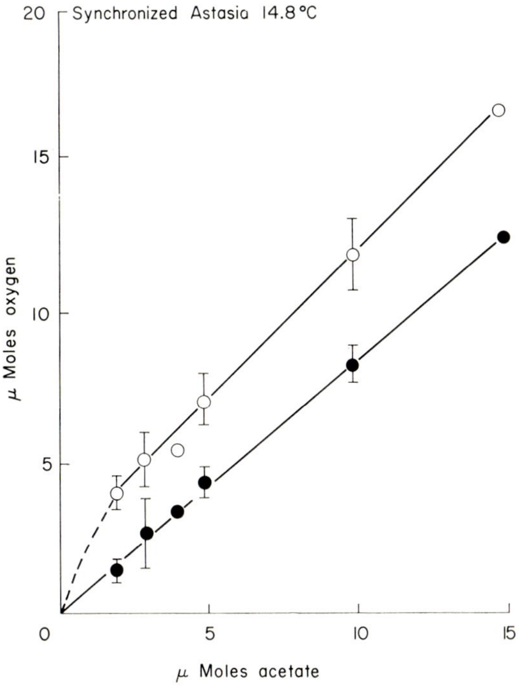

FIG. 6. The relationship between the oxygen consumed and the acetate utilized by synchronized *Astasia* sampled from the cold period and tested at 14.8°C. Open circles are the total oxygen consumptions; closed circles are the "corrected" oxygen consumptions, that is the extra oxygen consumptions due to the presence of acetate. (From Wilson, 1963.)

presence of a nitrogen source. It averages about 42% of the acetate available to the cells.

The percent of acetate oxidized by the substrate-dependent oxidation system of *Astasia* was also found to be independent of 2 other parameters, temperature and synchronization (Wilson, 1963).

Table II lists the corrected oxygen/acetate ratios for exponentially grown and synchronous *Astasia* cells. The ratios were relatively independent of temperature averaging 0.859 μmoles oxygen per μmole acetate over a temperature range of 15–28.5°C, equivalent to 43% oxidation of the acetate available to the cells (Wilson, 1963). It is interesting to note that the temperature independence of the corrected oxygen/substrate ratios had been noted by Winzler (1940) with respect to the oxidation of acetate by *Saccharomyces*. The report of Wilson (1963) stressed the significance of the temperature independence of the stoichiometry of the substrate dependent oxidation system. It was suggested that these

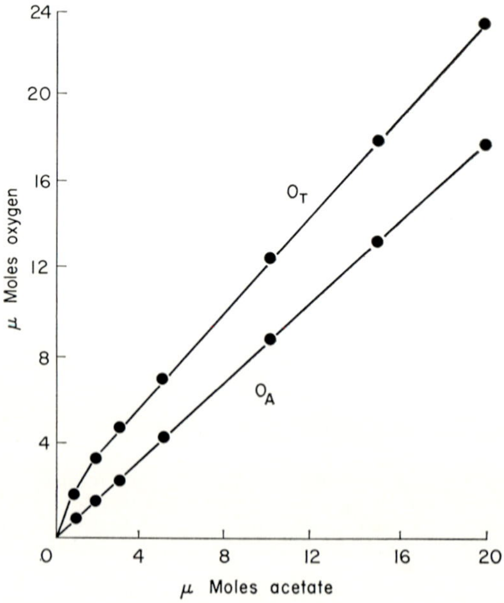

FIG. 7. Theoretical curve calculated from Eq. (1) for the relationship between the oxygen consumption and the acetate utilization of synchronized *Astasia* at 14.8°C. O_t are the total oxygen consumptions; O_a are the extra oxygen consumptions due to the presence of acetate. (From Wilson, 1963.)

ratios reflect a temperature compensated acetate regulating mechanism within *Astasia* and *Saccharomyces* and that this ratio "... may indicate the presence of a homeostatic cell mechanism ... basic to the control of cell metabolism." The simplest mechanism consistent with the results would be one in which the rate of partitioning of acetate, to processes utilizing it, would be limited by the rate of production of a product of the substrate dependent oxidation system, such as ATP. This would result in an oxygen/acetate ratio that would be compensated with respect to temperature, acetate concentration, and reaction rate. The stoichiometric relationships between the production and utilization of the limiting factor would determine the particular value of the oxygen/substrate ratio.

Considering the temperature independence of this phenomenon it is not surprising that the corrected oxygen/acetate ratios were unaffected by synchronization of the cells. Regardless, the results seem to exclude drastic changes in the effectiveness of substrate utilization and oxidation as playing a major role in the mechanism of synchronization of *Astasia*.

The corrected oxygen/acetate ratio is not an estimate of the percent of acetate oxidized during cell growth. During steady-state growth the amount of acetate oxidized by the cells will be more than is oxidized di-

TABLE II

Relation between Oxygen Consumption and Acetate Utilization of *Astasia*[a]

[a] (Adapted from Wilson and James, 1963.)

Temperature (°C)	(Corrected O_2/acetate[b] μmoles O_2/μmoles acetate)
Exponential growth	
28.5	0.846 ± 0.180 (6)
25.0	0.878 ± 0.080 (5)
20.0	0.032 ± 0.068 (4)
15.0	0.828 ± 0.175 (4)
Synchronous growth	
28.5	0.818 ± 0.094 (13)
14.8	0.852 ± 0.126 (19)
Mean	0.859

[a] (Adapted from Wilson, 1963.)
[b] Numbers in parentheses indicate the number of samples.

rectly via the substrate dependent respiration system since the acetate incorporated into the cell reserves and then oxidized via the reserve oxidation system must be included in the calculations. However, the oxygen consumed by the substrate dependent oxidation system appears to be a measure of the acetate taken up by the cells. The data indicate that acetate uptake is closely linked to the oxidation of a fixed proportion of it, as if an oxidation system is involved in some way with the permeability of the cells to acetate. This relationship between the corrected oxygen/acetate ratio and the acetate consumption of the cells has been confirmed in short-term studies of *Euglena* (Danforth and Wilson, 1961; Eshleman and Danforth, 1964), but long-term experiments on growing populations need to be performed.

If we assume for the moment that the data gained from the short-term experiments applies to growing cultures, the values of the corrected oxygen/acetate ratios, the endogenous respirations and the oxygen consumption rates of growing cells may be used to calculate estimates of the acetate uptake rates of synchronous and exponentially grown *Astasia*. These calculations are listed in the first column of Table III. They indi-

TABLE III

Estimated Acetate Uptake and Percent Acetate Oxidized by *Astasia*

Temperature (°C)	Acetate uptake (μmoles acetate/hr/10^6 cells)	Percent acetate oxidized
Exponential growth		
15	0.68	65
20	1.70	62
25	2.35	51
28.5	2.15	53
Synchronous growth		
14.8	1.18	51
28.5	2.58	50

cate that synchronized *Astasia* consumed acetate at faster rates than did *Astasia* grown exponentially at the cycle temperatures. Since other data showed that the rate of dry weight production of exponentially grown and synchronized *Astasia* were the same, it would seem that exponentially grown *Astasia* use acetate more effectively for cell syntheses than do their synchronized counterparts.

The data may also be used to estimate the percent of acetate oxidized during the growth of the cells, taking into account the carbon oxidized by the reserve oxidation system. These calculations are listed in the second column of Table III. With the exception of exponentially grown cells at 15°C, approximately ½ of the acetate utilized by *Astasia* was oxidized during exponential or synchronous growth. Thus, it seems unlikely that the stoichiometry of acetate utilization was drastically affected by the synchronization process.

V. CONCLUSION

We are now in a position to formulate one scheme by which limiting oxygen tensions might shift the production of respiratory machinery from the warm to the cold period of the synchronized cell cycle of *Astasia* and improve the synchronization of the cells, as suggested by James (1965) and to discuss some concepts and outline some future experiments which may unify the studies done on the oxidative assimilation, the intermediary metabolism and the growth of exponential and synchronized heterotrophic flagellate cells.

The preceding discussion suggests the hypothesis that the reduction in the rate of oxygen consumption forced upon the synchronized cells by steady-states of zero oxygen tension in the culture flasks accomplished by James (1965), have at the same time, limited the rate of acetate uptake of the cells, providing that the corrected oxygen/acetate ratio was not affected by oxygen tension. This limitation in the availability of substrate slows down the synthetic rates of the cells in the warm period. Cell division is not inhibited since the preparations for it were completed earlier in the cycle before zero oxygen tensions occurred. Thus, the daughter cells produced in the warm period are deficient in whatever cell machinery would have been produced in the latter part of the warm period, in this case, respiratory machinery. This lack is made up in the next cold period when oxygen tension no longer is limiting to the respiration rate of the cells.

An explanation of how this could lead to an improvement of the synchronization of the cells involves consideration of "intensity" and "capacity" factors which have been referred to before with respect to cell synchronization (Padilla and James, 1960). Alterations in metabolic activity may occur either from changes in the amount of metabolic machinery per cell, or the rates at which the machinery functions. The amount of machinery is the capacity factor, its unit rate of activity is the intensity factor and the one times the other yields the total metabolic activity of the particular process under consideration. Thus, if the time of the production of a factor important for cell division is shifted from the warm to the cold period its capacity (amount) would then increase during the cold period and the high incubation temperature of the following warm period would greatly increase the intensity (rate) at which it would operate, saturating the pathways necessary for cell division and causing the cells to divide simultaneously.

The above hypothesis predicts that cells exposed to steady-states of zero oxygen tension in the warm period will exhibit a decreased rate of acetate uptake and dry weight synthesis and a lowered rate of oxygen consumption when tested under conditions where oxygen is not limiting to the respiration of the cells. In addition, it follows from this that the acetate uptake, dry weight synthesis and oxygen consumption of the cells will be higher in the next cold period than that of cells incubated under nonlimiting oxygen conditions in the previous warm period, and that the next burst of cell division will be sharp.

The fulfillment of the above predictions in future experiments would not necessarily indicate that the total respiration of the cells was a capacity factor important to cell synchronization. The complex relationships between substrate utilization, cell syntheses, and cell respiration (touched upon in this chapter) are such that it is difficult to vary one of these parameters independntly of the others. Moreover, it is likely that changes in factors as gross as dry weight synthesis and respiration are merely reflections of crucial events occurring at specific metabolic sites.

One must be careful to avoid postulating causal roles for parameters solely because they can be easily measured, and attributing significance to concepts such as metabolic efficiency and intensity-capacity factors which are purposely defined broadly to provide a skeleton for future thought. Indeed, the major difficulty in applying the concept of intensity and capacity factors to problems of the control of cell division and metabolism is not that the approach is invalid but that the terms may be taken too literally and the true nature of the constituents that cause these effects be ignored. Consequently, studies on topics such as the ways in which mitochondria are replicated and activated should be the next order of business.

In summary, the experiments performed on synchronized *Astasia* lead to the conclusion that the respiration of Astasia is intimately geared to its division and synchronization. Some synchronous cultures increased in respiratory capacity during the cold period; the respiration was ordered in time over the synchronous cell cycle; it tended to rapidly approach values characteristic of cells grown exponentially at the cycle temperatures when expressed on a milligram basis; and the respiration rate per cell rose abruptly and then declined as cells in a well synchronized culture prepared to divide. These results seem sufficient to establish a relationship between the growth, respiration, and division of these cells, but specific metabolic events leading to these data remain to be elucidated. Similarly, the intensity and capacity factors active during synchronization are obscure, but, as Racker (1965) points out, "A good hypothesis is worth a few ugly little facts and a few hundred negative experiments."

REFERENCES

Barker, H. A. (1936). *J. Cellular Comp. Physiol.* **8,** 231.
Blum, J. J., and Padilla, G. M. (1962). *Exptl. Cell Res.* **28,** 512.
Blumenthal, H. J., Koffler, H., and Heath, E. C. (1957). *J. Cellular Comp. Physiol.* **50,** 471.
Clifton, C. E. (1963). *Ann. N.Y. Acad. Sci.* **102,** 655.
Danforth, W. F. (1961). *J. Protozool.* **8,** 152.
Danforth, W. F., and Wilson, B. W. (1961). *J. Gen. Microbiol.* **24,** 95.
Eshleman, J. N., and Danforth. W. F. (1964). *J. Protozool.* **11,** 394.
Gibor, A., and Granick, S. (1964). *Science* **145,** 890.
James, T. W. (1965). *Exptl. Cell Res.* **38,** 439.
Kaplan, N. O., and Cahn, R. D. (1962). *Proc. Natl. Acad. Sci. U.S.* **48,** 2123.
MacKenzie, C. G., Mackenzie, J. B., and Reiss, O. K. (1964). *Exptl. Cell Res.* **30,** 533.
Marzullo, G., and Danforth, W. F. (1964). *J. Gen. Microbiol.* **34,** 21.
Nass, M. M. K., and Nass, S. (1963). *J. Cell Biol.* **19,** 613.
Padilla, G. M., and James, T. W. (1960). *Exptl. Cell Res.* **20,** 401.
Racker, E. (1965). "Mechanisms in Bioenergetics," p. 17. Academic Press, New York.
Scopes, A. W., and Williamson, D. H. (1946). *Exptl. Cell Res.* **35,** 361.

Stern, H., and Kirk, P. (1948). *J. Gen. Physiol.* **31,** 243.
Wilson, B. W. (1963). *J. Cellular Comp. Physiol.* **62,** 49.
Wilson, B. W., and Danforth, W. F. (1958). *J. Gen. Microbiol.* **18,** 535.
Wilson, B. W., and James, T. W. (1963). *Exptl. Cell Res.* **32,** 305.
Wilson, B. W., Peterson, D. W., and Lilyblade, A. L. (1965). *Proc. Soc. Exptl. Biol. Med.* **119,** 104.
Winzler, R. J. (1940). *J. Cellular Comp. Physiol.* **15,** 343.
Wu, R., and Racker, E. (1963). *In* "Control Mechanisms in Respiration and Fermentation" (B. Wright, ed.), pp. 265–288. Ronald Press, New York.
Zeuthen, E. (1960). *Exptl. Cell Res.* **19,** 1.

CHAPTER 12

Synchrony of Cell Division in Root Meristems following Treatment with 5-Aminouracil*

Ellen Mattingly

I. INTRODUCTION

Partial synchrony of cell division exists in specialized tissues of many plants. Data about such naturally occurring synchronous division have recently been reviewed (Erickson, 1964). In the root meristem of flowering plants, several investigators have described a low degree of diurnal

* Research sponsored by the U. S. Atomic Energy Commission under contract with the Union Carbide Corporation.

rhythm of cell division; Jensen and Kavaljian (1958) found, for example, peaks of synchronized division at noon in *Allium* roots. These peaks were not sharply synchronized since all stages of the cell cycle were present and interphase cells were by far the most numerous. A more highly synchronized system would be useful in physiological experiments and biochemical determinations on roots, as Jensen and Kavaljian have pointed out. Therefore, a new technique for the induction of synchronized cell division by chemical treatment of the root with the thymine analogue, 5-aminouracil, has been of great interest.

The first published report, by Smith *et al.* (1963), concerning synchronized division following 5-aminouracil treatment of *Vicia* roots, showed that a high proportion of meristematic cells are in mitosis about 14 hours after removal of the roots from the chemical. The above authors suggested that this synchronous division depended on removal of the roots from the chemical to water or nutrient solution, and they ascribed the synchronized division to a prior inhibition of DNA synthesis during treatment with aminouracil. This total inhibition would result in the accumulation of cells at the beginning of the period of DNA synthesis. These cells, released from inhibition by transfer of the roots to water, would go through synthesis of DNA together and subsequently enter division together.

Other recent data (Jakob and Trosko, 1965) indicate that this view is an oversimplification, since cells in *Vicia* spontaneously recover the ability to divide in the presence of aminouracil, and levels of cell division above control values may be found in roots cultured continuously in the chemical. If cells in the process of DNA synthesis at the time of immersion of the roots ultimately divide in aminouracil, it appears that some DNA synthesis occurs during exposure of the roots to 5-aminouracil.

The basic physiological mechanisms responsible for the above effect in plants are still unknown, although some information is available concerning the action of 5-aminouracil in bacteria. In *E. coli*, 5-aminouracil serves as a precursor of nucleic acid cytosine and thymine but is not incorporated itself into DNA (Wacker *et al.*, 1960). Inhibition of the microorganism by the analogue is prevented by thymine (Dunn and Smith, 1955). On the basis of data available, it appears that the mechanism of action of 5-aminouracil is somewhat different in higher organisms. Some reports conflict rather sharply, however, and so it is impossible to reach any valid conclusions about the physiological mechanisms responsible for the induction of synchronized cell division. On the other hand, the technique does not appear to cause any serious or lasting overt effect on the root, other than the synchronization of cell division, which does not seem deleterious.

II. MITOTIC INHIBITION DURING TREATMENT WITH AMINOURACIL

A. The Normal Cell Cycle

The details of the cell cycle of the broad bean, *Vicia faba*, are known from the early labeling studies of Howard and Pelc (1951) with P^{32} and the more precise investigations of Evans and Scott (1964) with thymidine-H^3. The latter authors have shown that the average total cell cycle at 20°C is 19 hours. The greater part of this (about 17 hours) is spent in interphase, which may be further subdivided into G_1, a preliminary period during which DNA is not synthesized (4.9 hours); S, the phase of thymidine-H^3 incorporation, which is equated with DNA synthesis (7.5 hours); and G_2 (4.9 hours), the subsequent period during which there is again no evidence of thymidine-H^3 incorporation. The process of actual cell division (mitosis) occurs during the following 2 hours and consists of the typical prophase, metaphase, anaphase, and telophase stages.

B. Effect of 5-Aminouracil Treatment on Mitosis

Much of the work that has been done with 5-aminouracil has involved analyses of the effects of this chemical on the various stages of the cell cycle in the apical meristem. The original report by Duncan and Woods (1953) described the effects of aminouracil on root tips of *Allium cepa*. The most apparent cytological effect was a severe inhibition of cell division in the presence of various concentrations of aminouracil up to 4 mM (table I). In general, the higher the concentration of aminouracil, the fewer the cells seen in mitosis at the end of 24 hours. This severe inhibition

TABLE I

Effects of 24-Hour Treatment with Various Concentrations of 5-Aminouracil[a]

	ppm	Nuclei in mitosis (%)	Anaphases with Feulgen-negative regions (%)	Newly formed cells with indisputable micronuclei (%)
Control	—	8	0	0
5-Aminouracil	31	5	26	0
5-Aminouracil	63	5	88	1
5-Aminouracil	125	2	68	0
5-Aminouracil	250	1	82	0
5-Aminouracil	500	2	100	1

[a] From Duncan and Woods (1953).

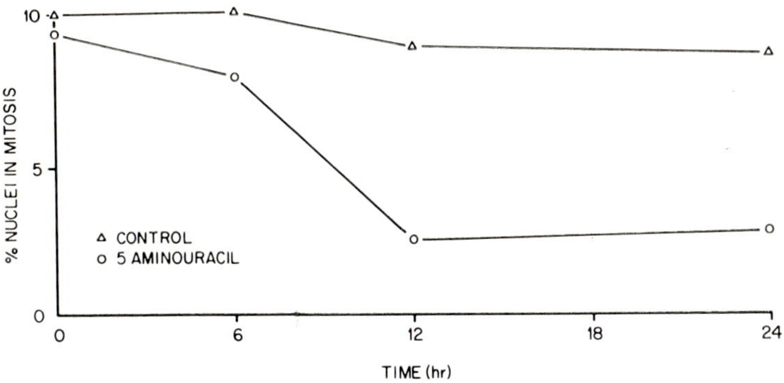

FIG. 1. Effect of duration of treatment with 5-aminouracil (100 ppm) on mitotic activity in *Allium cepa* root meristems. (Duncan and Woods, 1953.)

of division did not appear immediately after immersion of the roots in aminouracil, however, and the data show no sharp decline during the first 6 hours of treatment (Fig. 1).

These authors also showed that 5-aminouracil affects the structure of chromosomes with high frequency (second column, Table I). The structural changes (interstitial Feulgen-negative gaps) were especially noticeable at anaphase in roots treated for 24 hours. There was, however, no evidence of stickiness of the chromosomes.

Martinez-Pico and Duncan (1955) then investigated the cytological effects of aminouracil in *Vicia faba*. Although treatment with aminouracil severely depressed the mitotic index during the first 12 hours, cells in root tips of *Vicia* spontaneously recovered the ability to divide, even in the presence of the chemical. Jakob and Trosko (1965) confirmed the observations of spontaneous recovery, and Mattingly (unpublished data, 1965) found that after the recovery from the mitotic inhibition, cells in roots exposed to aminouracil for several days may go through repeated divisions.

C. THE EFFECT OF 5-AMINOURACIL ON INTERPHASE

Autoradiography following thymidine-H^3 treatment has been used to determine the effect of continuous treatment with aminouracil on the various stages of interphase. In control roots, cells that have incorporated thymidine-H^3 require an average time of 4.9 hours to appear in mitosis. Cells that are in S when aminouracil treatment is begun do not enter division.

S. Wolff and H. Luippold (unpublished data, 1964) studied the effect of 5-aminouracil on cells in G_2 at the beginning of treatment. They treated lateral roots of *Vicia faba* for 1 hour with thymidine-H^3 and then placed these roots in 6 mM 5-aminouracil. With 1-hour labeling, all cells in S, and those S cells that completed DNA synthesis and moved into G_2 during the hour were labeled. In control roots, labeled metaphase cells began to appear in quantity about 4 hours after treatment of the roots with thymidine-H^3. No change in mitotic index was observed. In 5-aminouracil-treated roots, the mitotic index fell to about zero within 9 hours after the beginning of 5-aminouracil treatment. In these roots, which had been treated with thymidine-H^3 for 1 hour before 5-aminouracil exposure, no labeled metaphase cells were seen during the first 9 hours. This indicated that the labeled cells in the early part of G_2 were blocked. This same experiment has been repeated by Mattingly (unpublished data, 1965), using 4 mM 5-aminouracil, and virtually identical results were obtained.

Prensky and Smith (1965) also studied the effect of 5-aminouracil on G_2. With only a slightly different experimental procedure, they studied the passage of G_2 cells into mitosis during the first 12 hours of 5-aminouracil treatment, and concluded on the basis of their data that it is highly probable that 5-aminouracil decreases the rate of progress of cells through the G_2 period.

III. RELATION OF THE AMINOURACIL EFFECT TO DNA SYNTHESIS

A. Effect of 5-Aminouracil Treatment on G_1 Cells

Mattingly (unpublished data, 1965) studied the effect of aminouracil on the progress of G_1 cells into S. Autoradiographs prepared from roots exposed simultaneously to aminouracil and thymidine-H^3 for 8 hours were scored for the number of cells able to incorporate thymidine-H^3 at intervals during this time (Fig. 2). During this first 8 hours, the mitotic index in 5-aminouracil-treated roots approached zero. In the same roots, however, the number of cells synthesizing DNA increased at the same rate as in control roots.

B. Decreased Rate of DNA Synthesis in S

Although G_1 cells continue to enter S in roots exposed to 5-aminouracil, and hence the total number of labeled nuclei increases with time, the incorporation of labeled DNA precursors, presumed to represent DNA synthesis, is depressed in the individual cells. Several investigators have studied this effect (Prensky and Smith, 1965; Jakob and Trosko, 1965; Mattingly, unpublished data, 1965), and all agree that this depression in DNA synthesis results in an extension of the S phase of the cycle. With such a system, cells that were in G_2, mitosis, G_1, and S at

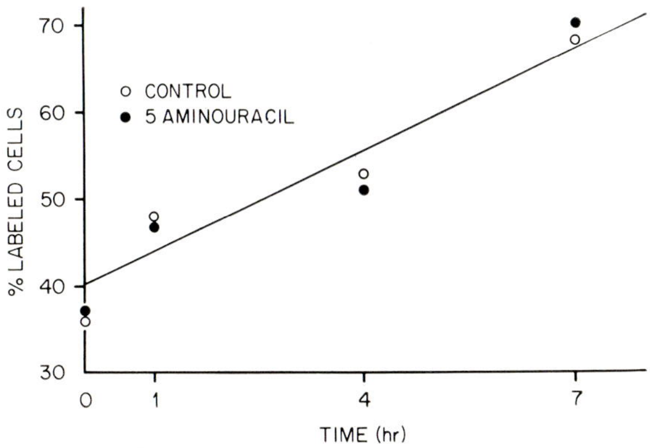

FIG. 2. Increase in number of labeled nuclei with increase in length of exposure to 5-aminouracil and thymidine-H^3. (Mattingly, unpublished data.)

the beginning of aminouracil treatment might be expected to be collected in *S* by a sufficiently long exposure to aminouracil.

In the presence of aminouracil, a highly synchronized group of cells in *S* is achieved. It appears, however, that there is an upper limit to the number of cells that can be labeled in the root apex. Howard and Pelc (1951), using P^{32} and autoradiography, found that after 24 hours of continuous exposure to the isotope, 75% of the cells in the *Vicia* meristem were labeled. One may conclude this is near the maximum. Mattingly exposed *Vicia* roots to thymidine-H^3 for 4 hours after a pretreatment with aminouracil for 12 hours. Although roots not treated with 5-aminouracil showed about 60% of the cells labeled during this time, the experimental roots had a labeling index of 82%, and in the subsequent peak of synchronized division, 95% of the metaphase figures were labeled. Brewen (1964) found that a high labeling index (greater than 70%) is seen in the synchronized division period following treatment of *Vicia* roots with 750 ppm 5-aminouracil. It seems, therefore, that a very synchronous period of DNA synthesis does exist in the presence of aminouracil; this does not necessarily indicate, however, that all cells are delayed to the same degree in *S*.

C. SYNCHRONIZED DIVISION RESULTING FROM 5-AMINOURACIL TREATMENT

Although Smith *et al.* (1963) reported that the removal of roots from 5-aminouracil is necessary for the appearance of a synchronous peak of cell division 14 hours later, it is now known that a wave of cell division appears in roots left continually in 5-aminouracil at about the same time as in roots transferred to water (Mattingly, unpublished data, see Fig. 3). The peak of cell division is not sharp, since in 5-aminouracil-

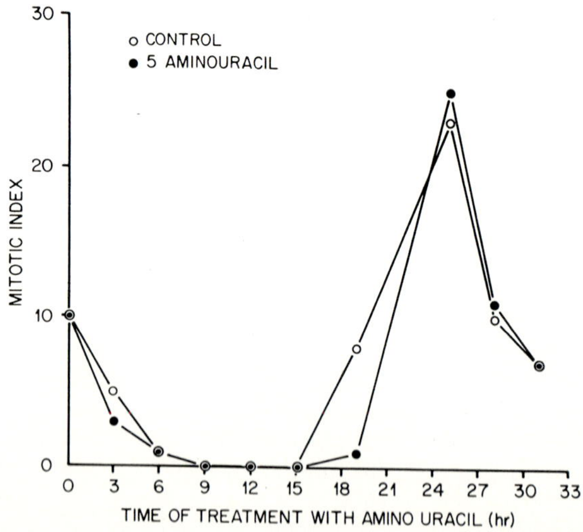

FIG. 3. Mitotic peak in presence of 5-aminouracil (500 ppm; 750 ppm).

treated roots only about 40% of the apical cells are in some stage of division at any one time. This mitotic index remains well above control values (10%) for several hours after the beginning of the wave of divisions.

D. Differential Effect According to Cell Stage at Beginning of Treatment

Mattingly investigated the relative delay of different stages of the cell cycle in their progress through S by determining the composition of the mitotic population at various intervals during the phase of synchronized division (Table II). Metaphase figures in roots pretreated with colchicine (0.2%) for 2 hours tend to accumulate even after removal of the roots from colchicine. Thus, roots pretreated with colchicine for 2 hours, exposed to thymidine-H^3 for 30 minutes, and then placed in aminouracil, showed the typical aminouracil-induced inhibition, but the mitotic index did not reach zero until 11 hours after the beginning of aminouracil treatment. In this type of experiment, the cells that were delayed in metaphase in the presence of aminouracil had all been in G_2 at the beginning of treatment; these cells ultimately formed restitution nuclei and would appear as polyploid cells when they next appeared in metaphase.

The rationale for this experiment was that the following populations of cells should be seen in the phase of synchronous division: (a) nonlabeled polyploid cells, which would result from those G_2 cells that had entered metaphase during the first 12 hours of aminouracil treatment; (b) labeled diploid cells, which would represent that group of cells that were

TABLE II

COMPOSITION OF CELL POPULATION AT INTERVALS OF THE SYNCHRONOUS DIVISION IN ROOTS PRETREATED WITH COLCHICINE (2-HOUR) AND SUBSEQUENTLY GROWN CONTINUOUSLY IN 5-AMINOURACIL[a]

Time in 5-aminouracil (hours)	Cells in interphase (%)	Cells in mitosis			Diploid metaphase cells		Tetraploid metaphase cells	
		Prophase (%)	Metaphase (%)	R[b] (%)	Labeled (%)	Unlabeled (%)	Labeled (%)	Unlabeled (%)
0	87	6	7	—	0	100	—	—
4	89	—	11	—	0	100	—	—
6	92	—	8	—	0	100	—	—
8	98	—	2	—	0	100	—	—
11	100	—	—	—	—	—	—	—
21	89	6	5	—	94	6	0	0
24	55	16	26	3	56	25	0	19
27	70	5	22	3	43	22	4	31
30	74	5	18	3	24	37	5	34
33	81	7	9	3	21	11	48	20

[a] Mattingly, unpublished data.
[b] R = restitution.

in S at the beginning of aminouracil treatment; (c) nonlabeled diploid cells, which would be derived chiefly from cells in G_1 at the beginning of treatment, with a possible contribution from very early G_2 cells.

When roots so treated were fixed at intervals during the synchronized division phase, an orderly progression of cells through metaphase was noted (Table II). Almost all of the first cells to enter mitosis were labeled diploid cells. These cells, according to the above rationale, were in S at the beginning of aminouracil treatment. This population of labeled diploid cells was rapidly joined by a group of cells with about equal proportions of nonlabeled diploid metaphases and nonlabeled polypoid cells. The nonlabeled diploid metaphases represent those cells that were in G_1 and mitosis at the beginning of treatment, with some also resulting from the block induced in early G_2. The nonlabeled polyploid cells were the result of the passage of G_2 cells through metaphase and restitution during the first 8 hours of treatment. During the time that the mitotic index remained high, the number of labeled diploid metaphase cells decreased, and among the unlabeled cells in metaphase, the ratio of polyploid metaphases to diploid metaphases increased.

It was concluded from such observations that cells that had been in G_2 at the beginning of aminouracil treatment and were delayed by pretreatment with colchicine, nevertheless progressed through G_1 and S at such a rate that they were able to enter mitosis with cells that had been in G_1 in the beginning of treatment.

During the later intervals of the synchronized division phase, an unexpected population of cells appeared in mitosis, i.e., labeled tetraploid metaphases. The only explanation for the appearance of this population of cells is that the labeled diploid cells that had gone through metaphase early in the synchronized phase had passed through a second comparatively brief interphase and again were entering mitosis. It appears from the data, that this interval might be as short as 12 hours.

From the reports that have appeared, it seems that although cells do accumulate in *S* in the presence of aminouracil, DNA synthesis is not suppressed to the same degree in all cells. Therefore, a relation between this relative suppression of DNA synthesis and the subsequent semisynchronous division is indicated. At present, there are no data available to suggest what physiological mechanisms may be responsible for the effects observed.

IV. STUDIES OF THE AMINOURACIL EFFECT IN SECTIONED MATERIAL

All possible cells in the apex may be labeled with thymidine-H^3 after a 12-hour pretreatment with 5-aminouracil (Mattingly, unpublished data, 1965). The subsequent phase of synchronized division, however, extends over at least 8 hours, generally reaches about 40% at best, and does not fall again to zero after the peak of synchronous division has passed. The broad phase of synchronized division may be explained as follows. Although at an interval early in the phase, i.e., 24 hours of aminouracil exposure, smears of root tips show a low degree of cell division, median longitudinal sections of such roots (Fig. 4) show that division is strictly localized in the cortical tissues proximal to the apex. The cells of the vascular cylinder and apex at this time are entirely in interphase, although the cortex may show mitotic indices as high as 70%.

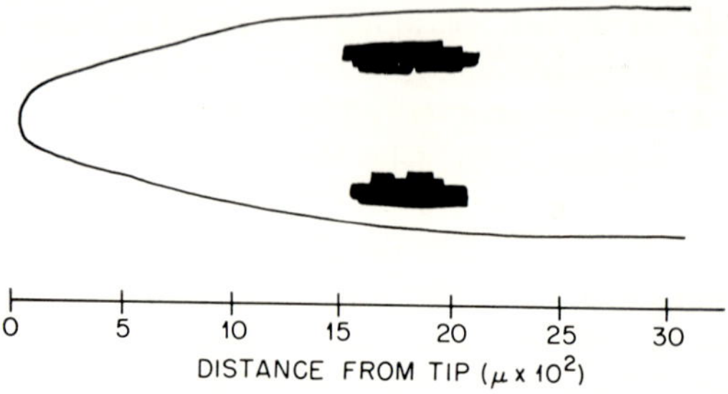

FIG. 4. Diagrammatic representation of root apex showing in black the areas in which division figures were seen after 24 hour of exposure to 5-aminouracil. (Mattingly, unpublished data.)

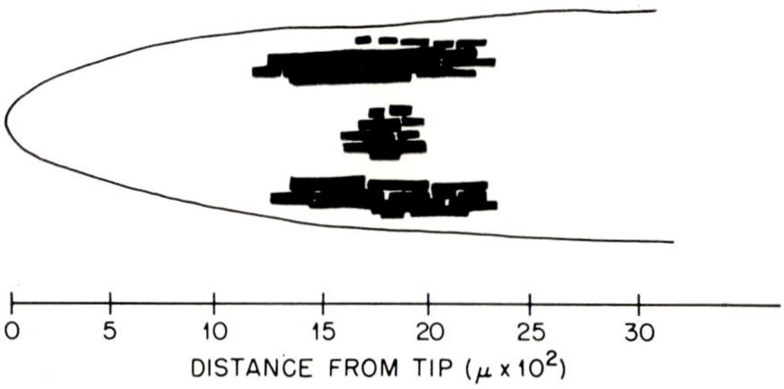

FIG. 5. Areas of high division frequency seen after 24-hour 5-aminouracil and 4-hour colchicine treatment. (Mattingly, unpublished data.)

If roots are treated with aminouracil for 24 hours and then placed in colchicine, the accumulation of metaphase figures by the colchicine increases the synchrony of cell division produced by the aminouracil. As the total mitotic index increases with time, longitudinal sections show that the small area of high division frequency previously seen in the cortex gradually moves toward the apex (Fig. 5). Cells that came early into metaphase recover from the colchicine inhibition and form reconstruction nuclei, while more distal cortical cells enter division. Ultimately, therefore, only the most apical cells of the meristem will be seen in metaphase.

As the more proximal cells of the cortex return to the restitution nuclei, numerous cells in the vascular cylinder begin to appear in prophase at the same level of the root in which cells of the cortex had previously entered mitosis. This small area of high division frequency spreads distally, and simultaneously cells of the protoderm and epidermis enter metaphase. This results eventually in the appearance of a vascular cylinder and epidermis having a high mitotic index surrounded by cortical cells that are apparently in tetraploid interphase (Fig. 6). Furthermore, at this time, levels of the root in which differentiated vascular elements may be distinguished show a layer of cells on the periphery of the

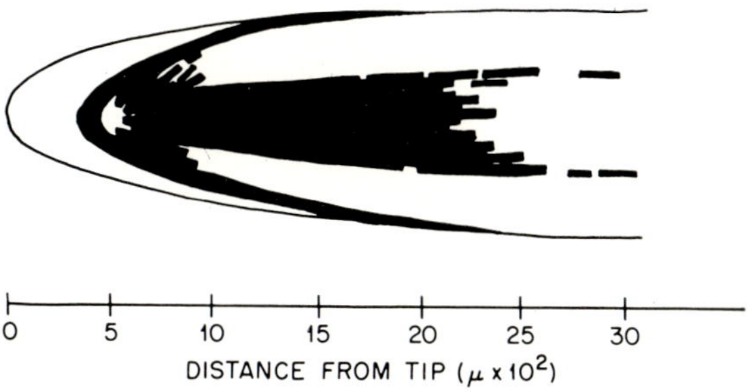

FIG. 6. Areas of high division frequency seen after 24-hour 5-aminouracil and 10-hour colchicine treatment. (Mattingly, unpublished data.)

vascular cylinder that contains cells in metaphase, although the surrounding tissues are entirely in interphase. These cells are thought to belong to the pericycle and therefore retain the capacity for cell division.

V. COLLATERAL EFFECTS PRODUCED BY AMINOURACIL

In addition to the induction of synchronous division, several observations have been made concerning other effects in roots after treatment with aminouracil. Duncan and Woods (1953) discussed an inhibition of root elongation following treatment of *Allium* roots with aminouracil. They found that this inhibition of elongation was independent of the repression of cell division produced by this chemical. Although the latter inhibition was unaffected by subsequent treatment of the roots with either thymine or folic acid, the inhibition of root elongation could be totally reversed in the presence of these chemicals.

Sormova *et al.* (1960) studied the effect of aminouracil on elongation of pea seedlings. They confirmed the previous report of an inhibitory effect and showed that treatment of roots with concentrations as low as 100 μg/ml caused a 44% inhibition of root elongation.

Finally, several investigators have studied the effect of 5-aminouracil on animal cells in tissue culture. Chu (personal communication) found that in Chinese hamster tissue cultures treated for 24 hours with growth medium containing 3×10^{-3} *M* 5-aminouracil, the mitotic index was reduced but not completely suppressed within the first 8 hours of exposure (Fig. 7). No recovery in the presence of the chemical was seen. When such cultures were removed from aminouracil at the end of 24 hours, a phase of synchronized division followed, reaching a peak at the eighth hour after removal. Similar results have been obtained by G. Brewen (unpublished data, 1964). This peak was of short duration, indicating a high degree of synchronization of cells that were capable of arriving at division. There was no evidence for a second peak 80 hours after inoculation, when these cells would have been expected to divide again. Instead, the synchrony appears to have decayed as the cells began to divide randomly at the normal control rates.

DNA synthesis in 5-aminouracil-treated hamster cultures was studied, both by autoradiography and by liquid-scintillation counting (Fig. 8). The data demonstrate that DNA synthesis was either blocked or slowed down by 5-aminouracil and that at least a portion of the cell population could recover from the chemical treatment and complete DNA synthesis before entering G_2 and division.

In comparison to the effects seen in *Vicia*, 5-aminouracil evidently affects animal cells in tissue culture quite differently.

For example, the peak of synchronized cell division following the re-

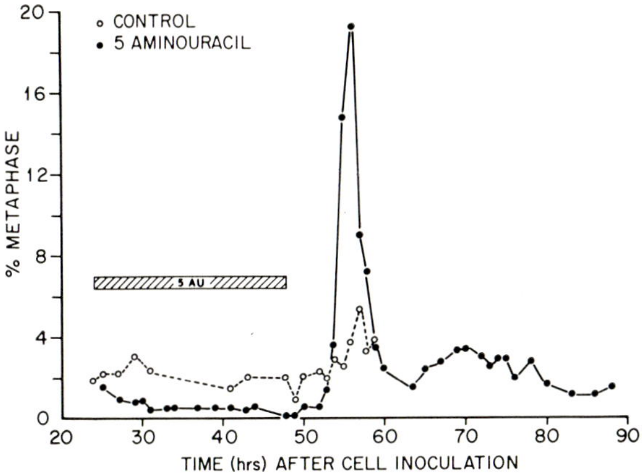

FIG. 7. The frequency of metaphases in Chinese hamster cell populations at various times in the presence of and after removal of 5-aminouracil. No colchicine was added to any sample. (E. H. Y. Chu, personal communication, 1965.)

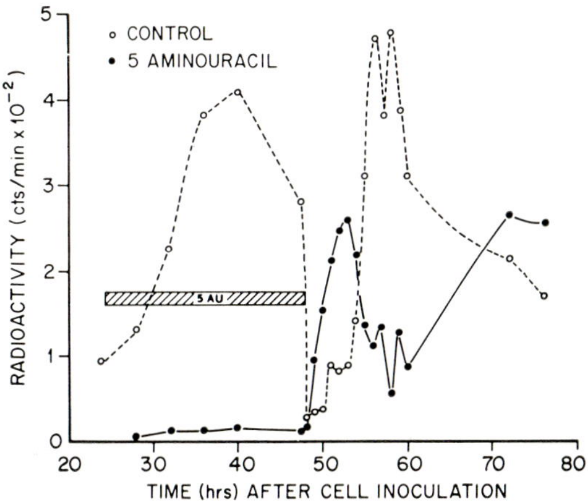

FIG. 8. Total DNA synthetic activity in Chinese hamster cell population in the presence and after the removal of 5-aminouracil. The activity was determined by liquid-scintillation counting of incorporated thymidine-H^3 immediately after 15-minute pulse labeling at each interval. (E. H. Y. Chu, personal communication, 1965.)

moval of hamster cells from aminouracil is exceedingly sharp in comparison with the type of synchronized cell division seen in *Vicia faba*. The explanation for this difference is that the population of hamster cells in tissue culture is relatively homogeneous, i.e., fibroblast cells. The distal 2 mm of the *Vicia* root tip contains several kinds of cells and tissues, whose average cell cycle times may differ considerably even in normal roots. As Mattingly has shown, the 5-aminouracil synchrony of division seen in root tip smears of *Vicia* is actually the result of very complex responses of the individual tissues present in the root tip.

REFERENCES

Brewen, G. (1964). *Genetics* **50,** 101.

Duncan, R. E., and Woods, P. S. (1953) *Chromosoma* **6,** 45.

Dunn, D. B., and Smith, J. D. (1955). *Nature* **175,** 336.

Erickson, R. O. (1964). *In* "Synchrony in Cell Division and Growth" (E. Zeuthen, ed.), pp. 11–37. Wiley (Interscience), New York.

Evans, H. J., and Scott, D. (1964). *Genetics* **49,** 17.

Howard, A., and Pelc, S. R. (1951). *Exptl. Cell Res.* **2,** 178.

Jakob, K., and Trosko, J. E. (1965). *Exptl. Cell Res.* **40,** 56.

Jensen, W. A., and Kavaljian, L. G. (1958). *Am. J. Botany* **45,** 365.

Martinez-Pico, L. M., and Duncan, R. E. (1955). *Cytologia* **20,** 378.

Prensky, W., and Smith, H. H. (1965). *J. Cell Biol.* **24,** 401.

Smith, H. H., Fussell, C. P., and Kugelman, B. H. (1963). *Science* **142,** 595.

Sormova, Z., Melichar, A., and Sorm, F. (1960). *Collection Czech. Chem. Commun.* **25,** 2889.

Wacker, A., Kirschfeld, S., Hartmann, D., and Weinblum, D. (1960). *J. Mol. Biol.* **2,** 69.

CHAPTER 13

The Physiology of Repetitively Synchronized *Tetrahymena**

G. M. Padilla, I. L. Cameron, and L. H. Elrod

I. INTRODUCTION

A combination of attributes, including ability to grow in pure culture, cell size, morphogenetic patterns of development, and responses to incubation temperature, has made *Tetrahymena pyriformis* a favorite organism for studies of cell division and growth. This ciliate was first synchronized by Scherbaum and Zeuthen in 1954 and has since been extensively studied. Not only was it one of the first cells to be successfully synchronized, but it is being used with ever increasing frequency in studies of cell division, morphogenesis, nutrition, and growth (Prescott,

* Research sponsored by the U. S. Atomic Energy Commission under contract with the Union Carbide Corporation.

1957; McDonald, 1958, 1962; Iverson and Giese, 1957; Thormar, 1962a,b; Elliott *et al.*, 1962; Kidder and Dewey, 1951; Holz, 1964; Summers, 1963; Stone and Prescott, 1964; Walker and Mitchison, 1957; Cameron and Prescott, 1961; Cameron, 1965). Recent comprehensive reviews of the literature on *Tetrahymena* include Zeuthen (1964), Prescott (1964), Holz (1964), Corliss (1954), and Elliott (1959).

Of necessity, this discussion of temperature-induced synchrony in *Tetrahymena* is limited primarily to the experimental approach first devised by James in his pioneer experiments leading to the synchronization of *Amoeba proteus* with a repetitive temperature cycle (James, 1954). These early studies were based on the recognition that certain cellular processes, e.g., cytokinesis, operate most effectively within a narrow temperature range, whereas other processes, largely comprising cytoplasmic functions, continue throughout the entire physiological temperature range of a cell. This is shown in the experimental observation that at low temperatures cells grow larger and have a reduced division rate. A consideration of the discontinuity in the temperature-dependence that is seen between the growth and division rates led James to postulate that a temperature program which alternates between 2 temperatures which favor nuclear events at one level and cytoplasmic functions at another would lead to an alignment of the individual cell cycles throughout a population of cells and thus yield repetitive synchrony as long as the temperature cycle was maintained. This hypothesis has been strengthened by the repetitive synchrony that has been achieved with a similar method in the flagellate *Astasia longa* (James, 1960; Padilla and James, 1960) and more recently in the ciliate *Tetrahymena pyriformis* HSM (Padilla and Cameron, 1964). Indeed this chapter is based on the successful synchronization of *Tetrahymena* by this method. In the formulation of such temperature cycles, cognizance has been taken of the overall behavior of a cell in response to widely different temperatures under transient and steady-state conditions (James and Read, 1957; Rao and Engelberg, Chapter 17) and availability of nutrients as they affect a cell's progress toward cell division (Summers, 1963; Christensson, 1959; Stone and Prescott, 1964; Scherbaum, 1956; Cameron, 1965; Cameron *et al.*, 1965; Hill and van Eys, 1964; Watanabe, 1963; Elliott *et al.*, 1962; and James, 1960).

This type of synchrony induction is fundamentally different from the method devised by Scherbaum and Zeuthen (1954). Repetitive synchrony is achieved not through introducing a low-temperature cytokinetic block, but by approximating a time-temperature interval which is required by the cells to complete a given function, say those reflected in growth, usually at a reduced temperature, and then shifting the temperature upwards, which by a yet undetermined alteration in biosynthetic regulatory mechanisms allows the cells to proceed at a faster rate toward cell division. To be sure, the exact makeup of the temperature cycle is empirically determined, but there has been an improvement in the degree and stability of the cell synchrony achieved, as evidenced particularly in the sequential studies with *Astasia longa* (Blum and Padilla, 1962; Padilla and James, 1964; Padilla and Cook, 1964).

Several of the physiological and biochemical features that characterize the repetitive synchrony in *Tetrahymena* will be examined in this chapter. The chapter will include an evaluation of the phase relationship between the DNA synthesis periods and the synchrony-inducing temperature cycle, and examination of the respiratory activity of the synchronized cells. The differential response to the synchronizing treatment that is seen in 2 strains of *Tetrahymena* will also receive brief consideration. The cytological and morphogenetic events that occur during synchrony are considered in Chapter 14 (Whitson *et al.*).

As stated by Zeuthen (1964, see the introduction there), "a certain degree of synchronization must always represent a balance between synchronizing factors imposed upon the system and a multitude of desynchronizing tendencies inherent in a population." Hence, synchronization of cell division must be approached with the view that even closely related cells will differ in their responsiveness to synchronizing factors. As discussed in this report, closely related cells, such as various strains within a given species, give different responses to a synchronizing treatment and hopefully offer a clue to the controlling factors underlying cytokinesis.

II. TEMPERATURE-INDUCED CELL SYNCHRONY IN *Tetrahymena*

A. Growth versus Temperature

The generalization that lower temperatures favor cytoplasmic growth rather than cell division is well established and has been studied experimentally in several ways, including (*a*) the relationship between generation time and temperature, (*b*) the relationship of cell size and generation time, and (*c*) the relationship between division index and temperature (Spear, 1928). *Tetrahymena pyriformis* HSM has a temperature optimum from 30°–34°C at which the generation time is minimal (2.25 to 2.5 hr; see Fig. 1). At higher and lower temperatures the generation time is lengthened. Similar responses have been reported for other strains of *Tetrahymena* (Prescott, 1957; Thormar, 1962a; Summers, 1963). With increasing generation time, the cells become larger (Summers, 1963; James and Read, 1957; Thormar, 1962b). The effects of lowered

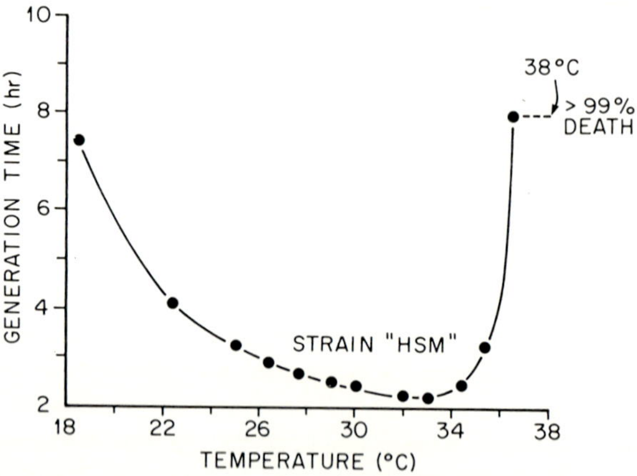

FIG. 1. A plot of incubation temperature against generation time for *Tetrahymena pyriformis* strain HSM (redrawn from Prescott, 1957).

temperature on *Tetrahymena's* division index have been considered at great length by Zeuthen (1964) and co-workers. They have examined the effects of a series of temperature shocks of various durations and temperature levels on the division index, and the degree of synchrony following such shocks. The effect of relatively short supraoptimal temperature shocks (in a temperature range which would eventually lead to death of the cells) is now well-known (Scherbaum and Zeuthen, 1954). Of interest are the studies of shifts to lowered temperatures (Zeuthen, 1964). In an experimental situation comparable to the synchrony procedure reported by Padilla and Cameron (1964), Zeuthen (1964) demonstrated that lowering of temperature for at least 2 hours to a temperature below 16°C results in an arrest of division; about 1.5 hours after the cells were returned to a higher temperature there was a rise in the division index, indicating the occurrence of a synchronous burst of division (Zeuthen, 1964, pp. 103, 121). However, this synchrony could be maintained for only a limited number of cycles following the return to optimal temperature (Zeuthen, 1964, p. 121), as is generally the case with ergodic cultures (see Engelberg and Hirsch, Chapter 2).

Indeed, the ability to maintain repetitive synchrony under a given synchrony treatment may be attributable to the strain of *Tetrahymena* used, for, as shown below, the HSM strain used in our studies yielded a repetitive type of synchrony but the GL-C strain failed to yield the same degree of synchrony (see Table V and Fig. 8).

B. Repetitive Cell Synchrony in Tetrahymena

Studies on repetitive cell synchrony have been greatly facilitated by the continuous counting technique developed by James and Anderson

(1963) in their studies on the synchrony of the flagellate *Astasia longa*. This method allows the continuous and immediate determination of population density changes of exponentially growing or synchronized cell cultures. The system involves the continuous and constant withdrawal of very small quantities of cells (0.3 ml/minute) that are counted with a modified Coulter counter and recorded on a strip chart. The unit operates with a minimum of attention and yields a series of long-term determinations on the developing synchrony patterns. This system allows biochemical determinations on a cell system during synchrony.

Figure 2 shows the degree of cell synchrony achieved with a 12-hour temperature cycle (9.5 hours at 12°C and 2.5 hours at 27°C). The tracings, constituting a series of cycles, demonstrate the repetitiveness that is characteristic of this synchronized system. In Zeuthen's low-temperature synchrony induction (1964, p. 105), as well as in ours, there is a

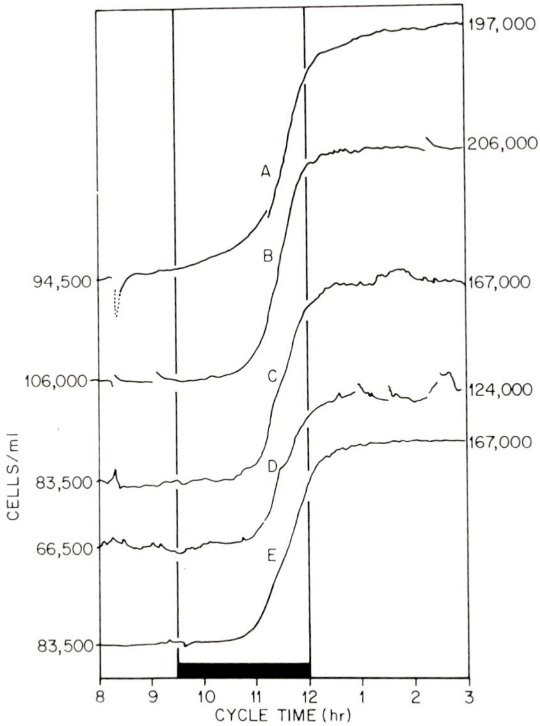

FIG. 2. Tracings of recorded, continuous cell counts of temperature-synchronized *Tetrahymena pyriformis* strain HSM. A series of 5 alternate bursts of division (A–E) is shown. The shaded portion indicates the duration of the warm period (2.5 hours at 27°C). Only portions of the cold period are shown (9.5 hours at 12°C). The initial and final population densities of the culture, sampled and counted by the technique of James and Anderson (1963), are given. (From Padilla and Cameron, 1964.)

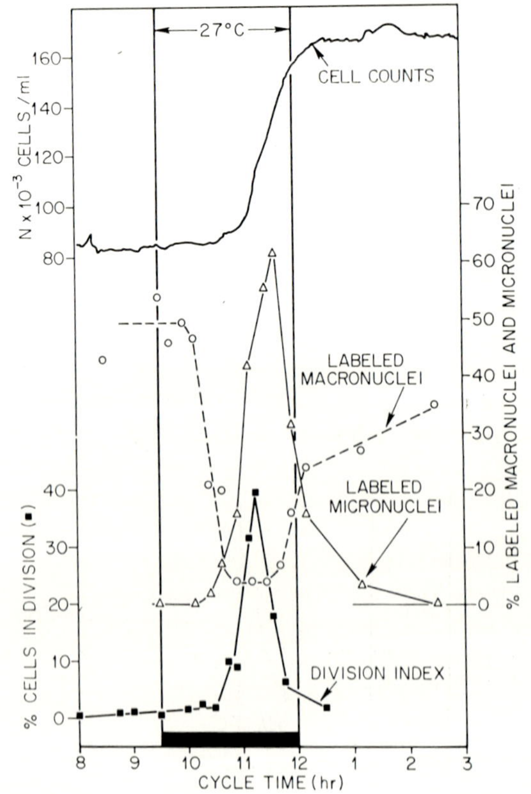

FIG. 3. Macronuclear and micronuclear thymidine-H^3 uptake in (pulse labeled for 5 minutes) synchronized *Tetrahymena pyriformis* strain HSM. Only portions of the 12-hour temperature cycle (2.5 hours at 27°C and 9.5 hours at 12°C) are shown. The shaded bar represents the duration of the warm period. The division index is also shown on this figure. From Padilla and Cameron (1964).

1.5-hour lag before the cells begin to divide and the division index is approximately 40% (Fig. 3).

Figure 3 shows the nuclear replication cycle in synchronized *Tetrahymena* as revealed by thymidine-H^3 incorporation and autoradiography. The cells actively incorporate the label into the macronucleus during the latter portions of the cold period. Thirty minutes after the cells have been exposed to the warm period, the number of cells in macronuclear DNA synthesis (also referred to as macronuclear *S* period) begins to decrease. The lowest level, approximately 4%, is reached in the next 45 minutes. Toward the end of the burst of division, the macronuclei resume incorporation of the isotope. Resumption of the cold period causes a reduction in the rate at which cells become labeled in the macronucleus.

Approximately 30 minutes after the uptake of macronuclear label decreases, the micronuclei begin to appear labeled, with over 60% being labeled in the middle of the synchronized burst of division. The peak in division index precedes the micronuclear maximum by approximately 30 minutes. Thus we have in this synchronized system a sequence of events that is comparable to the situation obtained in exponentially dividing cells; namely, the macronucleus is labeled at a time that excludes micronuclear labeling.

The results in Fig. 3 raise several questions regarding such a synchrony pattern: (*a*) is this the best level of synchronization that can be achieved; (*b*) how can the temporal pattern of macronuclear and micronuclear labeling be compared to the kinetics of labeling in exponentially dividing cells; and (*c*) can any significance be attributed to the sequence of macronuclear versus micronuclear labeling?

In answer to the first question, the individual generation times of *Tetrahymena* cells grown at constant temperature are rather widespread (Prescott, 1959). It is only when individual cells are isolated, by the techniques used by Stone and Cameron (1964), that one sees a higher degree of synchrony (Fig. 4). Such synchrony decays rapidly, for in the Stone and Cameron experiment (1964) the spread of the division burst had increased fivefold at the second division. Nevertheless, as is discussed by Whitson *et al.* (Chapter 14), the present level of synchrony in our

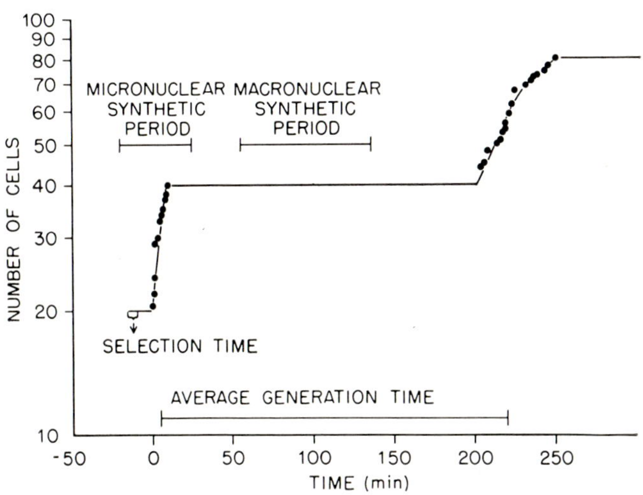

FIG. 4. Degree of synchrony of 20 cells selected in an early stage of division. Times required to collect dividing cells, subsequent separation, and the periods of micronuclear and macronuclear DNA synthesis are indicated (taken from Stone and Cameron, 1964).

repetitive method is sufficiently high to demonstrate an alignment of the morphogenetic events underlying stomatogenesis. Furthermore, the drug actinomycin D exerts its differential blockage of cell division during a short time interval (Whitson and Padilla, 1964).

In regard to the temporal pattern of labeling, the micronuclear DNA-synthetic period is undisturbed by the cell-synchrony-inducing temperature cycle (Fig. 4). Other workers (McDonald, 1962; Stone and Cameron, 1964; Cameron and Stone, 1964) have demonstrated that macro- and micronuclear DNA replication do not overlap temporally. With respect to the macronucleus, the thymidine-H^3 incorporation results in Fig. 3 show that the synchronous replication of DNA is not induced in this organelle.

In answer to the third question, the time differential between macronuclear and micronuclear labeling gives little information on the controlling factors that prevent simultaneous labeling of the 2 organelles. One can only say that, as is the case with exponentially dividing cells, the macronucleus appears to exert an inhibitory action on micronuclear DNA synthesis during its own DNA-replicating cycle.

C. Time Course of Cellular Nucleic Acid and Protein Synthesis, and Respiration Rate

Biochemical determinations of the time course of synthesis of DNA on a per cell basis confirm results achieved by autoradiography (Fig. 5). Most of the DNA is synthesized during the latter portions of the cold period and in the first hour of the warm period. It is during these periods that the largest percentage of macronuclei are labeled. These results are similar to those found in the flagellate *Astasia*, suggesting that cell synchronization is operating in a similar manner in 2 widely different species of protozoan cells (Padilla and Cook, 1964; Blum and Padilla, 1962). However, in contrast to *Astasia*, the predivision accumulation of cellular DNA in *Tetrahymena* precedes that of RNA. As expected, the DNA content of *Tetrahymena* fluctuates to the level of a 2 to 1 ratio for predivision: postdivision cells (35.5 to 16 $\mu g/10^6$ cells).

Figure 6 shows the changes of cellular protein content in synchronized cultures of *Tetrahymena* and the changes in the respiratory activity that occur during the warm period of synchrony. The respiratory determinations were performed with a Beckman oxygen analyzer. Samples from the parent culture were taken at close intervals during the cell cycle and placed in a calibrated 350 ml water-jacketed respirometer in which changes in the partial pressure of oxygen were measured with a Beckman oxygen electrode. This method allows one to determine the respiratory activity of cells under conditions which are essentially the same as

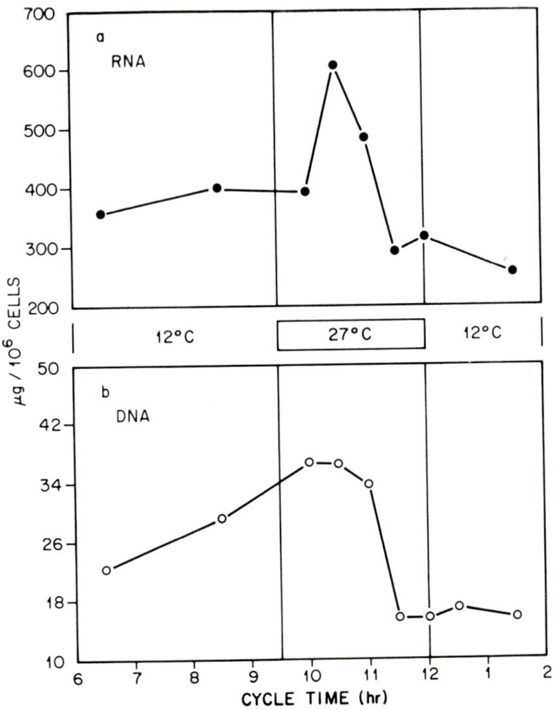

FIG. 5. Changes in the average cellular content of RNA (top) and DNA (bottom) in synchronized *Tetrahymena pyriformis* HSM. Nucleic acids were extracted by a modified Schmidt-Thannhauser (1945) procedure as described by Blum and Padilla (1962). The results are expressed as $\mu g/10^6$ cells and plotted against the time of 12-hour temperature synchrony cycle. Temperature changes are indicated. Ninety % of the cells regularly divide during the last hour of the warm period.

those of the parent culture. Table I summarizes a series of determinations on separate cycles. The Q_{O_2} at the onset of the warm period at a point just prior to cell separation and at the end of the warm period is indicated in Fig. 6.

The present study indicates that synchronized *Tetrahymena* exhibit a drop in the rate of respiration during cell division. The individual recordings from each of the runs, which contain more readings than the 3 shown in Fig. 6 and Table I, indicate that a drop in cellular respiration actually precedes the beginning of cell separation. These results agree with the earlier work of Zeuthen (1953), who found a drop or plateau in respiration rate at the time of division in single cell studies of *Tetrahymena*.

Tables II and III summarize certain biochemical changes in the cell cycle of *Tetrahymena* synchronized by our method. During the 9-hour

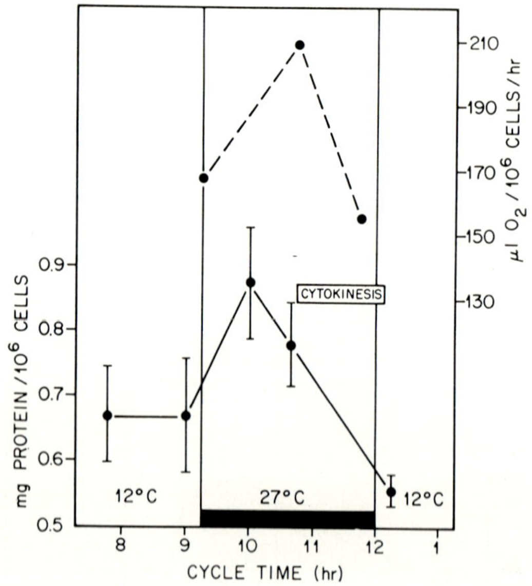

FIG. 6. Cellular respiratory rate (top) and protein content (bottom) of synchronized *Tetrahymena pyriformis* HSM. Respiratory rate is expressed as μliter/10^6 cells/hour as determined by a Beckman oxygen electrode (as described in text). The protein content was determined by the Lowry *et al.* (1951) method and is expressed in mg/10^6 cells [bars indicate standard deviation (S.D.)].

cold period, certain cellular activities continue while others are slowed or arrested altogether. The amount of carbohydrate and DNA in the cells increases about 100%; RNA, protein, and Q_{O_2} increase 33, 22, and 11%, respectively. Stomatogenesis is completely stopped. In the first 1.5 hours of the new warm period almost every cell undergoes stomatogenesis (see Whitson *et al.*, Chapter 14). RNA, protein, Q_{O_2}, and DNA increase 50, 31, 24, and 4%, respectively. Carbohydrates decrease 7%. During the 1-hour period when the cell number is increasing, RNA, DNA, carbohydrate, and number of mouths per cell drop almost in half. This is expected because of the partition of material to the daughter cells. On the other hand, protein content and Q_{O_2} do not halve, which suggests that they increase during cell division.

It would be futile to ascribe a cell division-triggering function to any of the above parameters of growth. In order to design experimental situations that indeed reveal the controlling mechanisms of cell division, one must consider the temporal relations between the various cellular functions. It would appear that repetitively synchronized *Tetrahymena* approach the model of a single cell, particularly with reference to the

TABLE I

FLUCTUATIONS IN THE RESPIRATORY ACTIVITY OF SYNCHRONIZED *Tetrahymena pyriformis* HSM[a]

Run no.	Oxygen consumption (μliters per 10^6 cells per hour)		
	Start of warm period	High	Low
1	175	235 (1.75)[b]	162 (2.50)[b]
2	180	238 (1.75)	195 (2.15)
3	170	270 (1.50)	150 (2.50)
4	162	182 (0.75)	162 (2.00)
5	160	190 (1.50)	155 (2.50)
6	162	182 (1.35)	122 (2.50)
7	160	169 (1.50)	140 (2.75)
Average ± S.E.	167 ± 2.8	209 ± 13.0	155 ± 7.5

[a] Synchronized by the one cold exposure per cycle method (a 12-hour cycle: 9.5 hours at 12°C and 2.5 hours at 27°C). Ninety % of the cells regularly divide during the last hour of the warm period.

[b] Figures in parentheses indicate time (hours) after onset of the warm period in which the high and low Q_{O_2} occurred.

TABLE II

CHANGES IN THE CELLULAR CONTENT OF PROTEIN, RNA, DNA, CARBOHYDRATE, CELLULAR RESPIRATION RATE, AND STOMATOGENESIS DURING THE CELL CYCLE OF SYNCHRONIZED *Tetrahymena pyriformis* HSM[a]

Cellular content	During the 9-hour cold period (12°C)	First 1.5 hours of new warm period (27°C)	Last hour of warm period during which cell number increases
Protein (mg/10^6 cells)	0.670 (+22)[c]	0.875 (+31)[c]	0.550 (−37)[c]
RNA (μg/10^6 cells)	400 (+33)	600 (+50)	300 (−50)
DNA (μg/10^6 cells)	34 (+110)	35.5 (+4)	16 (−55)
Carbohydrate[b] (mg/10^6 cells)	0.193 (+93)	0.179 (−7)	0.100 (−44)
Q_{O_2} (μliter/10^6 cells/hours)	168 (+11)	209 (+24)	155 (−26)
Number of mouths per cell	1 (0)	~2 (+100)	1 (−50)

[a] Synchronized by the one cold exposure per cycle method (a 12-hour cycle: 9.5 hours at 12°C and 2.5 hours at 27°C). Ninety % of the cells regularly divide during the last hour of the warm period.

[b] Data from Barber *et al.* (1965).

[c] Number in parentheses is the percent change which occurred during the indicated period of the synchronized cell cycle.

TABLE III

CARBOHYDRATE, PROTEIN, AND MITOCHONDRIAL CONCENTRATION OF *T. pyriformis* HSM[a]

Temperature period	Cycle time (hour)	No. of cells harvested[a] (× 10⁶)	Protein[b] (mg/10⁶ cells)	Carbohydrate[b] (mg/10⁶ cells)	Mitochondrial concentration[c] (no./unit surface area of the cells)
27°C warm period	9.5	107	0.710	0.193	11.4
	10.5	112	0.696	0.179	10.6
	11.5	162	0.469	0.105	8.0
	12.0				
12°C cold period	0.5	200	0.395	0.091	8.9
	1.5	200	0.395	0.085	9.5

[a] Population level rose from 70,000 to 130,000/ml in the last 1.5 hours of the warm period (total cycle time 12 hours).

[b] Data from Barber *et al.* (1965).

[c] These determinations were performed on separate cycles (Cameron, unpublished observations).

periodicity of micronuclear labeling. Changes in the biochemical profile also suggest that *Tetrahymena* adapts to a repetitive temperature cycle by undergoing oscillating phenomena that bear a close but not tightly coupled phase relationship to the temperature cycle. Scherbaum and Loefer (1964) have recently discussed the role of such oscillatory phenomena as controlling forces in cellular metabolism.

III. RELATION BETWEEN THE NUCLEAR DNA CYCLE AND THE DIVISION CYCLE IN MASS CULTURES OF SYNCHRONIZED *Tetrahymena pyriformis* HSM

Clearly it has now been demonstrated that *Tetrahymena pyriformis* strain HSM can be synchronized by a repetitive temperature cycle, and the synchrony can be maintained indefinitely (it has been maintained for more than a month at a time in this laboratory). Since micronuclear DNA synthesis takes place during cytokinesis, this synchrony method does not disturb the placement of the micronuclear DNA-synthetic period as compared with logarithmic growth conditions. The intimate connection between replication of micronuclear DNA and the division cycle suggests that either the cytoplasm or the micronucleus exerts a regulatory control over the other. Amicronucleate strains of *Tetrahymena pyriformis* undergo normal cytokinesis and growth. Moreover, the micronucleus in other micronucleated strains of *Tetrahymena pyriformis* has never been observed to synthesize DNA except during cyto-

TABLE IV

ANALYSIS OF THE NORMAL CELL CYCLE OF *Tetrahymena pyriformis* GROWN AT 29°C[a]

Strain	Medium	Generation time[b] (min)	DNA synthetic index at zero time (%)	Percentage of cell cycle occupied by each period[c]			
				G_1	S	G_2	D
HSM	1.5% proteose peptone + 0.1% liver extract	163	31	23.5	27.2	37.0	12.3
HSM	Defined medium[d] + 0.04% proteose peptone	225	33	41.6	34.8	12.6	11.0
GL-C	1.5% proteose peptone + 0.1% liver extract	135	52	14.1	36.3	34.8	14.8

[a] Unpublished data from Cameron, Padilla, and Wysinger.
[b] Generation time = mean culture doubling time.
[c] In reference to macronuclear S and cytokinesis.
[d] Defined medium as in Elliott *et al.* (1954).

kinesis. We must conclude therefore that the cytoplasm or macronucleus is dominant in its control and that the only course of the micronucleus in this species is to follow this control. The information which controls the micronucleus may indeed be stored in the cytoplasm as a result of some previous macronuclear message.

The macronuclear DNA replication period does not appear to be as intimately locked into the cell cycle as is the micronuclear DNA replication period (see Table IV and Zeuthen, 1964). One consistent fact evident from all these studies is that although the macronucleus undergoes an amitotic division (without the ordinary condensation of chromosomes as occurs in mitotically dividing cells), the macronucleus never replicates DNA during cytokinesis. This may, therefore, be thought of as macronuclear synchrony in the repetitive synchrony procedure. In this synchrony system macronuclear DNA replication and cytokinesis are mutually exclusive events in the cell cycle.

In an effort to better understand the relation between the nuclear replication cycle and division cycle in this synchrony regime, we extended the thymidine-H^3 labeling studies as follows: We incubated one portion of the cells in thymidine-H^3 from the beginning of the warm period and continued the incubation for 2 hours. Samples were periodically withdrawn for autoradiographic preparation. Two other portions were taken 45 minutes before the beginning of the cold period (at this time few macronuclei are in DNA synthesis, S phase; see Fig. 7, Curves II and III); both of these portions were also incubated constantly in thymidine-H^3. One portion was returned to the cold period at the nor-

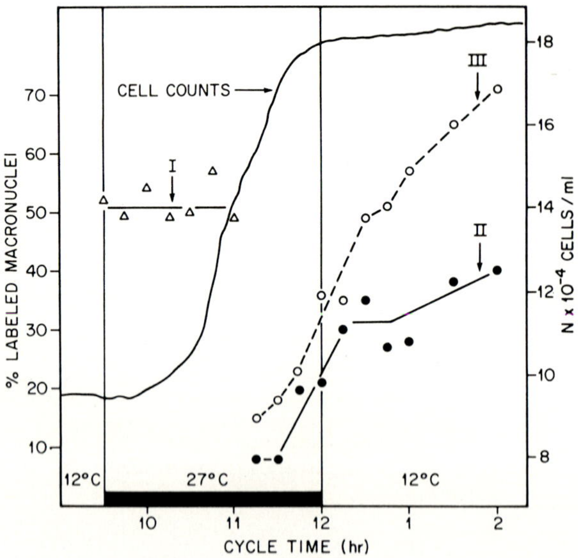

FIG. 7. Tracing of continuous recorded cell count of temperature-synchronized *Tetrahymena pyriformis* HSM according to the technique of James and Anderson (1963). Curves I, II, and III indicate the percentage of cells with labeled macronuclei when a portion of cells was removed from the parent culture (indicated by the first point of each curve) and exposed continuously to thymidine-H^3. Samples were taken for autoradiography at intervals and the results plotted. In Curve II the sample was returned to the cold (12°C) at the normal time, whereas in Curve III the sample remained at 27°C.

mal time (Fig. 7, Curve II), and the other was allowed to continue incubation at the warm temperature (Fig. 7, Curve III). Samples were periodically removed from both portions and autoradiographs prepared. The autoradiographs were analyzed for the percentage of cells showing macronuclear incorporation of thymidine-H^3.

When thymidine-H^3 is left in the culture, all of the cells that are in or that enter macronuclear DNA synthesis during the incubation period will be labeled. The results show that about 50% of the cells were in macronuclear DNA synthesis at the end of the cold and beginning of the warm period (Fig. 7, Curve I). If other cells had entered macronuclear DNA synthesis during this period, the percent would have increased. No evident increase in percent occurred. Therefore we may conclude that no cells or very few cells enter macronuclear *S* during the initial 2 hours of the warm period. Pulse labeling of cells with thymidine-H^3 (Fig. 3) showed that about 1 hour after the cells enter into the warm period the number of cells engaged in macronuclear DNA synthesis begins to decrease, falling from 50% and reaching its lowest level (about

5%) in the next 45 minutes. Taken together, these 2 findings (Fig. 3 and Fig. 7, Curve I) show that cells in the macronuclear *S* phase at the beginning of the warm period are either finishing their round of DNA synthesis or can no longer continue synthesis before cytokinesis. The results (Fig. 7, Curves II and III) demonstrate that cells begin to enter macronuclear *S* during the latter part of the warm period. At the end of the warm period 18 to 36% of the cells have initiated macronuclear *S* (Fig. 3 and Fig. 7, Curves II and III). The rate of entry of cells into macronuclear *S* can be determined from the slope of Curves II and III in Fig. 7. Clearly, imposition of the cold period slows the rate of entry into *S* but does not altogether stop initiation of macronuclear *S* by other cells. Pulsing with thymidine-H^3 during the cold period shows that cells in macronuclear *S* during the cold period continue DNA synthesis. Indeed, the results suggest that as many as 50% of the cells have finished macronuclear *S* during the cold period, since no new cell enters macronuclear *S* during the first part of the warm period. Because no or very few cells divide during the early warm period, we must conclude that even those cells which have finished *S* during the cold period still require at least 75–90 minutes of the warm period before they can divide. This suggests that some metabolic and/or morphological events (e.g., stomatogenesis) must occur before cell division begins. It also indicates that completion of macronuclear DNA synthesis is not the immediate "trigger" for cytokinesis.

IV. ATTEMPTS TO SYNCHRONIZE THE NUCLEAR DNA AND CELL DIVISION CYCLES IN THE GL-C STRAIN

At this point in the synchrony studies we were intrigued with the idea of synchronizing a strain of *Tetrahymena* which lacked a micronucleus. We believed that this might allow us to harvest large numbers of cells during cytokinesis, when few macronuclei and no micronuclei would be synthesizing DNA. One could then study the biochemistry of these cells, and examine such questions as the persistence of deoxyribonucleotide pools from one DNA cycle to another. Figure 8 shows the results of a synchrony experiment on the amicronucleate strain GL-C (kindly supplied by Dr. L. Rasmussen) performed as we had done for the micronucleate strain HSM. The synchrony bursts with the GL-C strain were never as good as with HSM. The percentage of cells in macronuclear *S* at the beginning of the warm period was about 50% (the same as HSM). At about 45 to 60 minutes later the number began to drop, but the drop was not as rapid as with the HSM strain. Instead of reaching a 5% level the GL-C strain dropped to only about 37%, although we have seen cases in which it dropped to 20%. Because of these results we did

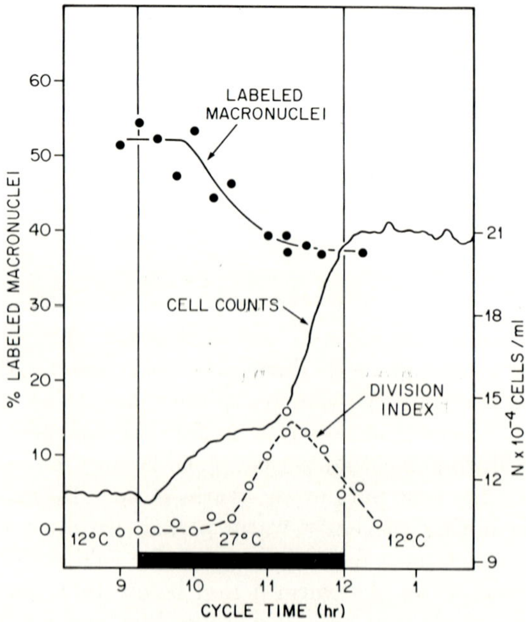

FIG. 8. Macronuclear thymidine-H³ uptake in (pulse labeled for 5 minutes) synchronized *Tetrahymena pyriformis* (amicronucleate strain GL-C). The division index and continuous recorded cell count are also shown. The temperature shifts are indicated on the figure. Earlier cycles of a 2.5-hour warm period followed by a 9.5-hour cold period showed the same pattern as these data.

not continue with biochemical studies as we had intended. Why were we unable to obtain as good a nuclear or division synchrony with the GL-C strain as with HSM? A plausible answer to this question came from analysis of the "normal" cell cycle of these 2 strains (Table IV) and from cell count curves (Fig. 8 and Table V).

Under the same culture conditions these 2 strains have similar generation times (Table IV; also see Thormar, 1962a; Prescott, 1957). In both strains the macronucleus begins DNA synthesis shortly after cytokinesis. The time between the end of macronuclear *S* and cytokinesis is shorter in GL-C than in HSM (47 versus 61 minutes). Also a larger percentage of the cell cycle is devoted to macronuclear DNA synthesis in the GL-C strain than in the HSM strain (36.3 versus 27.2%). The shorter interval between the end of macronuclear *S* and cytokinesis in the GL-C strain may partially account for the failure of the percentage of pulse-labeled cells in macronuclear *S* to decrease as rapidly as in strain HSM in relation to the synchronous burst of cell division.

That the GL-C strain lacks a micronucleus and that it requires a

TABLE V

ANALYSIS OF DIVISION KINETICS TAKEN FROM A CONTINUOUS RECORDING OF CELL COUNTS[a] OF SYNCHRONIZED *Tetrahymena pyriformis* STRAIN GL-C[b]

Synchrony cycle number	Warm period before increase in number of cells (hour)	Relative doubling first burst	Warm period plateau (hour)	Relative doubling during entire warm period	Relative doubling during entire cell cycle $\frac{N \text{ final}}{N \text{ initial}}$
6	0.50	1.08	1.40	1.18	1.27
7	0.33	1.34	1.60	1.81	1.89
8	0.33	1.12	1.70	1.30	1.55
9	0.33	—	—	—	—
10	0.50	1.12	1.50	1.33	1.50
11	0.33	1.22	1.60	1.84	1.87
12	0.50	1.10	1.50	1.40	1.70
13	0.40	1.12	1.70	1.57	1.80
14	0.30	1.08	1.60	1.10	1.30
15	0	1.29	1.50	1.29	1.29
Average	0.352	1.17	1.57	1.41	1.56

[a] By the technique of James and Anderson (1963).

[b] Using a 12-hour temperature cycle consisting of a 2.75-hour warm period and a 9.25-hour cold period. Earlier cycles of a 2.5-hour warm period followed by a 9.5-hour cold period showed the same pattern as these data.

greater part of its cell cycle to replicate macronuclear DNA could be reasons why neither the macronuclear replication cycle nor the division cycle was as synchronous as in HSM. Apparently the amicronucleate strain, because of its longer macronuclear *S* period, cannot respond in the same manner as does the micronuclear strain HSM to the synchrony-inducing regime. Perhaps in the GL-C strain not all of the cells have been able to initiate macronuclear DNA synthesis before the start of the next warm period, as was the case with the HSM strain. The temperature shock that the GL-C cells receive at the beginning of the warm period may be of sufficient severity to temporarily interfere with the initiation of macronuclear *S* in those cells which have not already initiated *S*. This could, therefore, result in a decreased degree of macronuclear and division synchrony. Another and perhaps better explanation may be that some cells divided early in the GL-C strain (Fig. 8, Table V), and that stomatogenesis may have been under way or complete in these cells at the time the warm period began. This early dividing subpopulation of cells could then perpetuate itself by finishing cytokinesis and DNA synthesis and being involved in stomatogenesis during the warm period. The imposition of the cold period could then "hold" the cells in stomatogenesis during the cold period, ready for almost immediate division at the start of the next warm period. Frankel (1962) indicates that stasis of stomatogenesis may occur under mild cold exposure, whereas heat shocks bring about resorption of primordial mouths. The cells which divide early in the warm period are presumably able to start macronuclear DNA replication during the middle of the warm period, and this would also help account for the fact that the macronuclear DNA labeling index did not decrease to the same extent as in the HSM strain.

The second and larger population of GL-C cells is apparently responding in the same way as did the cells of the HSM strain. That is, this larger population finishes division at the last of the warm period but is not able to start stomatogenesis until the next warm period. Clearly these speculations are subject to relatively easy experimental testing in the future.

It is evident from these experiments and the discussion that one must consider more than the relation between multiplication rate and temperature in the design of a synchrony procedure, even in closely related cells. Factors such as stomatogenesis, nuclear DNA replication cycle, as well as growth and cytokinesis, etc., must all be considered in the design of synchrony procedures.

V. SUMMARY AND CONCLUSIONS

A. Repetitive cell-division synchrony of *Tetrahymena pyriformis* HSM, growing in continuous culture on an enriched proteose peptone medium, has been induced with a 12-hour cycle consisting of 2.5 hours at 27°C and 9.5 hours at 12°C. More than 90% of the cells regularly divide in the last hour of the warm period.

B. Pulse labeling with thymidine-H^3 reveals that during early portions of the synchronous burst of division there is little macronuclear DNA synthesis. The micronucleus, on the other hand, actively incorporates thymidine-H^3 during cytokinesis.

C. The following biochemical measurements were made on these synchronized cells: the cellular content of protein, RNA, DNA, and carbohydrate, and also the cellular respiratory rate. The results suggest that an increase in DNA templates in one period will allow increased synthesis of RNA in the next period. Increased RNA (perhaps polysomes) would in turn allow more protein synthesis in the next period. Accumulation of carbohydrates (or energy reserves) in one period will free the cells' labors for increased respiration, stomatogenesis, and a generally increased synthetic activity in the next period.

Although the different cellular parameters change with different patterns in the cold and warm periods, a balance in chemical and structural elements of the cells appears to be maintained from one cell cycle to the text.

D. The thymidine-H^3 labeling data on the HSM strain of synchronized *Tetrahymena* permit the following statements:

1. Fifty % of the cells are in macronuclear *S* at the beginning of the warm period.
2. No or *very* few cells enter macronuclear *S* during the initial part of the warm period.
3. Cells in macronuclear *S* finish or stop *S* during the first hour of the warm period.
4. The first cells that divide at 1.5 hours after the beginning of the warm period begin their new macronuclear *S* phase during the last half hour of the warm period.
5. At the end of the warm period 18 to 36% of the cells have initiated macronuclear *S*.
6. The shift to the cold period (12°C) slows but does not stop the rate of entry of other cells into macronuclear *S*.
7. Cells in macronuclear *S* during the cold continue DNA synthesis during the cold.
8. At least 50% of the cells have finished macronuclear *S* during the cold period.
9. Even those cells which have finished macronuclear *S* during the cold period still require about 1.5 hours of warm before they divide.

E. Attempts to synchronize the nuclear DNA and cell division cycles of an amicronucleate strain of *Tetrahymena pyriformis* (GL-C) failed. An attempt was made to explain this failure on the basis of differences in the macronuclear DNA synthetic periods between the 2 strains and also because the GL-C strain lacks a micronucleus.

F. It is concluded that factors other than the relation between multiplication rate and temperature must be taken into account when designing synchrony procedures.

REFERENCES

Barber, A., Harris, W. W., and Padilla, G. M. (1965). *J. Cell Biol.* **27,** 281.

Blum, J. J., and Padilla, G. M. (1962). *Exptl. Cell Res.* **28,** 512.

Cameron, I. L. (1965). *J. Cell Biol.* **25,** 9.

Cameron, I. L., and Prescott, D. M. (1961). *Exptl. Cell Res.* **23,** 354.

Cameron, I. L., and Stone, G. E. (1934). *Exptl. Cell Res.* **36,** 510.

Cameron, I. L., Cline, G. B., Padilla, G. M., Miller, O. L., Jr., and Van Dreal, P. A. (1965). *J. Natl. Cancer Inst.* (in press).

Christensson, E. (1959). *Acta Physiol. Scand.* **45,** 339.

Corliss, J. O. (1954). *J. Protozool.* **1,** 156.

Elliott, A. M. (1959). *Ann. Rev. Microbiol.* **13,** 79.
Elliott, A. M., Brownell, L. E., and Gross, J. A. (1954). *J. Protozool.* **1,** 193.
Elliott, A. M., Kennedy, J. R., and Bak, I. J. (1962). *J. Cell Biol.* **12,** 515.
Frankel, J. (1962). *Compt. Rend. Trav. Lab. Carlsberg* **33,** 1.
Hill, D. L., and van Eys, J. (1964). *J. Tenn. Acad. Sci.* **39,** 117.
Holz, G. G., Jr. (1964). *In* "Biochemistry and Physiology of Protozoa" (S. H. Hutner, ed.), Vol. III, pp. 199–242. Academic Press, New York.
Iverson, R. M., and Giese, A. C. (1957). *Exptl. Cell Res.* **13,** 213.
James, T. W. (1954). Ph.D. Thesis, University of California (Berkeley).
James, T. W. (1960). *Ann. N.Y. Acad. Sci.* **90,** 550.
James, T. W., and Anderson, N. G. (1963). *Science* **142,** 1183.
James, T. W., and Read, C. P. (1957). *Exptl. Cell Res.* **13,** 510.
Kidder, G. W., and Dewey, V. C. (1951). *In* "Biochemistry and Physiology of Protozoa" (A. Lwoff, ed.), Vol. I, pp. 323–400. Academic Press, New York.
Lowry, O. H., Rosenbrough, N. J., Farr, A. L., and Randall, R. J. (1951). *J. Biol. Chem.* **193,** 265.
McDonald, B. B. (1958). *Biol. Bull.* **114,** 71.
McDonald, B. B. (1962). *J. Cell Biol.* **13,** 193.
Padilla, G. M., and Cameron, I. L. (1964). *J. Cellular Comp. Physiol.* **64,** 303.
Padilla, G. M., and Cook, J. R. (1964). *In* "Synchrony in Cell Division and Growth" (E. Zeuthen, ed.), pp. 521–535. Wiley, New York.
Padilla, G. M., and James, T. W. (1960). *Exptl. Cell Res.* **20,** 401.
Padilla, G. M., and James, T. W. (1964). *In* "Methods in Cell Physiology" (D. M. Prescott, ed.), pp. 141–158. Academic Press, New York.
Prescott, D. M. (1957). *J. Protozool.* **4,** 252.
Prescott, D. M. (1959). *Exptl. Cell Res.* **16,** 279.
Prescott, D. M. (1964). *In* "Synchrony in Cell Division and Growth" (E. Zeuthen, ed.), pp. 71–98. Wiley, New York.
Scherbaum, O. H. (1956). *Exptl. Cell Res.* **11,** 464.
Scherbaum, O. H., and Loefer, J. B. (1964). *In* "Biochemistry and Physiology of Protozoa" (S. H. Hutner, ed.), pp. 9–59. Academic Press, New York.
Scherbaum, O. H., and Zeuthen, E. (1954). *Exptl. Cell Res.* **6,** 221.
Schmidt, G., and Thannhauser, S. J. (1945). *J. Biol. Chem.* **161,** 83.
Spear, F. G. (1928). *Arch. Exptl. Zellforsch.* **7,** 484.
Stone, G. E., and Cameron, I. L. (1964). *In* "Methods in Cell Physiology" (D. M. Prescott, ed.), pp. 127–140. Academic Press, New York.
Stone, G. E., and Prescott, D. M. (1964). *J. Cell Biol.* **27,** 275.
Summers, L. G. (1963). *J. Protozool.* **10,** 228.
Thormar, H. (1962a). *Exptl. Cell Res.* **27,** 585.
Thormar, H. (1962b). *Exptl. Cell Res.* **28,** 269.
Walker, P. M. B., and Mitchison, J. M. (1957). *Exptl. Cell Res.* **13,** 167.
Watanabe, Y. (1963). *Japan. J. Med. Sci. Biol.* **16,** 107.
Whitson, G. L., and Padilla, G. M. (1964). *Exptl. Cell Res.* **36,** 667.
Zeuthen, E. (1953). *J. Embryol. Exptl. Morphol.* **1,** 239.
Zeuthen, E. (1964). *In* "Synchrony in Cell Division and Growth" (E. Zeuthen, ed.), pp. 99–158. Wiley, New York.

CHAPTER 14

Morphogenetic and Macromolecular Aspects of Synchronized *Tetrahymena*

G. L. Whitson,* G. M. Padilla, and W. D. Fisher

I. INTRODUCTION

Interest in ciliate morphogenesis stems from the early studies of Chatton and Lwoff (1935), who developed a silver staining technique which has made possible the study of the complex cortical architecture of ciliated protozoans. The development of techniques for induced cell division synchrony in axenically grown *Tetrahymena* by Scherbaum and Zeuthen (1954), made possible studies concerning the relationship of morphogenesis to cell division. Holz *et al.* (1957), were the first to observe that

* United States Public Health Service Fellow.

the synchronizing treatment blocked all cells in a very early stage of stomatogenesis (anarchic field stage) in *T. pyriformis* strain WH-6. A similar result was obtained by Williams and Scherbaum (1959), using the GL strain. Several papers have since appeared presenting a complete analysis of the effects of temperature shocks on morphogenesis (Frankel, 1962, 1964a,b, 1965). For a more detailed account of some of these and other studies, the reader is referred to an excellent review by Williams (1964).

Our interest is in the structural changes in *Tetrahymena*, particularly stomatogenesis and its relationship to cell division. Recent advances in cell biology, however, show the value of considering macromolecular changes and the patterns of their synthesis in relation to cell division and cell morphogenesis. The synthesis and synchronization of the DNA replicative periods, using different methods of cell synchronization, are discussed in Chapter 13. This chapter deals primarily with recent findings concerning RNA synthesis, its inhibition, and the nucleolar and ribosomal changes which accompany stomatogenesis and cell division during the repetitive synchrony of *Tetrahymena* by the procedure developed by Padilla and Cameron (1964).

II. TIMING OF MORPHOGENESIS

A. Normal Occurrence

The period of stomatogenesis was examined in cells of *Tetrahymena pyriformis* HSM synchronized by a repetitive temperature cycle (Whitson and Padilla, 1964a,b). Cells were subjected to a repetitive temperature regimen of 9½ hours at 12°C and 2½ hours at 27°C. Small aliquots of cells were taken at intervals throughout the entire cycle and fixed for cortical examination using the Chatton and Lwoff silver impregnation technique described by Corliss (1953). The time course of stomatogenesis during the cell cycle is shown by the histogram in Fig. 1. Stomatogenesis is confined to the warm period. Although none of the cells examined during the latter portion of the cold period had oral anlage, some of the cells immediately upon entry into the warm period began stomatogenesis. Also, 90% of the cells are in one stage or another of oral development just prior to the peak in cell division. Figure 2 is a sample of cells taken 1 hour after the beginning of the warm period. The oral anlage (OA) are visible at this time.

B. Chemical Inhibition

After establishing the temporal relationship of stomatogenesis during the synchronized cycle, the effects of an inhibitor (actinomycin D) were

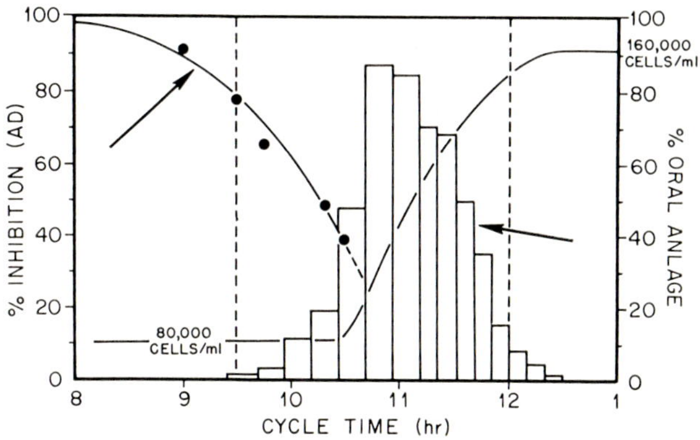

FIG. 1. A histogram of the percent frequency distribution of cells with oral anlage during portions of the synchronized cycle. The descending curve in this figure gives the percent inhibition of cell division. The ascending curve shows the normal doubling of the synchronized control population of cells. The temperatures during these portions of the synchronized cycle are as follows: 9.5 hours at 12°C; 2.5 hours at 27°C. (Courtesy of Academic Press.)

studied. Cells were exposed to varying concentrations of actinomycin in order to find out which concentration, if any, would effectively interfere with stomatogenesis or inhibit cell division* (Whitson and Padilla, 1964b). The inhibitor was added at a preselected time of 1 hour and 45 minutes before the beginning of the warm period. The number of cells which divided was determined by the ratio between the final and initial cell numbers. A concentration of 8 μg/ml of actinomycin D inhibited division in over 90% of the cells. In a more recent study, Whitson *et al.* (1965a) have shown that cells exposed to 10 μg/ml of actinomycin D do not divide during continuous exposure for 24 hours. None of the cells died during this prolonged treatment and complete reversal of this inhibition was observed once these cells were washed free of the actinomycin and placed in fresh medium. Complete reversal of this inhibition, however, was not observed in cells exposed for the same period of time to concentrations of 20 μg/ml or greater of actinomycin.

The effect of actinomycin D added at different times during the cell cycle was studied (Whitson and Padilla, 1964b) employing concentrations (8–10 μg/ml) which blocked cell division reversibly. When actinomycin was added 1.5 hours before the shift from the cold to the warm period, over 99% of the cells failed to divide, see Figs. 1 and 3 (Whitson and Padilla, 1964b). When the inhibitor was added 1 hour after the beginning of the warm period, only 39% of the cells were blocked.

Cells which were continuously exposed to actinomycin added 1.5 hour before the onset of the warm period showed no oral anlage throughout

* See Table I p. 294.

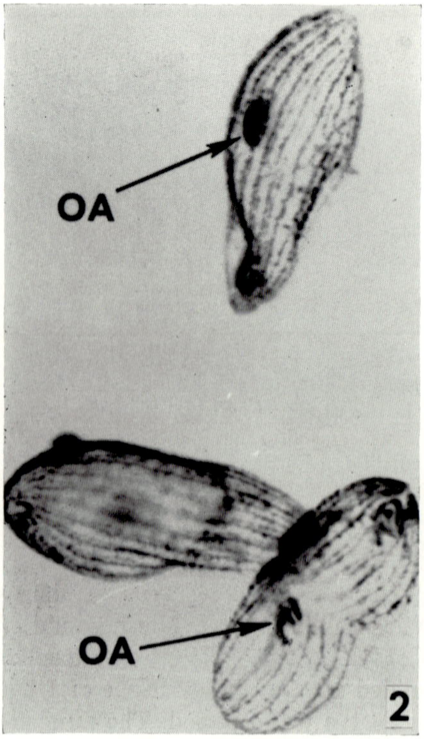

FIG. 2. A silver preparation of control cells 1 hour after the beginning of the warm period. Note the forming oral anlage (OA). (Courtesy of Academic Press.)

the remainder of the warm period. A diagrammatic representation of a comparison of controls and actinomycin-treated cells is shown in Fig. 4. Active morphogenesis (stomatogenesis) is obviously inhibited if actinomycin is added at a specific time during the latter portion of the cold period. For a discussion of possible implications of these results see Sections III, IV, VI, and VII.

C. Synchronization by Long Period Cold Exposure versus Short Duration Heat Shocks

Certain advantages of long exposures of *Tetrahymena* to cold temperatures as opposed to short-period heat shocks for synchronization are discussed in Chapter 13. For studies on stomatogenesis, the advantage is perhaps even more obvious. The long cold-exposure method shows 2 distinct periods in the repetitive synchronized cycle: one in which visual morphogenetic changes are absent and one in which visual morphogenesis is present. Morphogenetic studies by Holz *et al.* (1957), Williams

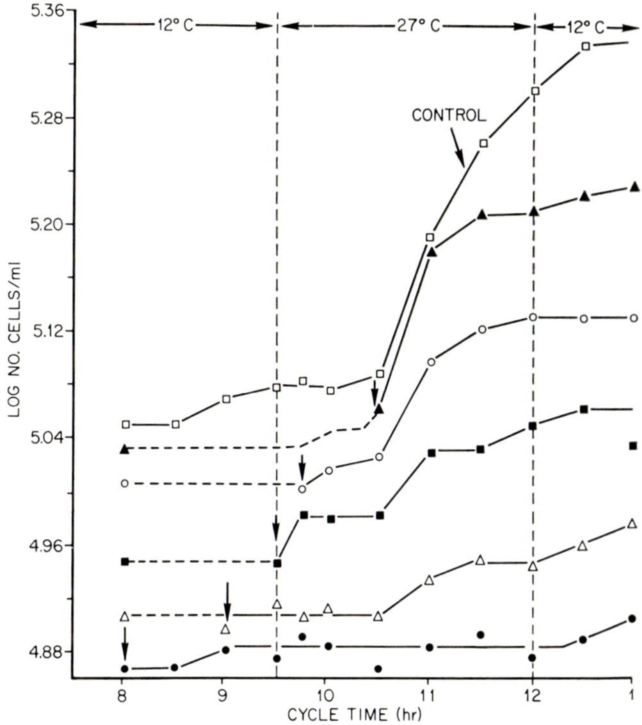

FIG. 3. The effect of actinomycin D on synchronized cell division and stomatogenesis in *Tetrahymena*. This figure shows the effect of the addition of 8 μg/ml actinomycin D at different times (as shown by the arrows) during the synchronized cycle. The curves, other than the control, have been displaced downward for clarity. The base line was the same in all cases (approximately 5.05 log units). (Courtesy of Academic Press.)

and Scherbaum (1959), and Frankel (1962) do not show complete separation of periods of morphogenesis in heat-shock synchronized *Tetrahymena*. Rather, there is a period of oral development extending to the anarchic field stage of stomatogenesis. All cells remain arrested for some time following the last heat shock until further reactivation of morphogenesis takes place. Therefore, inhibitor studies on morphogenesis in heat-shock synchronized *Tetrahymena* apply only to interference with processes occurring after oral anlage have already appeared, namely, from the point of morphogenetic arrest to morphogenetic reactivation and then on through cell division. It is virtually impossible to relate premorphogenetic macromolecular changes with morphogenesis in these cells which already have anlage present. Where the actual initial formation of anlage can be prevented by long exposures to low temperatures

TABLE I

THE EFFECT OF DIFFERENT CONCENTRATIONS OF ACTINOMYCIN D ON SYNCHRONIZED CELL DIVISION IN *Tetrahymena*[a,b]

Concentration (μg/ml)	N_I/ml	N_F/ml	$(N_F - N_I)/N_I$
16	128,700	129,000	0.01
8	131,560	137,930	0.05
4	129,675	141,140	0.09
1.6	128,790	231,675	0.80
0.16	132,610	241,020	0.82
Control	133,400	264,605	0.98

[a] N_I = Initial cell numbers.
[b] N_F = Final cell numbers.

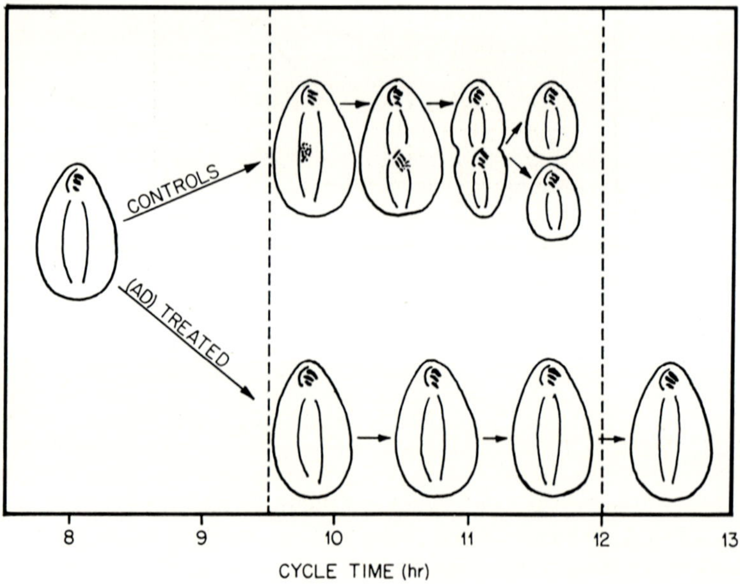

FIG. 4. Diagrammatic representation of the effect of actinomycin D (AD) on stomatogenesis.

(e.g., 9.5 hours at 12°C) it is, however, possible to examine cells during this period of premorphogenesis and to compare them with cells taken during the period when morphogenesis takes place. Sections IV and V deal in more detail with such comparative studies.

III. AUTORADIOGRAPHY AND OTHER TRACER STUDIES

By the use of pulse labeling and autoradiography, evidence has been obtained which indicates that cells actively engaged in morphogenesis

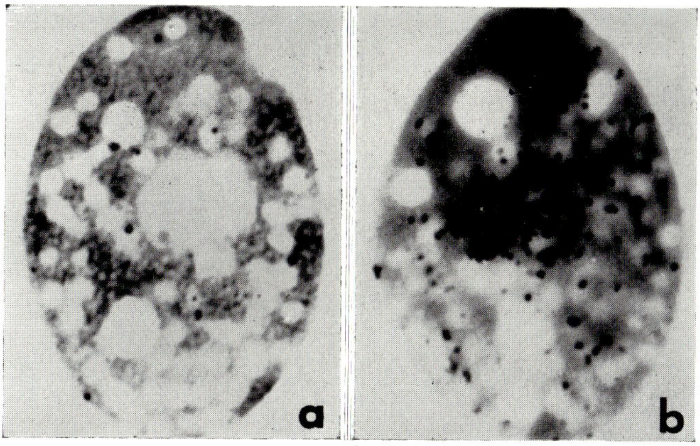

FIG. 5. Radioautograph of *Tetrahymena* pulse labeled for 10 minutes with cytidine-H^3. (*a*) cell treated with actinomycin D (*b*) control cell. (Courtesy of Academic Press.)

(stomatogenesis) are also actively synthesizing RNA (Whitson and Padilla, 1964a,b). On the other hand, cells inhibited by actinomycin showed very little synthesis of RNA when pulse labeled with cytidine-H^3 (Fig. 5a). Figure 5b is an autoradiograph of a cell from the control synchronized population which was pulse labeled for 10 minutes, 1 hour after the beginning of the warm period. The heavily labeled area is over the cell macronucleus. This label, which is removed by pretreating the cells with ribonuclease, indicates macronuclear synthesis of RNA and the presence of some cytoplasmic RNA. In a recent review concerning sites of RNA synthesis within cells, Prescott (1964) discusses the accumulated evidence which indicates that "the initial labeling of RNA with radioactive precursors takes place in the nucleus" in a variety of different cell types and that there is a time lag after which the radioactivity begins to move into the cytoplasm. Prescott (1964) also has shown that the time lag was from 10–30 minutes before labeled RNA began to appear within the cytoplasm in *Tetrahymena*. Our autoradiographic evidence also indicates that there is about a 10 minute lag before the radioactive RNA begins to appear in the cytoplasm (Fig. 5b).

Data from other isotope studies indicate that all RNA synthesis is not prevented in cells treated with actinomycin (Whitson *et al.*, 1964). Some unpublished results showing the kinetics of uptake of uridine-C^{14} are presented in Fig. 6. In this experiment, control synchronized cells and actinomycin-treated cells (exposed to 10 μg/ml of this drug 1.5 hours before the shift to the warm period) were used. Fifty μc/ml of uridine-C^{14} was added to each group of cells 1 hour after the beginning of the warm period. Duplicate samples of cells (10 ml each) from each group

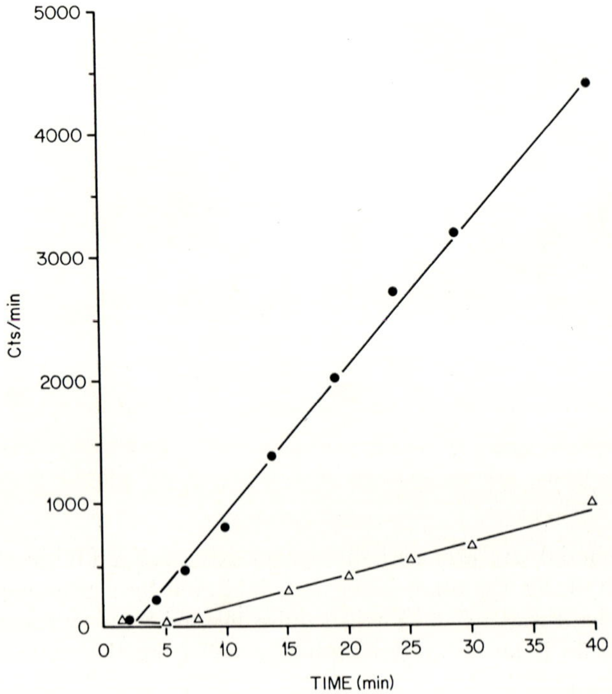

FIG. 6. Uptake of uridine-C^{14} by *Tetrahymena* upper curve control synchronized cells; lower curve actinomycin D treated cells.

were collected on Millipore filters, washed 3 times with cold 5% TCA, dried, and then counted in a Tracerlab gas-flow counter. The control synchronized cells showed a linear uptake with time of this label into cold TCA precipitable material. In actinomycin D treated cells there is an initial lag of uridine-C^{14} uptake and then a much reduced but linear uptake with time of this label into TCA insoluble material. Possible explanations for these results will be discussed in Section VII.

IV. BIOCHEMICAL STUDIES

A. Total RNA Measurements

The total RNA content in micrograms per 10^6 cells was determined for cells sampled late in the cold period and at different times during the warm period (Fig. 7). Although the results indicate a near doubling of the amount of RNA on a per cell basis just before the time when the number of cells undergoing stomatogenesis are at a peak, there appears to be also some RNA synthesis at least during the latter portion of the cold period. DNA synthesis is discussed in Chapter 13.

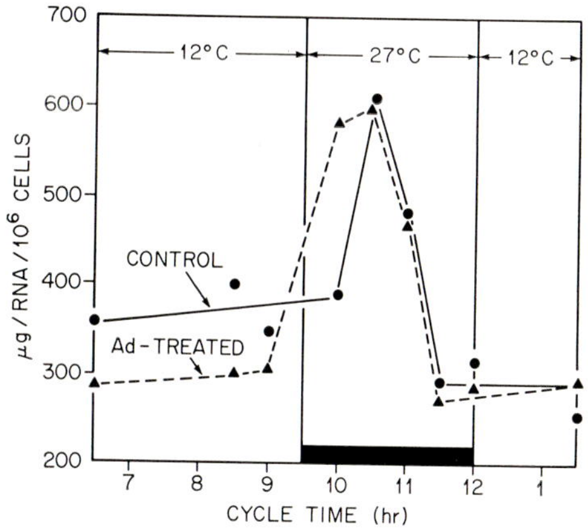

FIG. 7. RNA content of control synchronized and actinonycin D (AD) treated cells.

B. EFFECT OF ACTINOMYCIN ON TOTAL RNA

Whitson *et al.* (1964) have shown that no major differences exist in the total RNA content of control-synchronized cells and actinomycin D-treated cells (Fig. 7). The drop in RNA content in control synchronized cells is due to cell division: each daughter cell after division contains half of the RNA from the parent cell. The decrease in RNA in the non-dividing actinomycin-treated cell is not explainable in this way. It appears that although it is synthesized early in the warm period it is broken down by some mechanism after renewed synthesis has taken place.

V. NUCLEOLAR CHANGES DURING SYNCHRONY

The evidence for the nucleolar origin of most ribosomal RNA has been obtained by Perry (1960) and Perry *et al.* (1961) in studies on microbeam irradiation of the nucleolus in HeLa cells. Perry (1964) has also shown that nucleolar synthesis of RNA is inhibited by actinomycin. For a more thorough review of this subject, see Prescott (1964).

Changes in macronuclear cytology in *Tetrahymena* have been examined by Cameron *et al.* (1965a) with particular emphasis on nucleolar changes within cells grown in the following 3 ways: (*a*) normal log-phase cultures, (*b*) cultures synchronized by the Scherbaum and Zeuthen (1954) heat-shock method of synchronization, and (*c*) cells synchronized by the repetitive temperature cycling method of Padilla and Cameron (1964). They found that while cells given short-duration heat shocks showed fusion of nucleoli into large fusion bodies (Figs. 8a and b), cells examined at different times during the warm period during the Padilla and Cameron synchronizing procedure showed no large fusion bodies (Fig.

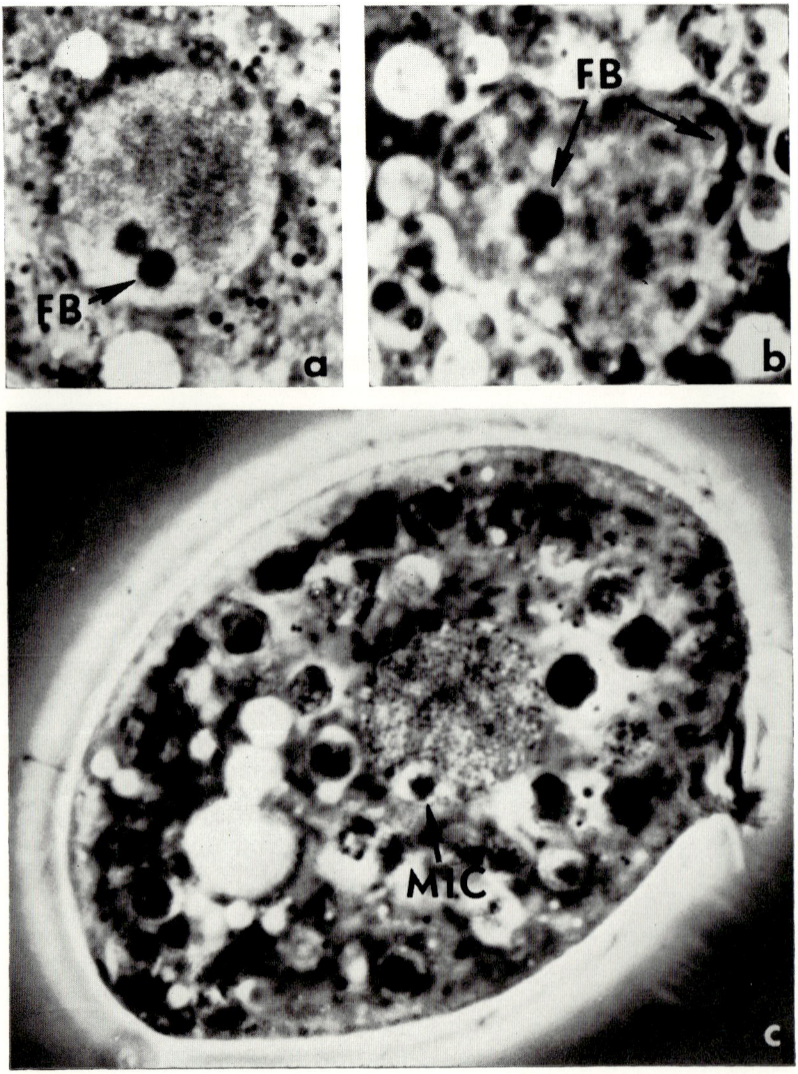

FIG. 8. Phase contrast micrographs of macronuclear morphology of heat-shocked cells a and b and cold synchronized cells c. FB fusion body, MIC micronucleus. (Courtesy of Dr. I. L. Cameron.)

8c). They further noted that the macronuclei of cells, synchronized by the latter procedure, more closely resembled macronuclei of log-phase cells than those of cells synchronized by the former method.

Whether or not the large fusion bodies, found only in heat-shock synchronized cells, result from the accumulation or greater synthesis of RNA is a matter of conjecture. It is known, however, that the amount of RNA in heat-shock synchronized *Tetrahymena* is much higher than that of log-phase cultures (Scherbaum *et al.*, 1959). Since no large fusion bodies were evident at the end of the cold period in cells synchronized by the Padilla and Cameron method, this suggests that no large abnormal amounts of nucleolar RNA accumulate during the cold treatment. Support for this contention comes from the evidence based on total cell RNA given in Section IV,B.

If, as suggested, large amounts of nucleolar RNA are accumulated in the form of large fusion bodies in heat-shocked cells, then possibly some RNA component is either being produced in an overabundance or is not being readily transferred to the cytoplasm in these cells. As indicated by Cameron *et al.* (1965a), it is only after several hours following the last heat-shock that fusion bodies or large nucleolar bodies in the macronucleus finally disappear. It is therefore reasonable to assume that heat shocks in some way elicit an abnormal nucleolar response in *Tetrahymena*—at least more so than in cells synchronized by the repetitive temperature cycling procedures outlined above.

VI. RIBOSOMAL STUDIES

Information gained from studies on a wide variety of different cellular systems suggest that polyribosomes are present in cells actively engaged in protein synthesis (Henney and Storck, 1964; Monroy and Tyler, 1963; Stafford *et al.*, 1964; and Cameron *et al.*, 1965b). Whitson *et al.* (1965a) have explored the possibility that different ribosomal patterns, particularly the formation of polysomes, may exist in relation to stomatogenesis and cell division in synchronized *Tetrahymena*. A comparison of ribosomal profiles of predivision warm period cells with actinomycin-treated cells showed a marked repression of polyribosomes by actinomycin. Using a biochemical method of synchronization, Cameron *et al.* (1965b) found that, while the polyribosome fraction was reduced in pyrimidine-starved *Tetrahymena* stalled in the G_1 period, there was an increase in polyribosomes following replacement of these essential metabolites.

Detailed descriptions of the methods used in obtaining ribosomes from *Tetrahymena* are given elsewhere (Whitson *et al.*, 1965a). Substantial amounts of polyribosomes are present at the end of the cold period (Fig. 9). In contrast, ribosome profiles of cells 1 hour after the beginning of the warm period show a decrease in 82 S ribosomes and the presence of trimers (148 S), Fig. 10. There is also a better resolution of the polyribosomes into individual peaks above 170 S. This has been interpreted to mean that there was either an improvement of the separation tech-

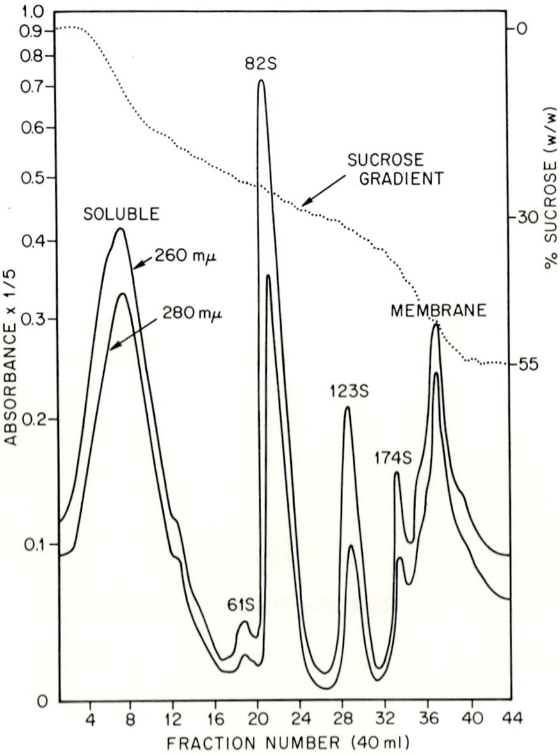

FIG. 9. Centrifugal separation on a sucrose gradient of ribosomes from synchronized *Tetrahymena* cells collected at the end of the cold period.

nique at this time or the existence of such ribosomal units at this point during the synchronized cycle. An electron microscopic preparation of ribosomes by Cameron *et al.* (1965b), taken from zonal ultracentrifuge fractions of log-phase *Tetrahymena*, show that, indeed, the clusters of ribosomes differ in each peak (Fig. 11). The 83 S fraction consists of single ribosomal units, the 125 S fraction of double units; the 148 S consists of trimers, and the 186 S fraction of tetramer units of ribosomes. Addition of actinomycin (10 μg/ml) to cells 1 hour before the shift from the cold period to the warm period, but lysed 1 hour after the warm period gave a somewhat different ribosomal profile (Fig. 12). There is a relative increase in 82 S ribosomes (single ribosomes) and a reduction in the higher multiple units of ribosomes. It is not known whether the increase in 82 S ribosomes is due to failure of mRNA synthesis or a breakdown of mRNA complexing with the multiple units of ribosomes in these actinomycin-treated cells.

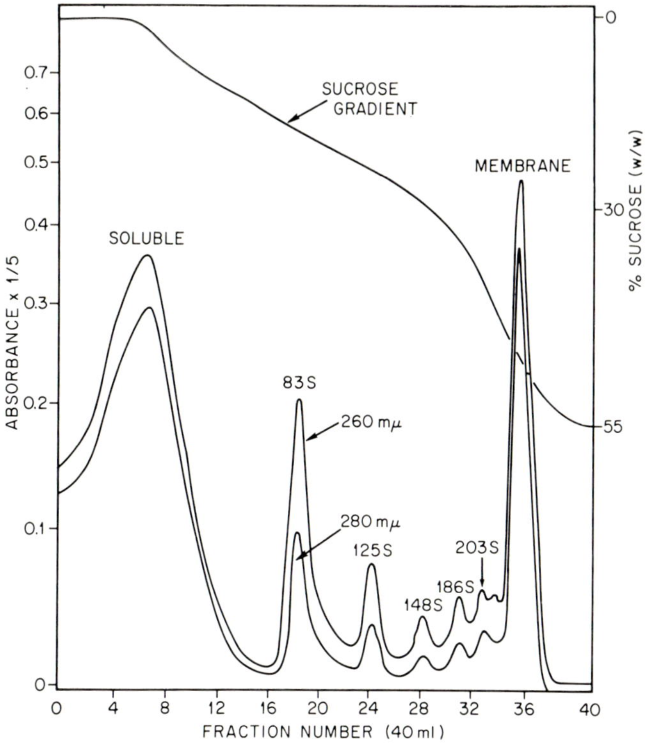

FIG. 10. Centrifugal separation of ribosomes from synchronized *Tetrahymena* cells collected 1 hour after the beginning of the warm period.

Other ribosomal studies have shown, however, that anomalous 70 S ribosomes are generated under actinomycin D treatment (Table II). These data were obtained by analyzing homogenates of *Tetrahymena* using the Beckman Model E analytic ultracentrifuge and will appear in more detail in another publication (Whitson *et al.*, 1965a). Note that 70 S ribosomes were also found in postdivision cells at a time when morphogenesis has just been completed. It may be that these changes in ribosome in actinomycin-treated and postdivision cells reflect a breakdown of polysomes in cells not actively engaged in protein synthesis. Whether these changes indicate the removal of protein or loss of mRNA at the ribosomal site is unknown.

VII. DISCUSSION

Tetrahymena is uniquely suited for the study of biochemical events associated with morphogenesis: It can be synchronized readily by dif-

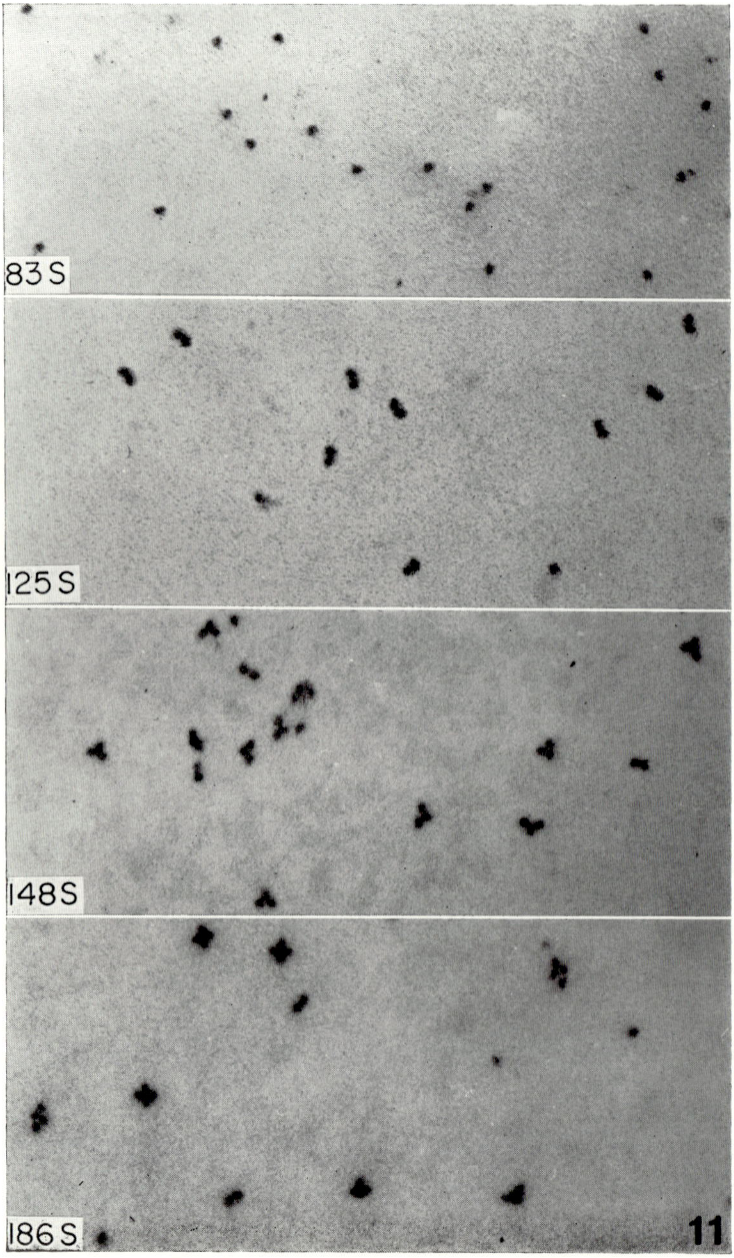

FIG. 11. Electron micrographs of ribosome fractions from zonal centrifuge run. (*a*) 82 peak monomer, (*b*) 125 peak (dimer), (*c*) 148 (trimer), and (*d*) 186 (tetramer). (Courtesy of Dr. I. L. Cameron.)

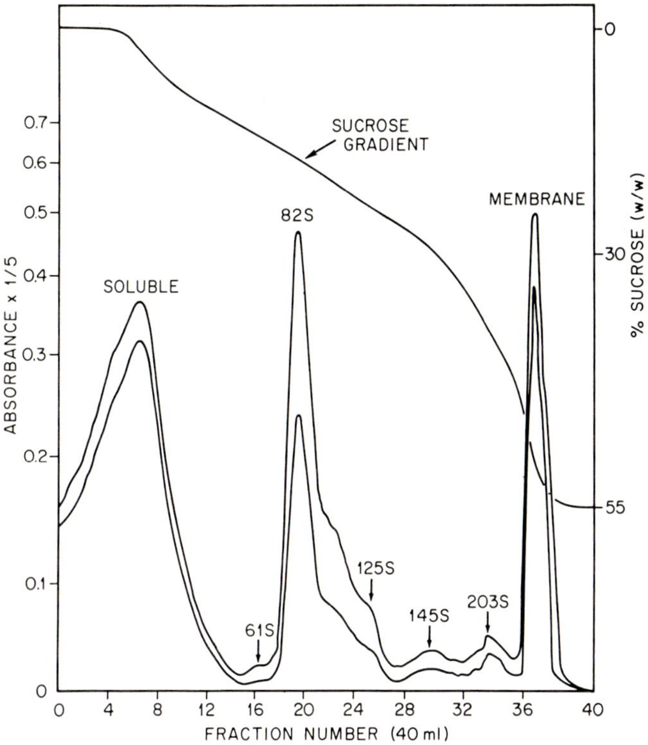

FIG. 12. Centrifugal separation of ribosomes from actinomycin treated *Tetrahymena* cells collected 1 hour after the beginning of the warm period.

TABLE II

A COMPARISON OF $[S_0]$ VALUES OF RIBOSOMES FROM CELLS AT DIFFERENT TIMES DURING THE SYNCHRONIZED CYCLE[a]

End of cold period:	82.8	—	57.2	29.5
Predivision warm period:	(83.7) 81.8	—	(61.8)	(33.1)
Postdivision beginning of cold period:	82.1	70.0	57.9	29.0
Actinomycin D-treated cells:	(81.5) 81.7	(71.5)	(59.3) 63.6	(32.4)

[a] Numbers in parentheses indicate measurements performed on material recovered from the 82 S fraction of the zonal run.

ferent means and has a complicated cortical morphology. One such cortical structure is the new cell mouth which begins to form some time prior to cell division (Frankel, 1962). As yet many of the underlying biochemical events which lead to the formation of such complex cell structures are poorly understood.

Most recent evidence suggests that synthesis of new organelles requires the translation of genetic information into structural proteins. Very little, if anything, is known, however, concerning the kinds and the number of steps involved in such processes. Some of the studies mentioned in this chapter were begun with this in mind. For instance, the recent finding that actinomycin inhibits DNA-dependent RNA synthesis (Reich *et al.*, 1961) made it possible to study the relation of RNA to stomatogenesis and cell division in synchronized *Tetrahymena* (Frankel, 1965; Whitson and Padilla, 1964a,b; Whitson *et al.*, 1965a; Nachtwey and Dickinson, 1964). All available data indicate that some newly synthesized RNA is needed for cells to undergo complete oral development; the addition of actinomycin D prevents oral development, presumably by blocking this RNA synthesis. Results obtained by Frankel (1965) and by Whitson and Padilla (1964a,b) suggest that cells which are in the latter stages of oral development and just beginning cell division are no longer sensitive to the addition of actinomycin. It is possible that at this time further oral development and cell division require no further synthesis of RNA. If some mRNA is needed for completion of these events, then it must already be present and stable—at least until the synthesis of proteins associated with further development is completed. It appears, however, that the fraction of RNA which mediates protein synthesis for stomatogenesis and cell division is not in the form of stable templates existing from one cell division to the next.

Evidence from autoradiographic and isotope studies presented in this chapter indicate that actinomycin (10 μg/ml) does not interfere with the major part of RNA synthesis in synchronized cells. In other findings, Whitson *et al.* (1965b) have found that a high concentration of actinomycin (50 μg/ml) added to cells in radioactive medium containing uracil-C^{14} prevents further incorporation of this isotope following addition of this drug. This result as well as other evidence obtained from different cells (Perry, 1964; Franklin, 1963; Levy, 1963) indicates that actinomycin in high concentrations not only inhibits nearly all RNA synthesis but possibly prevents transport of nuclear RNA to the cytoplasm.

Spectrophotometric determinations of the total RNA content of actinomycin (10 μg/ml) treated cells also indicate that only a small fraction of the total RNA synthesis is inhibited in synchronized cells. These results indicate that this concentration of actinomycin D is not likely to inhibit ribosomal RNA synthesis. Perry (1964) and others have shown, however, that low concentrations of actinomycin D does selectively block nucleolar RNA synthesis. The nucleolar origin of ribosomal RNA has already been discussed (see Section V). Some evidence presented by Georgiev and Mantieva (1963) indicates that mRNA, or at least RNA with base ratios similar to DNA, may be routed via the nucleolus to the cytoplasm in certain cells. Perhaps actinomycin inhibition of the nucleolus results in the prevention of mRNA transfer in this way.

Ribosomal studies, particularly those mentioned in Section VI, indicate that polyribosomes are present in cells actively synthesizing protein. An apparent relationship in polyribosome changes with morphogenesis and cell division exists in *Tetrahymena*. An earlier study by Plesner (1961) was made on single 70 S ribosomes in heat-shock synchronized *Tetrahymena*. His study did not mention polysomes but pertained to changes in single ribosomes. It was suggested that changes noted in the single ribosomes reflected protein changes associated with these units prior to cell division. From his observations, Plesner (1961) suggested that the loss of protein from single 70 S ribosomes after division indicated that a "division protein" was being formed at these sites prior to cell division. Worthy of note, however, is the fact that no "division-directing" proteins have yet been isolated from this organism.

VIII. CONCLUSION

There is evidence that RNA plays an important role in morphogenesis and cell division. Interference with the synthesis of certain RNA by the direct action of an inhibitor (actinomycin) indicates that renewed synthesis of RNA (possibly mRNA) is required some time prior to and during early morphogenesis. Information from the study of polyribosomes in synchronized *Tetrahymena* indicates that a breakdown of polyribosomes occurs in cells which are not actively engaged in protein synthesis and morphogenesis. Whether this breakdown of polyribosomes reflects loss of mRNA is not known. More information concerning this and other underlying biochemical events are now needed for a better understanding of the biosynthetic regulation of morphogenesis and cell division.

ACKNOWLEDGMENT

This research was sponsored jointly by the United States Atomic Energy Commission under contract with the Union Carbide Corporation and the United States Public Health Service (Fellowship 1 F2 Ca-21, 160-01 from the National Cancer Institute).

REFERENCES

Cameron, I. L., Padilla, G. M., and Miller, O. L., Jr. 1965a). *J. Protozool.* (in press).

Cameron, I. L., Cline, G. B., Padilla, G. M., Miller, O. L., Jr., and Van Dreal, P. A. (1965b). *J. Natl. Cancer Inst.* (in press).

Chatton, E., and Lwoff, A. (1935). *Arch. Zool. Exptl. Gen.* **77,** 1–453.

Corliss, J. O. (1953). *Stain Technol.* **28,** 97.

Frankel, J. (1962). *Compt. Rend. Trav. Lab. Carlsberg* **33,** 1.

Frankel, J. (1964a). *J. Exptl. Zool.* **155,** 403.

Frankel, J. (1964b). *Exptl. Cell Res.* **35,** 25 and 349.

Frankel, J. (1965). Communicated *Tetrahymena* Information Exchange Group.

Franklin, R. M. (1963). *Biochim. Biophys. Acta* **72,** 555.

Georgiev, G. P., and Mantieva, U. L. (1963). *Biochemistry (USSR) (English Transl.)* **27,** 805.

Henney, H. R., Jr., and Storck, R. (1964). *Proc. Natl. Acad. Sci. U.S.* **51,** 1050.

Holz, C. G., Scherbaum, O. M., and Williams, N. E. (1957). *Exptl. Cell Res.* **13,** 618.

Kidder, G. W., and U. C. Dewey (1951). *In* "Biochemistry and Physiology of Protozoa" (A. Lwoff, ed.), Vol. I, pp. 323–400. Academic Press, New York.

Levy, H. B. (1963). *Proc. Soc. Exptl. Biol. Med.* **113,** 886.

Monroy, A., and Tyler, A. (1963). *Arch. Biochem. Biophys.* **103,** 431.

Nachtwey, D. S., and Dickinson, W. J. (1964). *J. Protozool.* **11,** Suppl., 24.

Padilla, G. M., and Cameron, I. L. (1964). *J. Cellular Comp. Physiol.* **64,** 303.

Perry, R. P. (1960). *Exptl. Cell. Res.* **20,** 216.

Perry, R. P. (1964). *Natl. Cancer Inst. Monograph* **14,** 73.

Perry, R. P., Hell, A., and Errera, M. (1961). *Biochim. Biophys. Acta* **49,** 47.

Plesner, P. (1961). *Cold Spring Harbor Symp. Quant. Biol.* **26,** 159.

Prescott, D. M. (1964). *In* "Progress in Nucleic Acid Research and Molecular Biology" (J. N. Davidson and W. E. Cohn, eds.), Vol. 3, p. 33–58. Academic Press, New York.

Reich, E., Franklin, R. M., Shatkin, A. J., and Tatum, E. L. (1961). *Science* **134,** 556.

Scherbaum, O. H., and Zeuthen, E. (1954). *Exptl. Cell Res.* **6,** 221.

Scherbaum, O. H., Louderback, D. L., and Jahn, T. L. (1959). *Exptl. Cell Res.* **18,** 150.

Schmidt, G., and Thannhauser, S. J. (1945). *J. Biol. Chem.* **161,** 83.

Stafford, D. W., Sofer, W. H., and Iverson, R. M. (1964). *Proc. Natl. Acad. Sci. U.S.* **52,** 313.

Whitson, G. L., and Padilla, G. M. (1964a). *Am. Zoologist* **4,** 313.

Whitson, G. L., and Padilla, G. M. (1964b). *Exptl. Cell Res.* **36,** 667.

Whitson, G. L., Padilla, G. M., and Elrod, L. H. (1964). *J. Cell Biol.* **23,** 102A.

Whitson, G. L., Padilla, G. M., and Fisher, W. D. (1965a). *Exptl. Cell Res.* (in press).

Whitson, G. L., Padilla, G. M., and Elrod, L. H. (1965b). *Abstr. 9th Meeting Biophys. Soc., San Francisco* p. 107.

Williams, N. E. (1964). *In* "Synchrony in Cell Division and Growth" (E. Zeuthen, ed.), pp. 159–175. Wiley (Interscience), New York.

Williams, N. E., and Scherbaum, O. H. (1959). *J. Embryol. Exptl. Morphol.* **7,** 241.

Note added in proof: The following publication is relevant to the above text. Watanabe, Y. and Ikeda, M. (1965). Isolation and characterization of the division protein in *Tetrahymena pyriformis*. *Exptl. Cell Res.* **39,** 443.

CHAPTER 15

Behavior of Sulfhydryl Groups in Synchronous Division

Katsuma Dan

I. INTRODUCTION

Any person will readily admit that the term synchrony has a much wider meaning than simultaneity of events. Although simultaneity is a part of synchrony, it is obvious that sheer simultaneity such as widespread demolition by an earthquake is not synchrony. For the concept of synchrony, the presence of some kind of rhythmicity is implied and biological synchrony means coincidence of the phases of the cycle of rhythmicity among different individuals. Since this is the way the author feels about the term, particular attention in the following account will be paid to the SH cycle, which underlies the phenomenon of cell division.

II. SH CYCLES IN SEA URCHIN EGGS

A. Fluctuation of SH Groups Soluble in TCA

As examples of synchrony of division, artificially inseminated egg cells have been considered as the most easily available material for nearly a century and Agrell's recent article (1964) has collected many beautiful instances. However, Rapkine (1931) was the first person who succeeded in disclosing rhythmic biochemical reactions which lie at the base of division synchrony in sea urchin eggs. What he found was an orderly sequence of changes, correlated with the division cycle, in the amounts of SH in a protein extractable from sea urchin eggs by 25% TCA. The sequence is such that the quantity of SH per egg is relatively high before and immediately following fertilization; this, however, decreases gradually until the streak stage, when a minimum is reached (see Fig. 1). Then the SH value rises again to a maximum, at a stage which Rapkine identified as the metaphase. After this, however, the SH value immediately begins to drop even before the first cleavage sets in (Fig. 1).

B. SH $\rightleftharpoons$ S—S Interchange in KCl-Soluble Protein of Sea Urchin Eggs

In the laboratory of the Tokyo Metropolitan University, after confirming Rapkine's results (Sakai and Dan, 1959), we substituted 0.6 *M* KCl for 25% TCA as the extraction medium, because of the possibility that the encountered substance might be a protein. If so, since a protein soluble in such a high concentration of TCA must be a protein of an exceptional or aberrant nature, the application of knowledge derived from more ordinary proteins seemed extremely dangerous or almost impossible. As a matter of fact, Rapkine himself never thought that the substance of his discovery, which carried sulfhydryl groups and was soluble in TCA, could be a protein. Hence, he assumed that the substance must be glutathione. But this suggestion turned out to be entirely erroneous (Sakai and Dan, 1959).

Fortunately, the protein extracted with 0.6 *M* KCl by our group (Sakai, 1960) also shows an SH cycle similar to that of the TCA-soluble protein of Rapkine (Fig. 2). Moreover, while only one cleavage cycle was dealt with by Rapkine, we have traced the SH fluctuation through

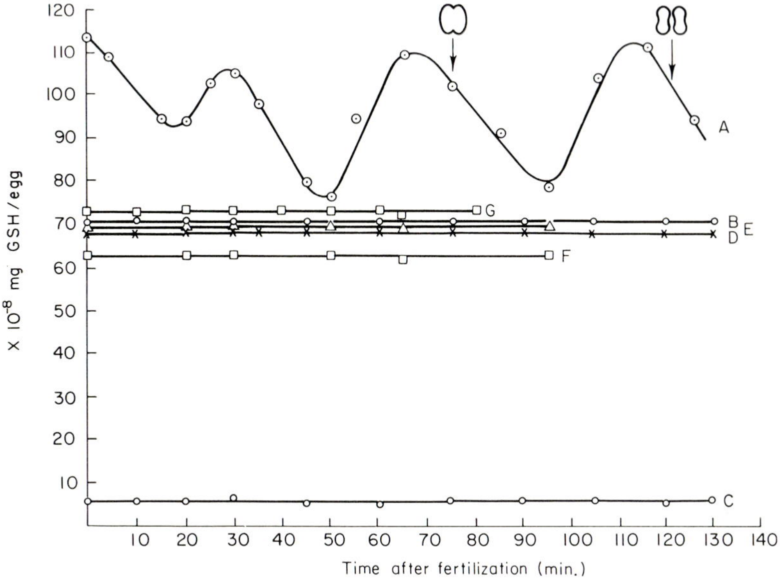

FIG. 1. TCA-soluble sulfhydryl groups of the eggs of the sea urchin, *Pseudocentrotus depressus*. A: nondialyzable sulfhydryl. B and C: reduced (GSH) and oxidized glutathione (G—S—S—G) as measured by iodimetry. D, E, F, G: GSH recovered and measured by different methods. (Courtesy of *Exptl. Cell Res.*)

several cleavages, establishing beyond doubt the cyclic nature of the change (Ikeda, 1965). Since, however, we had concluded that the sulfhydryl groups in question represented protein SH rather than that of glutathione as Rapkine had originally suggested, the necessity for more critical analysis of the nature of the sulfhydryl change became apparent. Some of the questions raised are as follows.

(*a*) The first point to be considered is whether or not proteins extracted with KCl at different developmental stages of the sea urchin are of the same kind. In other words, the mere fact that the same extraction medium (0.6 *M* KCl) is used throughout may not adequately guarantee the identity of the proteins extracted. For example, if a situation exists where there are 2 kinds of proteins extractable with KCl, one with a high and the other with a low SH-content, and if the proportion of the 2 types regularly changes in connection with cell division, the SH amounts of the mixture will show a cyclic fluctuation. But this possibility has been ruled out by an immunological test (Dan, unpublished data, 1964). Precipitation lines in agar diffusion tests between KCl-soluble protein derived from different stages and the antisera for both unfertilized and fertilized KCl protein all gave continuous lines.

(*b*) The next point is whether or not the SH fluctuation may be due to folding and unfolding of the protein molecules in a reversible denaturation. To check this point, SH is measured in a completely denatured state of the protein.

For this purpose, the protein is denatured in acetone or in guanidine hydrochloride and the flocculum is stained with mercury orange (Bennett and Watts, 1958; Sakai, 1960). In preliminary studies, the specificity and stoichiometry of the staining reaction have been scrutinized. After staining the flocculum and removing excess dye with pure acetone, the medium is changed to acetone containing 0.06 *M* HCl to release the bound dye stuff (Ikeda, 1965). The elution is quite thorough, bleaching the flocculum pure white, so that the eluted dye in the supernate can be measured photometrically at 470 $m\mu$. For determining the amount of protein, the method of Lowry *et al.* (1951) was adopted to match the probes against standards, the nitrogen values of which were obtained by the micro-Kjeldahl method.

When SH is expressed by a weight ratio, SH/protein-N, and plotted against developmental stages, the curve again follows a course very similar to the original curve of Rapkine, which was calculated on the basis of SH in terms of glutathione per egg. This experiment assures that the apparent increase and decrease of SH in the KCl-soluble protein is not due to unfolding and folding of the molecules but represent real or net increase and decrease in the SH amount against protein nitrogen (Fig. 2).

(*c*) The last point is to find out from where sulfhydryl groups are derived when their amounts increase, and how they disappear when SH values diminish. This is a relatively simple question to answer because if the total amount of SH is measured after a thorough reduction of the S—S bridges of the denatured protein into SH by BAL, the total SH does not fluctuate with development (Fig. 2B′). This indicates that since the total SH remains constant, the apparent increase of sulfhydryl groups is brought about by the reduction of disulfide bonds of the same protein and the decrease is due to a reaction in the reverse direction. Furthermore, since the observed fluctuation of SH can easily be blocked by parachloromercuribenzoate (PCMB) while the sum of the sluggish and masked SH remains always unchanged, it is obvious that such compensatory or coupled oxidation and reduction between sulfhydryl and disulfide bonds must be occurring only among free reactive groups of SH.

As was mentioned above, Rapkine suggested that the maximum of the SH curve corresponds to the metaphase of the nuclear division. However, as the measurements were repeated many times, it became more certain that the position of the maximum of SH/protein-N value occurred later than the metaphase. But until the present time, it has not been possible

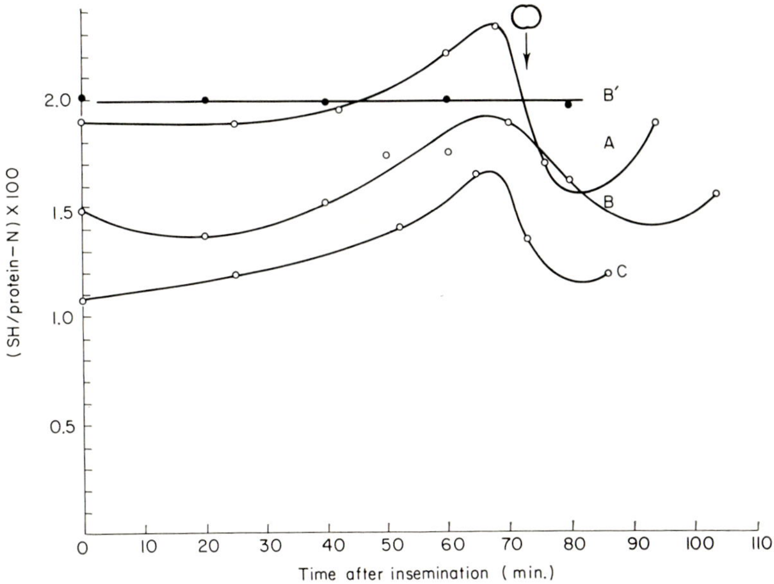

FIG. 2. Fluctuation in the value of SH/protein-N of KCl-soluble protein of sea urchin eggs. A: *Anthocidaris crassispina.* B: *Hemicentrotus pulcherrimus.* *C*: *Pseudocentrotus depressus.* B′: *Hemicentrotus,* SH + S—S. (Courtesy of *J. Biophys. Biochem. Cytol.*)

to ascertain whether or not the maximum really coincides with the beginning of cleavage. Since the ascending part of the SH curve is gradual while the descending part is steep (Dan, 1963), even under the circumstance in which a coincidence does exist, averaging the values of many cells tends to pull the peak toward the earlier side. Therefore, to settle this point, a method other than chemical analysis is needed; this is now being planned.

Now that we know the nature of the sulfhydryl of the KCl protein, let us go back to the main line of the argument. By the logic proposed in the Introduction, such a rhythm of SH fluctuation of the KCl protein, as has been described, can well offer a basis for the synchrony observed in the cleavage of sea urchin eggs. In this case, however, since the cycle is set off by artificial insemination, the impetus for synchrony is imparted by an environmental hazard planned by the investigator. But the realization of later events—particularly synchronous division—is made possible by the fact that the SH change goes through an orderly sequence and the cleavage process is tied to a definite phase of this change. As will be shown later, this statement holds true for synchronization of *Tetra-*

hymena, except for the point that the temperature treatment replaces insemination.

C. Contractility of the KCl Protein

Defining the role played by the SH changes of the KCl protein in cell division is an extremely difficult problem which has not yet been solved. When the new extraction method by 0.6 *M* KCl was started, we had the rather naive idea that 0.6 *M* KCl would probably extract some kind of contractile protein resembling myosin or actomyosin. But this optimism was soon disappointed because a thread model made by pulling this protein out of acetone into air failed to shorten on application of ATP even after adjusting the pH and magnesium concentration. Interestingly enough, however, Sakai (1962a) found that the thread model so prepared can be made to shorten and elongate reversibly by 2 other means.

(*a*) Shortening by polyvalent cations and reversal by metal chelators such as ethylenediaminetetraacetic acid (EDTA). In this case, cobalt and iron are 2 exceptions which make the thread shorten only irreversibly.

(*b*) Shortening by cystine and also by oxidized forms of glutathione and ascorbic acid and elongation by cysteine and reduced forms of glutathione and ascorbic acid.

If the amounts of SH are measured in the contracted and dilated conditions of the thread by staining it with mercury orange, it can be shown that the contraction is accompanied by oxidation of sulfhydryl groups of the thread and elongation is brought about by reduction of the disulfide bonding. This finding immediately explains why cobalt and iron ions are 2 exceptions. Since these 2 ions are well known to catalyze the oxidation of sulfhydryl groups, they must cause an irreversible formation of disulfide bridges, and consequently, an irreversible contraction of the thread (Sakai, 1962a).

A series of measurements which Sakai (1962b) conducted is very illuminating for understanding the interrelationship between the 2 conditions pointed out above. He prepared samples of KCl proteins at various stages of development, from the unfertilized condition until after the completion of the first cleavage. Then thread models were made from the protein samples and their degrees of contraction were measured under the influence of a constant concentration of cadmium chloride ($CdCl_2$) as the metal ion. The degree of contraction changed from stage to stage, in spite of the fact that the conditions for causing contraction were kept constant. An intermediate grade of contraction characterizes protein obtained from eggs before and immediately after fertilization (Fig. 3A). Contractility rises to a maximum several minutes before cleavage, after which it falls rather abruptly with the advance of the cleavage furrow.

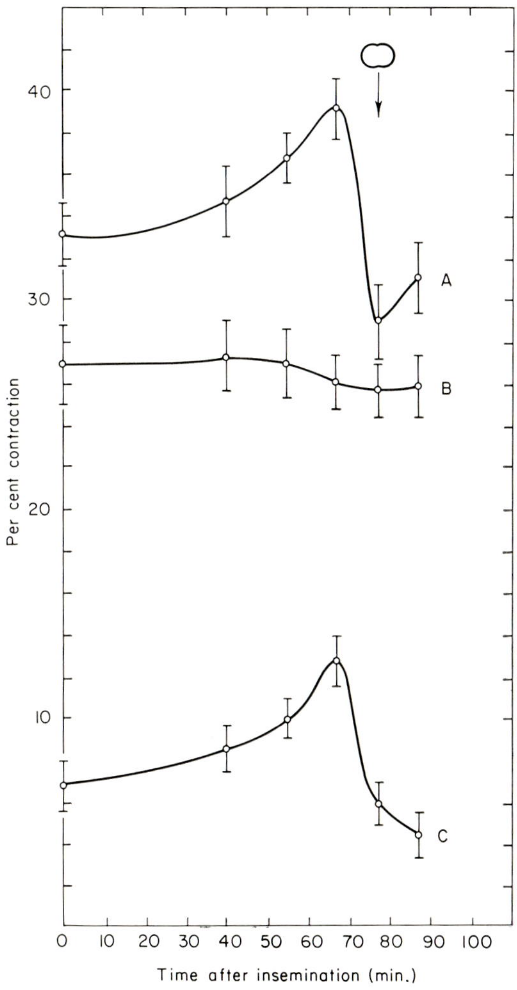

FIG. 3. Change in contractility of thread model during first cleavage cycle (*Hemicentorus pulcherrimus*). A: contraction induced by $CdCl_2$. B: $CdCl_2$ contraction after blocking SH. C: contraction induced by cystine. (Courtesy of *J. Gen. Physiol.*)

In other words, the curve of contractility resembles the curve of SH values described earlier (see Fig. 2).

Next, on the same samples, the degree of contraction is tested in a similar fashion but by using cystine as the agent. The curve runs quite parallel to the cadmium curve except that the magnitude of contraction is much smaller (Fig. 3C). Finally, using the same samples again, Sakai blocked the sulfhydryl groups of the thread by PCMB before adding cadmium chloride. Although the contraction is of a fair magnitude, being just short of the degree of contraction of the unblocked threads, the degree of contraction this time does not fluctuate according to the stage from which the samples are taken; i.e., the curve is flat and horizontal (Fig. 3B). This is interpreted to mean that the contraction of the thread after blockage of the free sulfhydryl groups may represent a contraction caused by the charge effect of cadmium cations on the presumably negatively charged protein. This, in turn, means that the curve of contractility of unblocked threads (A) must be the sum of the basic contraction due to the charge effect plus the contraction resulting from the oxidation of sulfhydryl groups. As a matter of fact, if the curve of cystine-induced contraction (C) is superimposed on the curve of blocked thread (B), the original curve (A) can almost be reconstructed.

If our interpretation is accepted, 2 important conclusions emerge. (*a*) Cadmium ions, besides making the thread contract by their charge-neutralizing effect, are simultaneously oxidizing the sulfhydryl groups of the thread. (*b*) The extent of the contraction due to the oxidation of SH groups (C) must depend on the amount of free sulfhydryl groups in the thread model available for oxidation. In fact, if the SH values of the KCl protein taken from various stages are plotted against the degrees of contraction of the respective stages, a linear relationship is quite evident, showing that the qualitative impression so far obtained—that the curves of SH and of contractility run parallel to each other—can be extended into a quantiative proportionality (Sakai, 1962b).

D. Water-Soluble Protein of Sea Urchin Eggs

Kane and Hersh (1959) have described a water-soluble protein obtained from sea urchin eggs by adding calcium chloride (0.05 *M*) to the supernate of the water homogenate of sea urchin eggs. Sakai (1963a) obtained a similar fraction by half saturating the supernate with ammonium sulfate. These 2 fractions will collectively be called "water-soluble protein" in the following account.

Sakai (1963a) prepared the water-soluble protein fractions from eggs at various stages of development and measured the SH content. A finding of extreme importance is that the SH of the water-soluble fractions persistently behaves in a fashion exactly reciprocal to that of the SH of the KCl-soluble fraction: i.e., the SH value is rather low in the initial stage and then rises to a maximum at the streak stage. Several minutes before cleavage, the curve dips to a rather low minimum, after which it goes up steeply with the formation of the cleavage furrow (Fig. 4).

What this reciprocity of the SH change between the 2 fractions signifies is a fascinating problem. The reciprocity alone might already be

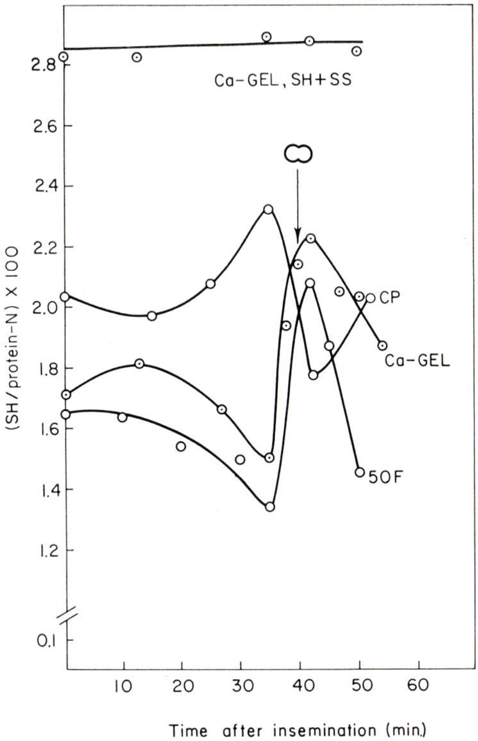

FIG. 4. Reciprocal fluctuation of amounts of SH in KCl- and water-soluble proteins of the eggs of *Anthocidaris crassispina*. CP: KCl-protein. Ca-Gel: water-soluble protein precipitated by $CaCl_2$. 50F: water-soluble protein precipitated by half-saturation with NH_4SO_4.

alluding to a close relationship between them. Sakai, in fact, succeeded in inducing *in vitro,* under appropriate conditions, reciprocal reduction and oxidation between the SH of the 2 fractions. Naturally, this reaction proceeds in such a manner that the more oxidized member at the start is reduced and the more reduced partner is oxidized, regardless of which partner is the KCl- or water-soluble fraction. Sakai, at present, tends to think that there must be electron transfer between the 2 fractions which is mediated by an enzyme, as he will report in detail in a forthcoming paper.

An earlier part of this article has referred to reversible oxidation and reduction between the sulfhydryl and disulfide of the KCl protein. Although the situation in the SH and S—S of the water-soluble fraction has not yet been analyzed, this mutuality might follow the cycle represented by the following scheme: when SH increases and S—S decreases in the

KCl protein, the change may probably be realized via the S—S and SH of the water-soluble fraction.

$$\begin{array}{ccc} SH_{KCl} & \rightleftharpoons & S\text{—}S_{KCl} \\ \uparrow\downarrow & & \uparrow\downarrow \\ S\text{—}S_{H_2O} & \rightleftharpoons & SH_{H_2O} \end{array}$$

This somewhat hypothetical scheme finds support in the experiment dealing with thread contractility. It was mentioned previously that the thread model made of KCl-soluble protein could be made to shorten by polyvalent cations, cystine, oxidized forms of glutathione, and ascorbic acid. Interestingly enough, however, a water-soluble fraction with a high S—S content can replace these reagents. This point is very important biologically, because although these substances can induce contraction *in vitro,* it is impossible to imagine that cations or cystine, etc., will suddenly make their appearance within the living cell at the time of cell division. As long as the water-soluble fraction can do the job, such a difficulty is completely avoided.

Sakai (1962b) proceeded to investigate the efficiency of water-soluble fractions prepared from various stages in inducing contraction; he tested these on a thread model made from a single sample of KCl protein. If the water-soluble proteins taken from different stages are compared, their efficiency in causing contraction varies in an inverse proportion to their sulfhydryl content, or in direct proportion to the amount of disulfide in the water-soluble protein used. Putting the stories of the 2 fractions together, the richer the KCl protein is in SH, the larger its capacity of contraction will be; the lower the SH or the higher the S—S content of the water-soluble protein is, the greater its efficiency in causing the KCl protein to contract will be. Furthermore, the best combination of the 2 factors to insure the maximal contraction is realized a few minutes before every cleavage. The author feels that this SH $\rightleftharpoons$ SS mechanism can perhaps be compared to a spring, the 2 ends of which are pulled apart before each cleavage and released with the onset of division. The energy necessary to stretch the SH-spring may be derived from metabolic activities, while the contraction may be more or less spontaneous, since it is possible to obtain reciprocal oxidation and reduction *in vitro* without providing an energy source.

E. The Rhythm of the SH $\rightleftharpoons$ S—S Cycle and the Rhythm of Cleavage

In order to push the analysis of the cleavage mechanism a step further, attempts were made to arrest the rhythm or a part of it. It must be made clear here that since the work in connection with the water-soluble protein is still in progress, the available data concern the sulfhydryl behavior of the KCl-soluble protein only.

It has been known for over 60 years that the cleavage of sea urchin eggs can be reversibly suppressed by a mixture of 0.6 volume percent of ether in seawater (Wilson, 1901). By measuring SH/protein-N of the KCl protein of the eggs during this treatment, it is revealed that the SH cycle is stopped by the treatment, and the level of SH remains stationary while the eggs are left in the ether-seawater mixture. Once they are taken out of the mixture, the SH cycle is resumed. When the remainder of the cycle is run through, division follows the usual SH maximum (Sakai, 1963b).

In fertilized eggs of the sea urchin *Anthocidaris crassispina* which have been kept at 26°C, a temperature shock of 34C° for various durations brings about the same result (Fig. 5) (Ikeda, 1965). A feature common to both cases is the fact that in spite of the absence of cytokinesis, the nucleus continues to divide so that multinucleate cells result which divide, on recovery, into more than 2 blastomeres. In the case of ether, cytokinesis is prevented as long as the eggs are kept in the solution, while with heat shock, the duration of suppression is roughly proportional to the length of the initial heat treatment.

Another method inhibiting cytokinesis is ultraviolet irradiation. The telophase of the first cleavage is chosen by Ikeda (1965) as the ap-

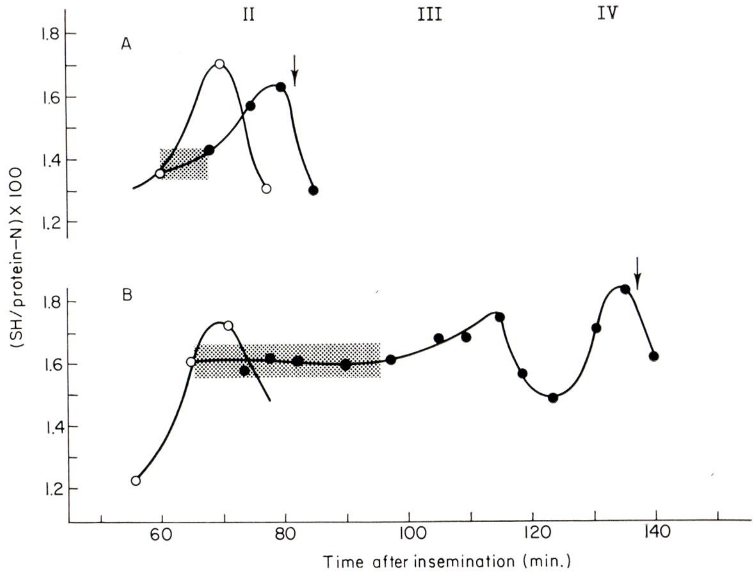

FIG. 5. Effect of heat treatment (34°C) on the pattern of SH change in the eggs of *Anthocidaris crassispina*. Open circles: control eggs. Filled circles: heat-treated eggs. Period of heat shock is shaded. Arrow indicates division in treated eggs. (Courtesy of M. Ikeda.)

propriate time to apply the irradiation, in order to avoid the complication of interfering with monaster formation or syngamy. Another reason for choosing this stage is that since telophase occurs later than the so-called critical stage for the first division (Swann, 1957), irradiation at this stage allows the first division to take place quite normally, only affecting the second and later divisions. Therefore, by taking the time of the first division as zero time, the timing of later division after recovery can be made conveniently.

Ikeda follows the changing SH level in the KCl protein of irradiated eggs after the first cleavage (which is not affected by the irradiation). Figure 6 shows clearly that the sulfhydryl content fluctuates strictly in pace with that of normal, nonirradiated eggs. This is quite striking in view of the fact that the experimental batch has missed a few division cycles, and consequently no morphological change is occurring which can be observed microscopically. In strong contrast to the result following ether and heat treatment, which stop cytokinesis without interfering with the nuclear division, the irradiated eggs invariably go into the 4-cell stage when they recover, regardless of the duration of the cleavage in-

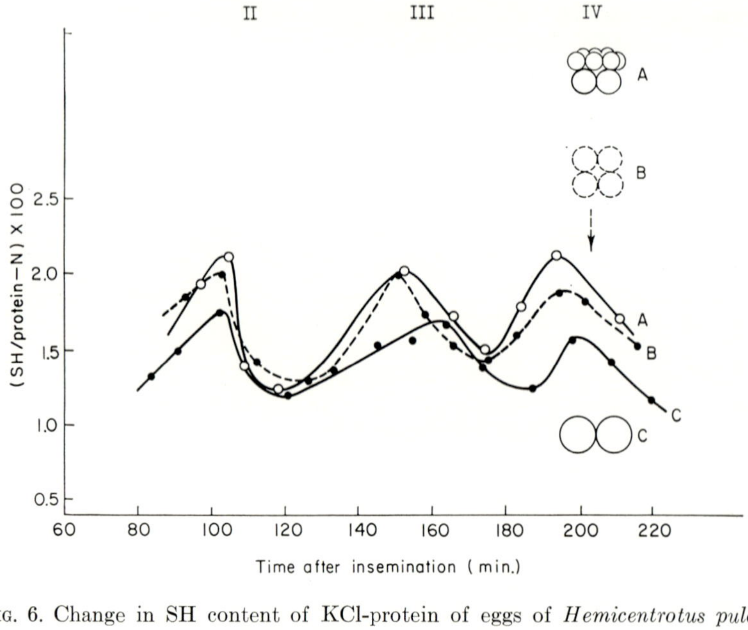

FIG. 6. Change in SH content of KCl-protein of eggs of *Hemicentrotus pulcherrimus* in which cleavage was suppressed by ultraviolet irradiation. A: control eggs; measurements begun after first cleavage, subsequent cleavage times indicated by Roman numerals at top of figure. B: irradiated eggs omitting two cleavages. C: irradiated eggs with three cleavages suppressed. (Courtesy of M. Ikeda.)

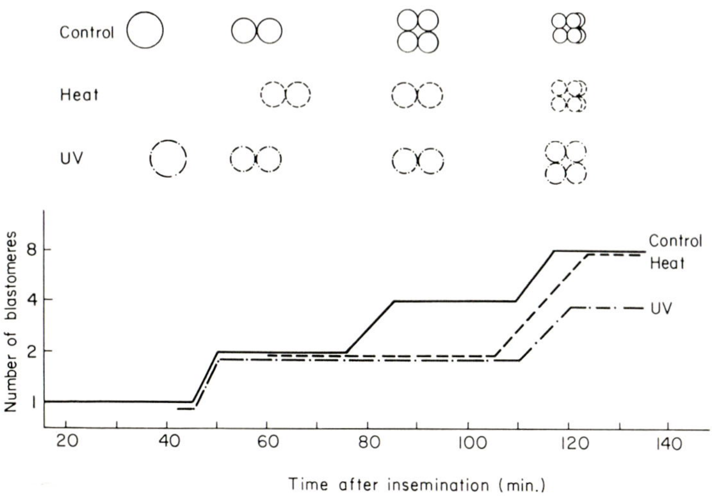

FIG. 7. Comparison of cleavage patterns after heat shock and ultraviolet irradiation. (Courtesy of M. Ikeda.)

hibition. This indicates that under the inhibition by irradiation, the nucleus must have remained quiescent although the cytoplasmic SH cycle has been going on. The situations under heat shock and irradiation are diagrammatically shown in Fig. 7.

These considerations indicate that if the cytoplasmic SH cycle and the nuclear cycle exist separately, neither of them alone can induce cytokinesis; a division takes place only when the cytoplasmic and nuclear cycles are synchronized.

F. THE RHYTHM OF SH ⇌ S—S CYCLE AND THE RHYTHM OF DIFFERENTIATION

During the ultraviolet irradiation experiments, precocious formation of the micromeres attracted our attention. In cleaving eggs which have skipped one nuclear cycle, the experimental batch only reaches the 8-cell stage when the control batch goes from the 8-cell stage (8 equally sized blastomeres) into the 16-cell stage (8 mesomeres + 4 macromeres + 4 micromeres). When this happens in the eggs of *Clypeaster japonicus*, a cake urchin, most of the irradiated eggs form 4 macromeres and 4 micromeres instead of 8 equally sized cells (Fig. 8). This formation of micromeres at the 8-cell stage is precocious as far as the number of blastomeres is concerned, but since the SH clock has been going, such seemingly precocious micromere formation occurs accurately in relation to the fourth maximum of SH, as found in the control batch.

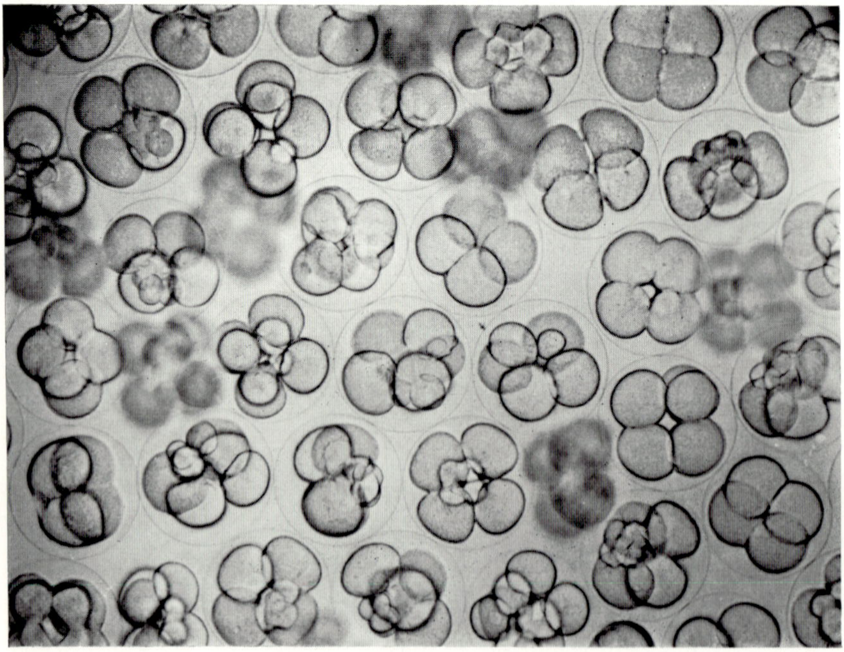

FIG. 8. Micromere formation at the 8-cell stage in ultraviolet irradiated eggs of *Clypeaster japonicus*. (Courtesy of M. Ikeda.)

This indicates that the timing of the micromere formation is governed by the SH clock, not by the number of divisions the nucleus has gone through. Since micromere formation is the first step in segregating cells with a definite function (spicule formation), this stage can also be considered as the first step in cellular differentiation. The possibility thus arises that the SH clock may be setting the pace for other differentiation phenomena. This possibility is supported by various other observations (Dan, 1960) which indicate that the steps of sea urchin development advance with the steps of cleavage: e.g., stepwise enlargement of the blastocoel in connection with first 7 divisions in *Mespilia globulus*, association of cilia formation, and hatching with the tenth cleavage in many sea urchin forms, etc. The only modification of the previous statement to be made is to replace the term "cleavage" with the term "SH cycle." This change is justifiable because in the normal development, the cleavage and the SH fluctuation proceed together.

As a result, it appears that the SH $\rightleftharpoons$ S—S cycle is achieving a dual task. The SH $\rightleftharpoons$ S—S interchange of the KCl-soluble fraction, coupled with that of the water-soluble fraction must be contributing directly to the performance of cytokinesis when it is synchronized with the right phase of the nuclear activity. But at the same time, the chronological accuracy of the SH fluctuation must also be setting the tempo for the progress of division. This may be the main reason why synchrony can last among an artificially inseminated batch of eggs for as long as 10 or more division cycles before the necessity arises for synthesis of different substances among the blastomeres (Agrell, 1964). Furthermore, because of the presence of such an accurate clock system within the cell, it is not surprising that the steps of differentiation are also geared into this system, the SH clock serving here as the pace maker for cellular differentiation.

III. SH CYCLE IN SYNCHRONIZED *Tetrahymena* CELLS

A. KCl- AND WATER-SOLUBLE PROTEIN FRACTIONS OF *Tetrahymena pyriformis*

Ikeda and Watanabe (1965) subjected *Tetrahymena pyriformis*, strain W, to 8 alternate changes of temperature between 26°C and 34°C for 30 minutes each. The cells of this strain divide synchronously (more than 80%) 70 minutes after the end of the heat treatment.

Samples of 200 ml are taken at different stages during the 70 minute interval. The cells are washed 3 times by centrifugation with cold inorganic medium and once with cold distilled water; the packed cells are immediately homogenized after addition of 9 times volume of distilled water. After the homogenate has been kept standing on ice for 3 hours, it is centrifuged 15,000 rpm for 15 minutes. The supernate is dialyzed and centrifuged. The supernate of the last centrifugation is used as water-soluble protein fraction. The residue is washed twice in distilled water and dissolved in an amount of 0.6 *M* KCl equal to 5 times the initial volume of the packed cells. After 3 hours of standing, it is centrifuged and the supernate is dialyzed against 0.6 *M* KCl to give the KCl-soluble protein fraction.

The changes in the amounts of sulfhydryl groups in the 2 protein fractions are measured by the mercury orange method. The result (Fig. 9) clearly shows that the sulfhydryl groups of the 2 fractions fluctuate in a reciprocal fashion, and the maximum of the SH of the KCl-soluble and the minimum of the SH of the water-soluble fraction practically coincide with the beginning of the synchronous division. In other words, the behavior of the SH groups of these 2 fractions of *Tetrahymena* is exactly the same as that of sea urchin eggs.

Special attention should be paid to the flexion points of the 2 curves at 45 minutes, because this stage is no other than the so-called "physio-

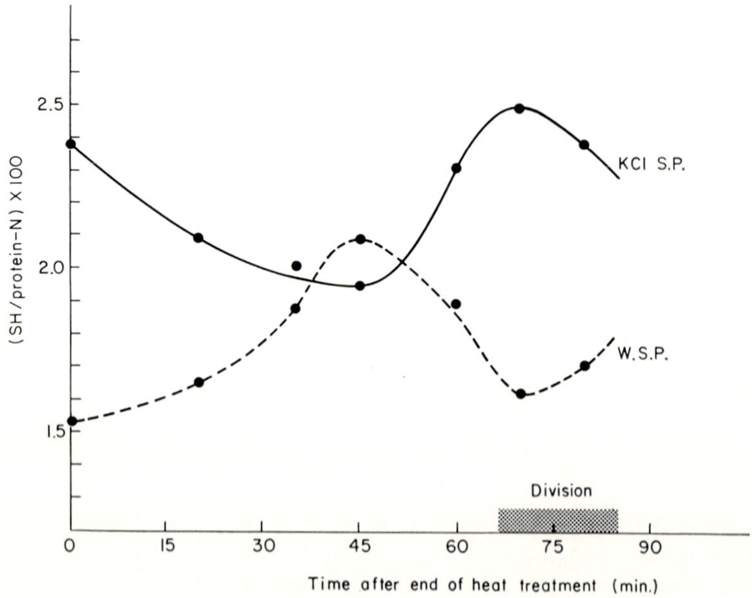

FIG. 9. Changes in amounts of SH of 0.6 *M* KCl-soluble and water-soluble proteins of *Tetrahymena pyriformis* during the recovery period. Division period is shaded. (Courtesy of M. Ikeda and Y. Watanabe.)

logical transition point" of Zeuthen (1964). In connection with division, this point is often referred to as "the point of no return" because if inhibitors of protein synthesis are added after this time, they allow one division to take place although later divisions are suppressed, while if they are applied before this time, no division can ever occur. At any rate, the fact that similar SH fluctuation is found in both sea urchin eggs and synchronized *Tetrahymena* cells is a point of the utmost importance, suggesting that the SH $\rightleftharpoons$ S—S mechanism must lie deep in the roots of the division phenomenon.

B. SH Behavior when Division Is Suppressed

In order to confirm the *sine qua non* nature of the SH cycle in the division of *Tetrahymena*, Ikeda and Watanabe (1965) proceeded to block division by sodium fluoride. Immediately after the termination of the temperature treatment, the cells are transferred to 30 m*M* sodium fluoride, which is sufficient to suppress their division completely. As shown in Fig. 10, in spite of the presence of sodium fluoride from the beginning, the SH curves of both fractions proceed along the same course as the control until the transition point. But after that the curves fail to

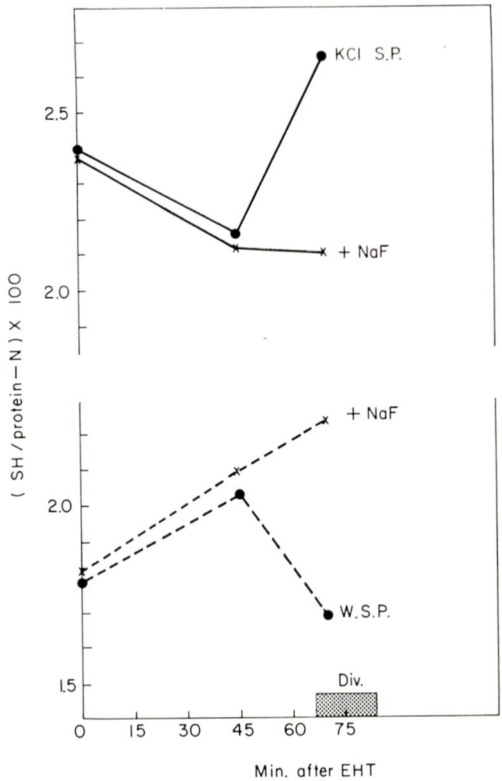

FIG. 10. Comparison of the changes in SH with and without NaF as an inhibitor of cell division in *Tetrahymena*. (Courtesy of Y. Watanabe and M. Ikeda.)

show usual flexion, continuing more or less along the same course. This means that the effect of inhibition reveals itself only after the critical point. We are tempted to think that the reversal of SH $\rightleftharpoons$ S—S after the transition point may be a factor directly contributing to the ensuing division.

C. Subfractionation of the KCl- and Water-Soluble Proteins

Watanabe and Ikeda (1965a) separated the 2 proteins further into subfractions by the use of a DEAE cellulose column, and obtained 3 subfractions from the KCl-soluble protein and 10 subfractions from the water-soluble protein, as distinguished by their optical density at 280 mμ.

Measuring the SH content by mercury orange staining, they have found that the SH groups are concentrated in one (No. 2) of the 3 sub-

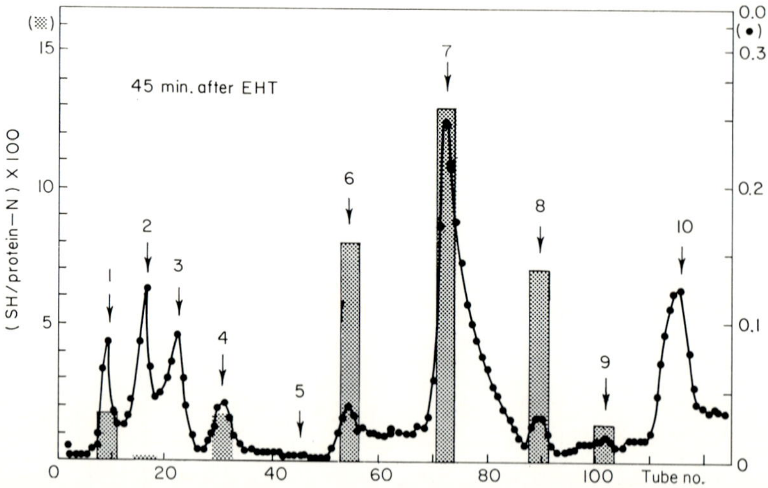

FIG. 11. Optical density profile (right ordinate) of 10 subfractions of the water-soluble protein of *Tetrahymena* at 45 minutes after the end of heat shock, as separated by DEAE cellulose chromatography. Shaded columns are SH/protein-N (left ordinate). (Courtesy of Y. Watanabe and M. Ikeda.)

fractions of the KCl-soluble protein and in 3 (Nos. 6, 7, 8) of the 10 subfractions of the water-soluble protein. Although the general trends of the SH changes of these subfractions are the same as those of the original preparations, the fluctuation is more acute, indicating that these are the components chiefly responsible for the SH changes of the overall preparations.

For reasons which will become clear later, the 10 subfractions of the water-soluble protein will be described more fully. Figure 11 is a chromatogram of the water-soluble protein taken at 45 minutes after the end of the heat treatment, when the SH content of this protein is at its maximum. The continuous lines in the figure represent the optical density profile, and the shaded columns show the SH/protein-N of the respective peaks. It is evident that Peak 7 is particularly rich in SH, to an amount about 6 times that of the whole preparation. A high rate of incorporation of methionine-H^3 into this subfraction, found in a labeling experiment of the log-phase cells, further supplements this conclusion.

D. Significance in Cell Division of Subfraction 7 of the Water-Soluble Protein

Watanabe and Ikeda (1965a,b) set up several criteria for diagnosing the "division protein" postulated by Zeuthen (see Rasmussen and Zeuthen, 1962) and scrutinized the 13 subfractions with respect to their qualifications for this role. Five features have been tentatively selected; some of these represent duplicate assurance for a single point, and some of them are only collaries to others. In any case, these points are adopted only for convenience because the range of selection is limited to testable cases within the capacity of the experimental arrangement. The criteria are as follows:

(1) Synthesis of the division protein as indicated by chromatographic separation should take place around 20–40 minutes before the forthcoming division.

(2) Point (1) can be reaffirmed by following the incorporation pattern after pulse labeling by radioactive amino acids.

(3) Suppression of the synthesis or dysfunction of the product should occur under conditions which inhibit division.

(4) As a corollary to (3), the product or its function, in *Tetrahymena*, should be particularly sensitive to heat shock.

(5) Although this is not a deciding point, it is imaginable that the synthesis of the division protein might be compensatorily enhanced over other kinds of protein around the transition point, since the formation of the division protein has been under a continued suppression through the temperature treatment.

The results of their examinations of the 13 subfractions are tabulated in Table 1.

As far as the criteria here selected are concerned, Subfraction 7 of the

TABLE I

QUALIFICATION OF 13 SUBFRACTIONS FOR "DIVISION PROTEIN" BY 5 CRITERIA

Subfractions (No.)	Criterion (1)	Criterion (2)	Criterion (3)	Criterion (4)	Criterion (5)
KCl soluble					
1		+			
2	−				
3	+		+		
Water soluble					
1					
2					
3					
4		+			
5		+			
6		+			
7	+	+	+	+	+
8					
9					
10	+				

water-soluble protein satisfies all 5 conditions, whereas none of the others fulfill more than 2. Obviously much more thorough studies will be required before the substance appearing as Peak 7 can be identified as the division protein. However, it is at least possible to state that one protein subfraction is now obtainable, the sulfhydryl groups of which fluctuate in a characteristic cyclic fashion and which seems to have an especially close relation to synchronous division in *T. pyriformis*.

IV. CONCLUDING REMARKS

Throughout the foregoing account, striking similarities between the SH cycles involved in the synchronous divisions of sea urchin eggs and of the protozoan, *Tetrahymena*, have been reported. However, the comparison is admittedly rather superficial.

In the former case, the nature of the SH change was viewed chiefly from the standpoint of the contractility of the KCl-soluble protein and its cyclic change by developmental stages and the situation was mainly analyzed to see whether or not this protein could contribute to the process of cytokinesis. As a result, the protein is tentatively considered to be present in the eggs in a more or less constant amount—hence, the possibility that this protein is being synthesized is ignored. This simplification may not be so arbitrary as it might seem, because the sea urchin egg is not a growing system, most of the substances needed for its early development having been bestowed upon it by the mother organism during oogenesis.

In the case of *Tetrahymena*, on the other hand, the explanation of the SH fluctuation is mainly sought for in protein synthesis, particularly that of Subfraction 7 of the water-soluble protein, but no analysis has been made concerning the concrete part which this subfraction would be playing in the fission of the protozoan. When this is worked out, it will not only enrich the now practically empty concept of "division protein," but will also tie together the now separate data accumulated in sea urchin eggs and the ciliate.

Obviously, one of the links connecting the 2 fields of study would be to obtain information about the localization of the KCl-soluble and water-soluble proteins within the cell. An attempt toward this end applying the immunostaining technique to sea urchin eggs is now in progress.

REFERENCES

Agrell, I. (1964). *In* "Synchrony in Cell Division and Growth" (E. Zeuthen, ed.), p. 39. Wiley (Interscience), New York.

Bennett, H. S., and Watts, R. M. (1958). *In* "General Cytochemical Methods" (J. F. Danielli, ed.), Vol. 1, p. 317. Academic Press, New York.

Dan, K. (1960). *Intern. Rev. Cytol.* **9,** 321.

Dan, K. (1963). *In* "Cell Growth and Cell Division" (R. J. C. Harris, ed.), p. 261. Academic Press, New York.

Ikeda, M. (1965). *Exptl. Cell Res.* **40,** 282.

Ikeda, M., and Watanabe, Y. (1965). *Exptl. Cell Res.* **39,** 584.

Kane, R. E., and Hersh, R. T. (1959). *Exptl. Cell Res.* **16,** 59.

Lowry, O. H., Rosenbough, N. J., Farr, A. L., and Randall, R. J. (1951). *J. Biol. Chem.* **193,** 265.

Rapkine, L. (1931). *Ann. Physiol. Physicochim. Biol.* **7,** 382.

Rasmussen, L. and Zeuthen, E. (1962). *Comp. Rend. Trav. Lab. Carlsberg* **32,** 333.

Sakai, H. (1960). *J. Biophys. Biochem. Cytol.* **8,** 609.

Sakai, H. (1962a). *J. Gen. Physiol.* **45,** 11.

Sakai, H. (1962b). *J. Gen. Physiol.* **45,** 427.

Sakai, H. (1963a). *Seitai No Kagaku* **14,** 185.

Sakai, H. (1963b). *Exptl. Cell Res.* **32,** 391.

Sakai, H., and Dan, K. (1959). *Exptl. Cell Res.* **16,** 24.

Swann, M. M. (1957). *Cancer Res.* **17,** 727.

Watanabe, Y., and Ikeda, M. (1965a). *Exptl. Cell Res.* **39,** 443.

Watanabe, Y., and Ikeda, M. (1965b). *Exptl. Cell Res.* **39,** 464.

Wilson, E. B. (1901). *Arch. Entwicklungsmech. Organ.* **13,** 353.

Zeuthen, E. (1964). *In* "Synchrony in Cell Division and Growth" (E. Zeuthen, ed.), p. 99. Wiley (Interscience), New York.

CHAPTER 16

A Method of Making Synchronous Cell Cultures by Density Gradient Centrifugation

J. M. Mitchison and W. S. Vincent

We have recently developed a simple method of preparing synchronous cultures which should prove useful in studying the biochemistry of the cell cycle (Mitchison and Vincent, 1965). Cells from a normal asynchronous culture in exponential growth are concentrated and layered as a thick suspension on top of a linear gradient of sucrose in a centrifuge tube. The gradient is then centrifuged until the cell layer has moved about two thirds of the way down the tube. The small cells, at the beginning of the cell cycle, move more slowly down the tube and can be separated off the top of the layer with a syringe. When these cells are put into fresh medium, they grow up as a synchronous culture. This technique is the same as that which has been widely used in recent years for separating different sizes of macromolecules, except that the times and speeds of centrifugation are very much less. The main function of the sucrose gradient is to stabilize the column of liquid in the centrifuge tube, and it does not act as an equilibrium gradient.

This method has been used by us on *Schizosaccharomyces pombe, Saccharomyces cerevisiae,* and *Escherichia coli,* and by Sinclair and Bishop (1965) on strain L mouse cells. Figures 1 to 5 show the increase in cell numbers in synchronous cultures of these organisms and, with the fission yeast and the mouse cells, the mitotic index or its equivalent. The details of the method are given in the original papers.

There are at least 2 disadvantages of this technique. The degree of synchrony is not very high because of the imperfect size separation in the

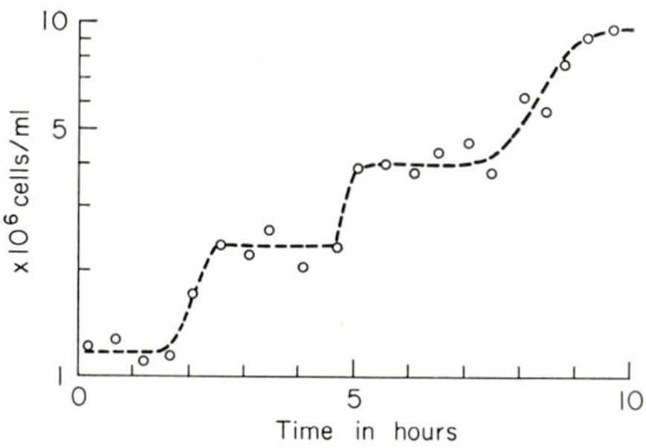

FIG. 1. Cell numbers in synchronous culture of *S. pombe*. 10–40% gradient in 80 ml tube loaded with 3×10^9 cells from exponential culture (3.2×10^6 cells/ml). Centrifuged at 500 g for 10 minutes at 25°C. Top sample suspended in fresh medium. (From Mitchison and Vincent, 1965.)

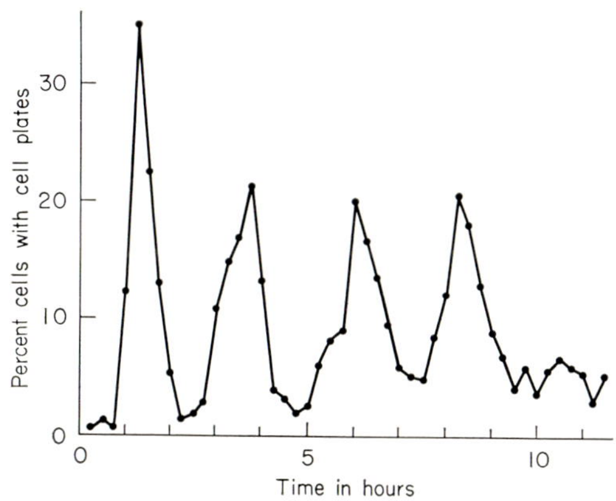

FIG. 2. Percentage of cells with cell plates in synchronous culture of *S. pombe* (equivalent to mitotic index). 10–40% gradient in 80 ml tube loaded with 3×10^9 cells from exponential culture (2×10^6 cells/ml). Centrifuged at 500 g for 9 minutes at 25°C. Top sample (4% of cells loaded) suspended in fresh medium. (From Mitchison and Vincent, 1965.)

gradient. There may also be a lag before the synchronous culture attains its full growth rate (e.g., in Fig. 4) which is presumably due to concentrating the cells in the harvesting. On the other hand, the method has certain merits. It is easy, quick, and requires little apparatus apart from a centrifuge, and a gradient machine which is simple to make. It uses growing cells and, although they may be subjected to some physiological shock during harvesting, this is small compared to some of the other methods of achieving synchrony. It gives a yield which is large enough

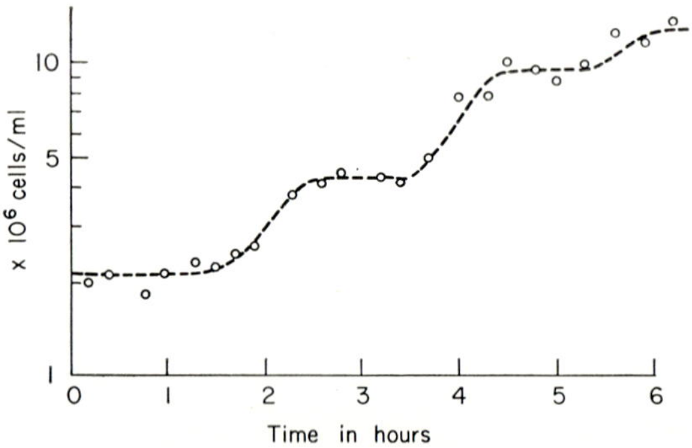

FIG. 3. Cell numbers in synchronous culture of *S. cerevisiae*. 10–40% gradient in 80 ml tube loaded with 4×10^9 cells from exponential culture (2.9×10^6 cells/ml). Top sample (3% of cells loaded) suspended in fresh medium. (From Mitchison and Vincent, 1965.)

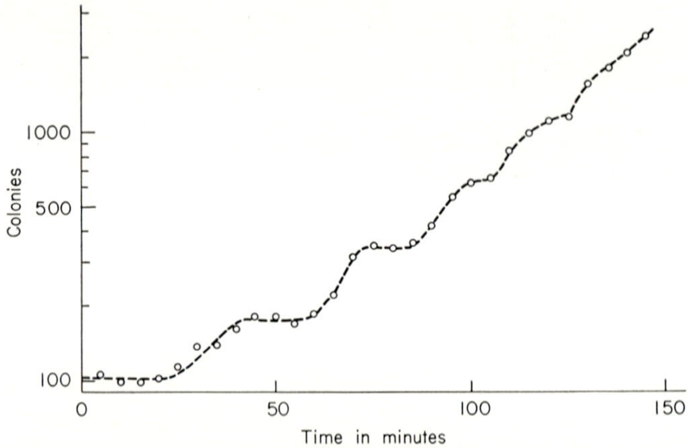

FIG. 4. Colony counts from synchronous culture of *E. coli* K12λ F⁻. 10—40% gradient in 15 ml tube loaded with 2×10^{10} cells from exponential culture (6×10^8 cells/ml). Centrifuged at 2,500 g for 20 minutes at 12°C. Top sample (2% of cells loaded) suspended in fresh medium. (From Mitchison and Vincent, 1965.)

for biochemical analysis. We have started synchronous cultures of *S. pombe* with 10^9 cells from one gradient, and it would be quite easy to scale this up by using several tubes. The practical upper limit is set by the size of the initial culture which can be handled in harvesting, since the recovery of cells from the gradient is only about 5% of the cells which are loaded. Finally, the method has the considerable merit that it works over a wide range of cells—from *E. coli* to mammalian cells.

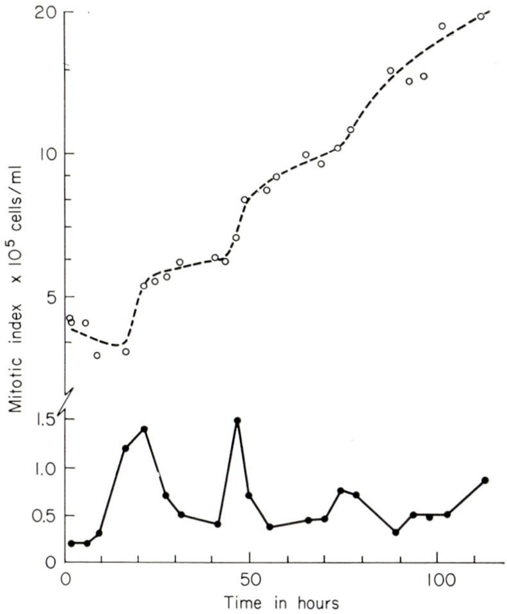

FIG. 5. Cell numbers and mitotic index from synchronous culture of Strain L mouse cells. 5–15% gradient in 15 ml tube loaded with 2×10^8 cells from exponential culture. Centrifuged at 80 g for 6 minutes at 18°C. Top sample (4% of cells loaded) freed from sucrose and suspended in 18 ml used medium, 2 ml fresh medium, 1 drop of trypsin solution (50μg/ml) and gassed with 5% carbon dioxide in air. Samples (1 ml) removed periodically for counting. After sampling, 1 ml fresh medium added and culture regassed. (From Sinclair and Bishop, 1965.)

There seems no reason why it should not work with any nonmotile growing cell which can survive in free suspension. If cells suffer osmotic damage in sucrose, an effective gradient could be made with macromolecules such as protein or dextran.

REFERENCES

Mitchison, J. M., and Vincent, W. S. (1965). *Nature* **205,** 987.

Sinclair, R., and Bishop, D. H. L. (1965). *Nature* **205,** 1272.

CHAPTER 17

Effects of Temperature on the Mitotic Cycle of Normal and Synchronized Mammalian Cells

Potu N. Rao and Joseph Engelberg

I. INTRODUCTION

The marked delaying effect of extremes of temperature on cell division in vertebrate cells (Spear, 1929; Lewis, 1933; Barber and Callan,

1943) and the expectation of differential temperature sensitivities of various synthetic processes of the division cycle has led to the use of temperature shocks for the induction of division synchrony in tissue culture cells (Gey *et al.*, 1954; Wildy and Newton, 1958). This procedure for synchronizing mammalian cells has been successful in some cases (Newton and Wildy, 1959; Newton, 1964) but not in others (Miura and Utakoji, 1961). At the inception of the studies to be reported here, our primary aim was to make a detailed study as to the effects of temperature on different phases of the mitotic cycle of cells in tissue culture. We hoped to attain a better understanding both of the phenomenon of temperature-induced synchrony and of the integration of the events which make up the mitotic cycle of the mammalian cell.

Cinemicrographic studies of human amnion cells in monolayer cultures immediately following a temperature shift had revealed a characteristic response in the duration of each of the phases (G_1, S, G_2, M)* of the mitotic cycle (Sisken, 1963). These studies were extended in our laboratory on HeLa cells† grown in suspension. The pattern of cellular proliferation following a sudden shift from 37°C to a new temperature was followed both during the period of transition immediately after the shift as well as after the steady-state (pure exponential growth) was reached. We will report here (*a*) on the durations of G_1, S, G_2, M, and Tg (generation time) at different temperatures during the period of transition and in the steady-state, (*b*) on the rate with which different phases of the mitotic cycle reach the steady-state, and, (*c*) on the effects of low temperature on synchronized (by the excess thymidine technique) and unsynchronized HeLa cells. We were unable, however, by the use of temperature shocks, to induce significant degrees of synchronization in our HeLa cell cultures.

* Following the terminology of Howard and Pelc (1953) 3 phases of the interphase period may be identified by labeling techniques: (*a*) the period of DNA synthesis (S); (*b*) the pre-DNA synthetic period (G_1) that extends from the end of telophase to the initiation of DNA synthesis; and (*c*) the post-DNA synthetic period (G_2) commencing from the completion of DNA synthesis and ending with the onset of prophase. M refers to mitosis.

† This clone was isolated for suspension culture by Dr. N. Salzman of the National Institutes of Health. Stock suspension cultures of these cells were maintained in steady-state growth at 37°C in Eagle's minimum essential medium supplemented with Eagle's nonessential amino acids, pyruvate and 5% calf serum. The pH of the culture was maintained at 7.4 by passing air containing 2.7% CO_2 through the culture flasks. Cell concentrations were maintained between 1.5 and 3.5 $\times 10^5$ cells/ml by daily dilution of the growth medium.

II. EFFECTS OF TEMPERATURE ON THE DURATION OF THE MITOTIC CYCLE AND OF MITOSIS

A. Growth Rate as a Function of Temperature

The *generation time* is the time required for a culture to double its cell number. This was measured at various temperatures from 33°–40°C, a range in which the cells could be maintained in steady-state growth. Generation times were obtained both by direct cell counts (hemocytometer) and from the rate of accumulation of mitoses in cultures to which Colcemid had been added. Results from both methods were in good agreement. Measurements were made over a 4-day period after the cultures appeared to be in the steady-state at each new temperature. The reciprocal of generation time ($1/Tg$), hereafter referred to as *growth rate,* is plotted as a function of temperature in Fig. 1. The quasilinear nature of the plot on both sides of the maximum is noteworthy.

Steady-state growth could not be maintained at temperatures of 32°C and below, or temperatures of 41°C and above. Following a shift to 32°C the cells divided at least once over a period of 72 hours after which no measurable increase in cell number was observed. After a shift to 41°C cell division ceased almost immediately.

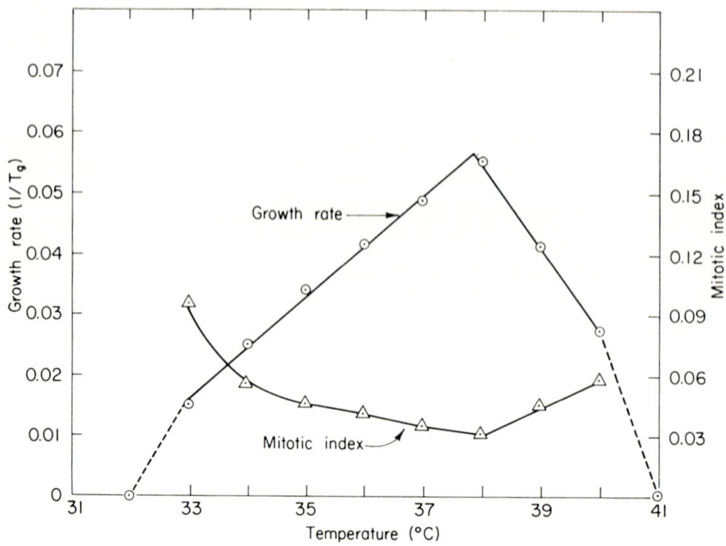

FIG. 1. Growth rate (reciprocal of the generation time) and mitotic index versus temperature for the HeLa cell in steady-state (exponential) growth.

The temperature-growth rate pattern of the HeLa cells is similar to that observed in a variety of biological systems (Woodruff and Baitsell, 1911; Ephrussi, 1926; Mitchell, 1929; Smith, 1940; Johnson and Lewin, 1946; Phelps, 1946; Brown, 1951; Hinshelwood, 1952; Mucibabic, 1956; Prescott, 1957; Scherbaum, 1957; Zeuthen, 1958, 1964; Thormar, 1962). It has been customary to plot growth-rate-temperature data in the form of an Arrhenius Plot, a plot of the logarithm of the growth rate versus the reciprocal of the absolute temperature. This representation derives from the field of chemical kinetics where, for simple chemical reactions, such a plot usually falls on a straight line (Glasstone, 1946). Johnson and Lewin (1946), for example, assumed that the temperature dependence of the growth rate of *E. coli* at suboptimal temperatures was due to the effect of temperature on a rate-limiting enzyme system. An Arrhenius Plot of their data is linear on both sides of the growth rate maximum. A similar plot for *Tetrahymena pyriformis,* however, is not linear on either side of the maximum (Thormar, 1962). An Arrhenius Plot of our data on HeLa cells is likewise nonlinear. However, we have found that the simple plot of growth rate versus temperature to be relatively linear (Fig. 1).

What is the reason for the maximum in the plot of growth rate versus temperature? It is reasonable to assume an increase in growth rate with temperature since chemical reaction rates increase with temperature. This then would explain the rising part of the plot. Why, however, does the growth rate fall when a critical temperature (in our case 38°C) is exceeded? It has been argued that this may be due to the denaturation of enzymes at higher temperatures. However, the fact that this critical temperature is approximately the same for many even distantly related organisms suggests that a more fundamental phenomenon may be at stake. Drost-Hansen (1964) has made the appealing suggestion that higher order phase transitions in the state of the intracellular water may account for precipitous changes in biological parameters with temperature. This hypothesis is interesting because it centers on a substance which is present in large quantity in all biological systems. Moreover the interaction of water with biological macromolecules plays an important role in the specification of cell structure and cell function. One should bear in mind the possibility, however, that biological temperature effects may reflect changes in the state of the cell lipids. These would influence the activity of membranous organelles—mitochondria, the cell membrane, the nuclear membrane, the endoplasmic reticulum, etc.—and would therefore be expected to have a profound effect on the functioning of these organelles. Investigations on the micellar state of lipid membranes *in vitro* do indeed exhibit the consequences of startling changes in molecular properties with temperature (Thompson, 1964).

B. Mitotic Index as a Function of Temperature during Steady-State Growth

1. *Determination of Mitotic Index*

The *mitotic index* (MI)—the fraction of cells in mitosis—was scored on slides prepared by the following procedure: By use of an automatic sampling device 1 ml samples of cell suspension were collected in a tube containing 2 ml of a hypotonic fixative (a 2% solution of 3 to 1 absolute ethanol-glacial acetic acid). The hypotonic fixative served both to expand the cells and to fix them. The cells were left in this fixative for anywhere from 10 minutes to 24 hours. This simple procedure was effective in yielding easily recognizable mitoses for scoring. After centrifugation the cells were resuspended in undiluted fixative and allowed to stand for 10 minutes. This completed fixation and facilitated, subsequently, the even spreading of the cells on a glass slide. The cells were centrifuged again and resuspended in the last drop of the supernatant fluid. This concentrated drop of cells was placed on a clean, wet slide and quickly dried under a fan. The cells were stained with a drop of aceto-orcein, covered by a coverslip, and sealed with wax.

2. *Mitotic Index in Steady-State Cultures*

For cultures in steady-state growth at 37°C the mitotic index averaged 0.03, and mitosis occupied about 4% of the generation time. The shortest duration for mitosis (0.8 hour) was obtained at 38°C. When cells were grown above and below 38°C the duration of mitosis and the mitotic index (see Fig. 1) increased. In short, growth rate varied inversely with mitotic index, the growth rate being a maximum when the mitotic index was a minimum.

III. LIFE CYCLE ANALYSIS OF HeLa CELLS AT DIFFERENT TEMPERATURES

Measurements of the durations of G_1, S, G_2, and M (mitosis) were carried out in 2 sets of experiments: (*a*) During the transient, that is, immediately after the cells were transferred from the routine growth temperature of 37°C to a new temperature. (*b*) In the steady-state, that is, after the cells had been growing for a period of 5 or more days at a given temperature. At the time at which the observations were begun Colcemid (0.25 μg/ml) as a mitotic block, and thymidine-H^3 (0.05 μc/ml, 0.36 C/mM, New England Nuclear Co.) as a specific DNA label were added to the HeLa cell suspension culture and samples taken at regular intervals. The samples were processed for radioautography as described in Section II,B,1. Unstained slide preparations were dipped in

Kodak NTB-3 nuclear track emulsion, developed 40 hours later and stained through the emulsion with toluidine blue (Prescott, 1964). The slides were scored for the number of labeled and unlabeled interphases and mitoses. The durations of G_1, S, G_2, M, and Tg were obtained from these data by the method of Puck and Steffen (1963) (see also Puck *et al.*, 1964). The determination of generation times by this method was more accurate than that obtained from cell counts using a hemocytometer. In the HeLa cell the Colcemid block at metaphase was absolute and no cell disintegration could be seen up to 60 hours. Unlike the HeLa cell, the mouse fibroblast cell (strain L) was blocked by Colcemid at metaphase for only 10 or 12 hours following which the cell reentered interphase without undergoing division and formed a multinucleate cell (Rao and Engelberg, unpublished data, 1964). In partially hepatectomized rats injected with colchicine, the regenerating liver cells started recovering about 12 hours after the injection of the drug by forming multinucleate cells (Brues and Jackson, 1937). Similarly a Colcemid block was also temporary in chinese hamster cells and led to the formation of multinucleate cells (Stubblefield, 1964). Do all rodent cells have this response to Colcemid? This question is under investigation.

A. Variation of Growth Rate during the Period of Transition

Whenever a physical system is suddenly shifted from one set of constant environmental conditions to another set of constant environmental conditions, there follows a period of transition during which the parameters descriptive of the system vary with time and asymptotically approach their steady-state values. In this and in the next section we will describe the variation with time of the duration of various phases of the mitotic cycle of the HeLa cell following a sudden shift from 37°C to a new temperature.

With a shift to a lower temperature there was an immediate decrease in growth rate, the greater the temperature drop the greater the decrease in growth rate. The growth rate then continued to decrease during the period of transition until it reached a steady-state value at the new temperature. With a temperature shift from 37°C to 34°C, for example, the growth rate dropped immediately from 0.046 to 0.031 $hour^{-1}$; with time the growth rate then approached the value 0.022 $hour^{-1}$ at 34°C. The duration of the period of transition varied with the magnitude of the temperature shift—the greater the shift the longer the period of transition. The presence of a similar transition period in the kinetics of reproduction of *Tetrahymena* was observed by Phelps (1946) and Prescott (1957).

We have detected a remarkable asymmetry in the pattern of proliferation at the very beginning of the transient period. When the temperature of a HeLa cell culture growing at 37°C was *lowered*, cell division continued, though at a lower rate—the growth curve did not manifest a lag period. When the temperature of a culture growing in the steady-state at a lower temperature, e.g., 34°C, was *raised* to 37°C, however, there followed a lag period of about 4 hours at the higher temperature during which no cell division took place. Thus cells in mitosis and G_2 at the lower temperature were not initially able to proceed through mitosis at the higher temperature. This phenomenon remains to be explained.

B. DURATIONS OF G_1, S, G_2, AND M IMMEDIATELY AFTER A CHANGE IN TEMPERATURE

Life cycle analyses were carried out on a culture of HeLa cells following a temperature shift from a steady-state temperature of 37°C to a new temperature. Immediately after the shift mitosis showed the greatest relative increase in duration of all the phases, followed by G_1, S, and G_2 in a decreasing order (Fig. 2A; Table I column "tr"). In other words, the immediate impact of the temperature shift was greatest on mitosis and G_1. An immediate and relatively large increase in the duration of G_1 following a temperature shift has been observed by Sisken (1963) on human amnion cells. Thus, at the beginning of the transient, G_1 shows a relatively large increase (Fig. 2A). With time S and G_2 catch up with G_1 (Fig. 2B). As with the generation time the durations of the individual phases of the mitotic cycle increased with time during the transient period until they approached a steady-state value. By contrast Sisken's (1963) studies on human amnion cells using time-lapse cinemicrography did not demonstrate any progressive increase in the duration of the mitotic phases with time at the suboptimal temperature beyond the initial change immediately following the temperature shift. The discrepancy between his results and ours may be due to the differences in cell types. A collaborative effort is being made by our laboratories to find the cause of this discrepancy.

C. STEADY-STATE DURATIONS OF G_1, S, G_2, AND M

Determinations of the durations of G_1, S, G_2, and M were carried out after cells were grown 5 or more days at a given temperature. Steady-state values are shown in comparison with those of the transient period in Table I. The pattern of response immediately following a sudden change in temperature (Fig. 2A) was different from that obtained in the steady-state (Fig. 2B). At low temperatures, there was an enormous increase in the duration of mitosis and of metaphase. In a culture at 33°C mitosis averaged 13 hours as compared to 0.9 hours in a 37°C culture.

TABLE I

DURATION OF G_1, S, G_2, AND M DURING THE TRANSIENT PERIOD AND IN THE STEADY-STATE

Phase of the mitotic cycle	33°C			34°C			36°C			37°C	38°C			40°C		
	tr[a]	ss[b]	tr/ss[c]	tr	ss	tr/ss	tr	ss	tr/ss	ss	tr	ss	tr/ss	tr	ss	tr/ss
G_1																
Hours	17.5	26.0	0.67	13.5	16.0	0.85	10.3	13.0	0.80	10.4	6.7	7.5	0.90	10.1	15.0	0.67
%[d]	48.6	35.4		42.0	35.8		44.2	50.5		48.0	36.4	39.1		42.2	44.5	
S																
Hours	12.0	22.4	0.54	12.8	14.8	0.86	8.4	7.4	1.12	7.0	7.6	7.6	1.00	9.3	11.2	0.83
%	33.4	30.3		39.6	33.0		35.8	28.6		32.0	41.3	39.5		38.6	33.2	
G_2																
Hours	3.9	12.2	0.32	4.3	10.5	0.41	3.7	3.9	0.95	3.5	3.3	3.3	1.00	3.2	5.0	0.64
%	10.8	16.6		13.2	23.4		15.7	15.1		16.0	17.9	17.2		13.3	14.9	
M																
Hours	2.6	13.0	0.20	1.7	3.5	0.49	1.1	1.5	0.72	0.9	0.8	0.8	1.00	1.4	2.5	0.56
%	7.2	17.7		5.2	7.8		4.3	5.8		4.0	4.4	4.2		5.9	7.4	
Tg																
Hours	36.0	73.6	0.49	32.3	44.8	0.73	23.5	25.8	0.91	21.8	18.4	19.2	0.96	24.0	33.7	0.72

[a] tr = Duration immediately following a shift from 37°C to the new temperature

[b] ss = Duration in the steady-state.

[c] tr/ss = The ratio tr/ss.

[d] % = Percentage of each phase of the generation time.

This is in line with findings of other workers that mitosis is the most temperature sensitive part of the division cycle (de Wildeman, 1891; Blakeslee and Cartledge, 1927; Barber and Callan, 1943; Makino and Nakahara, 1953; Chevremont-Comhaire and Chevremont, 1956; Savage and Evans, 1959; Evans and Savage, 1959; Sisken, 1963; Rao and Engelberg, 1965).

D. DISCUSSION

In interpreting these results it is well to consider what is meant by the G_1, S, and G_2 periods. Only the duration of S is an actual measure of the duration of a specific cellular process, namely DNA synthesis. The duration of G_1 is merely the time interval from the beginning of interphase to the beginning of DNA synthesis. When the duration of G_1 in-

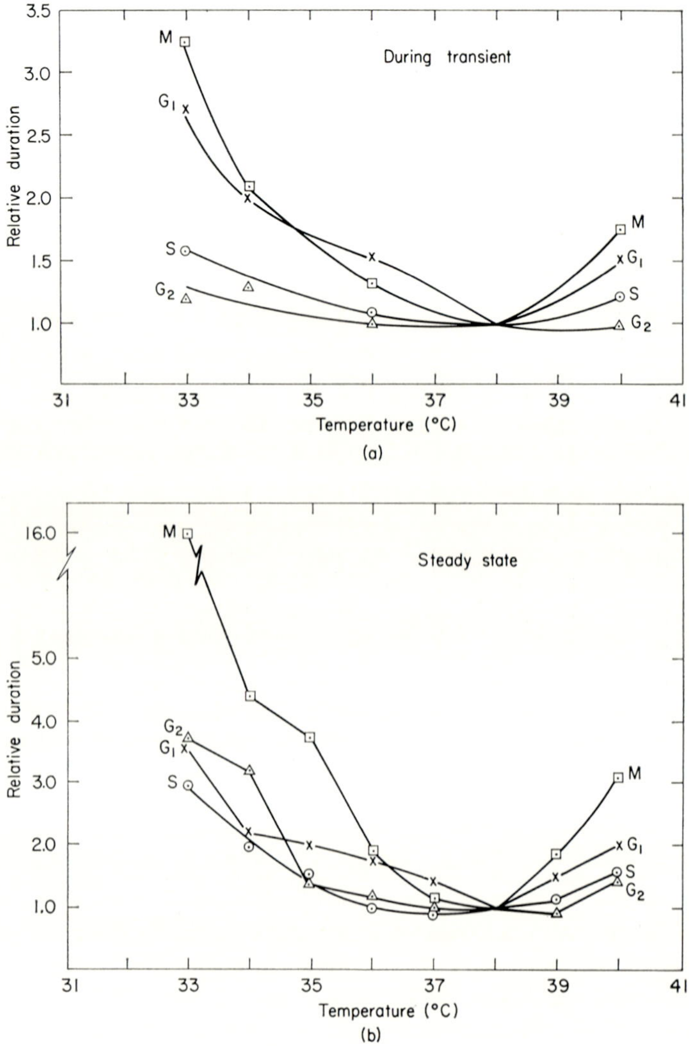

FIG. 2. (A) Relative (to the duration at 38°C) durations of G_1, S, G_2, and M immediately after a shift from steady-state growth at 37°C to a new temperature. (B) Relative (to the duration at 38°C) durations of G_1, S, G_2, and M in the steady-state.

creases this is only an indication that the initiation of DNA synthesis is delayed; it is not a general statement about other processes which take place in early interphase.

In general it can be said that as the temperature of steady-state growth in these cells was changed from 38°C, Tg, G_1, S, and G_2 were increased in duration. That is to say, the initiation of DNA synthesis was delayed, and the average rate of DNA synthesis decreased. When the temperature was lowered there was an immediate, relatively large increase in G_1 signifying a delay in the initiation of DNA synthesis. In the steady-state G_1, S, and G_2 showed comparable increases in duration. It is obvious that since all phases of the mitotic cycle increased their duration during the transient period that the generation time likewise increased during this period.

These findings may be interpreted as follows. The rate of a chemical reaction is a function (*a*) of the temperature, and (*b*) of the concentration of the reactants. Two different effects of temperature upon the life cycle of the cell may therefore be anticipated. Immediately after the temperature shift the concentration of reactants will not have appreciably changed. Hence, the major (direct) effect will be the direct effect of temperature upon chemical reaction rates. This would account for the drop in growth rate from 0.046 to 0.031 hr^{-1} when the temperature was shifted from 37°C to 34°C, for example. The second (indirect) effect becomes noticeable after the culture has been at the lower temperature for an appreciable time. The compositions of cells growing in the steady-state at 37°C and 34°C are undoubtedly different even when the cells are genetically the same. One would expect cell size and the number of cell constituents (enzymes, ribosomes, etc.) per cell to be different at the 2 temperatures. The second effect, then, is a consequence of changes in the concentration of chemical reactants resulting from the altered life history of the cell at the new temperature. These interpretations are testable and require further investigation.

IV. MITOTIC ACCUMULATION AT LOWER TEMPERATURES

In the range from 26°C through 31°C the cells could not be grown in the steady state; here, however, the cells manifested an interesting transient during which the rate of cell division approached zero while the mitotic index increased with time.

The results of a typical experiment, in which a HeLa cell culture was subjected to a temperature shift from 37°C to 29°C and maintained at the lower temperature, are presented in Fig. 3 (the curve marked "control"). Here the mitotic index remained constant for about 12 hours after the shift, then increased with time in a linear fashion. A maximum mitotic index of 0.36 was reached at about 60 hours followed by a decline in mitotic index. When Colcemid was added to the culture at the time of the temperature shift no lag was observed and the mitotic index increased immediately in a linear fashion (Fig. 3, curve: with Colcemid). The lag in the control signifies that during this period the number of cells entering mitosis equals those leaving. Subsequently, the mitotic

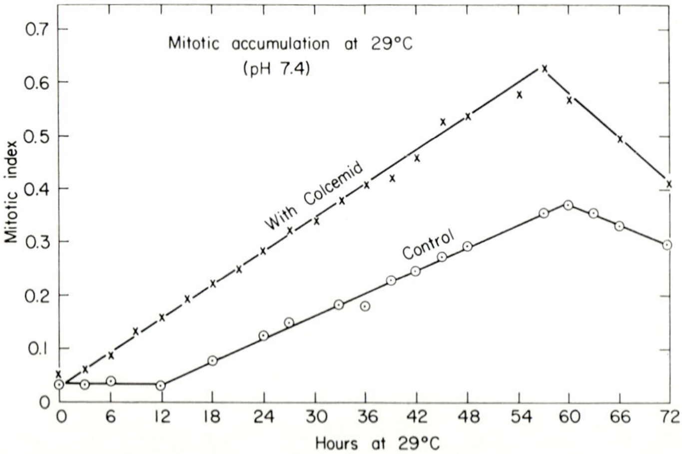

FIG. 3. Mitotic accumulation of HeLa cells after a shift in temperature at time zero from 37° to 29°C. Colcemid was not added to the "control" culture.

TABLE II

MITOTIC ACCUMULATION IN HELA CELLS FOLLOWING A TRANSFER FROM 37°C TO A LOWER TEMPERATURE

Temperature (°C)	Time at which mitotic index started increasing ("control" culture) (hours)	Mitotic index after 60 hours at new temperature ("control" culture)	Rate at which cells enter mitosis in %/hour (Colcemid culture)	Rate of mitotic accumulation relative to 37°C (Colcemid culture)
37° (control)	∞	0.03	4.60	1.00
31°	18.0	0.29	1.27	0.28
30°	10.5	0.30	1.13	0.25
29°	11.0	0.36	1.08	0.23
28°	12.0	0.31	0.90	0.19
27°	18.0	0.21	0.68	0.15
26°	24.0	0.11	0.42	0.09

index increased with time when the rate of entry into mitosis exceeded the rate of cell division. After the lag the lines for the "control" and "Colcemid-treated" cultures do not run parallel because some of the cells in the "control" culture were dividing at a progressively decreasing rate.

Data obtained from similar studies over the range from 26°–31°C are shown in Table II. In this temperature range the lag period varied from 10.5–24 hours. The maximum mitotic index (without Colcemid treatment) at each temperature (Fig. 4) was reached at about 60 hours at pH 7.4. At 29°C the greatest mitotic accumulation was observed. The

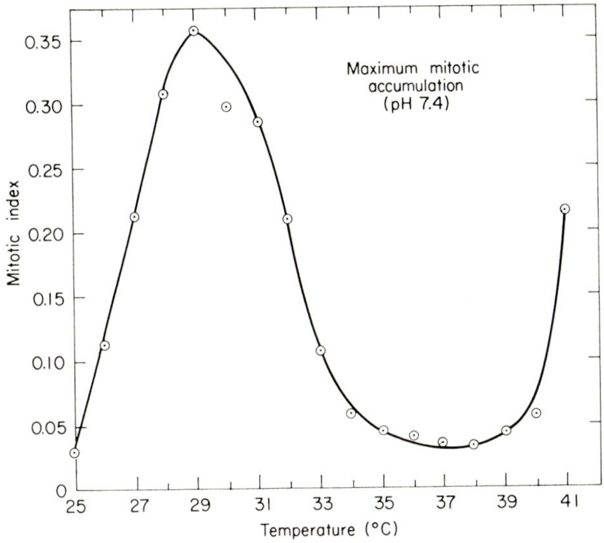

FIG. 4. Maximum mitotic accumulation at pH 7.4 at low temperatures (25°C through 32°C). The mitotic indices of HeLa cell cultures in steady-state growth are given for comparison (33°C through 40°C).

degree of mitotic accumulation was influenced by pH. A record high of 44 mitoses (almost all were in metaphase) per 100 cells was observed at 29°C when the pH was raised to 7.7.

Cells in metaphase at these subnormal temperatures had normal appearance, manifesting intact spindles and well defined and aligned chromosomes. This was in contrast to cells treated with Colcemid in which the spindles were disorganized and the chromosomes scattered. When cold-accumulated mitoses were lysed with 0.1 *M* citric acid (Sanford *et al.*, 1951) intact mitotic figures were obtained. Our study reveals the possibility of using subnormal temperatures as a means of obtaining large numbers of mitotic figures for use in analytical studies on the mitotic apparatus of mammalian cells. Mitoses accumulated in the cold, however, did not divide normally when returned to 37°C.

The difference in the rate of mitotic accumulation in experiments on HeLa cells with and without Colcemid treatment (Fig. 3) was used to estimate the increase in cell number in the "control" culture by use of the following equation:

$$\Delta n/n_0 = (m^{\text{colc}} - m)/(1 + m)$$

where Δn is the total number of cells dividing between time 0 and t; n_0 is the number of cells at time 0; m^{colc} is the mitotic index at time t with Colcemid and m is the mitotic index at time t in the control. Values

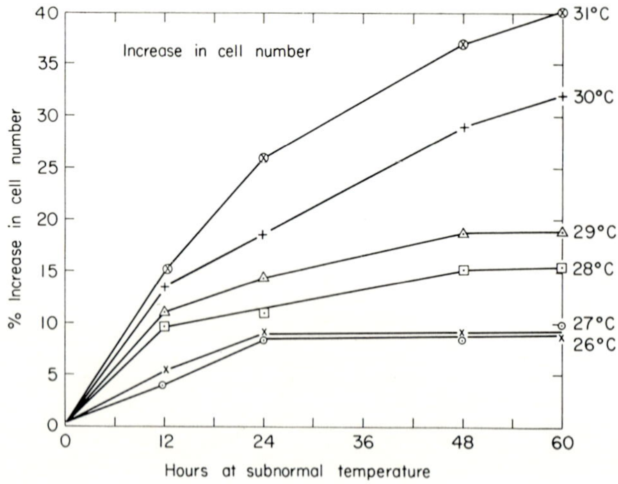

FIG. 5. Increase in cell number following a temperature shift from 37°C to a lower temperature.

calculated by means of the above equation were in good agreement with hemocytometer cell counts. The increase in cell number immediately following a shift from 37°C to temperatures between 26°C and 31°C is shown in Fig. 5. The initial rate of cell division increased with temperature. The time at which cell division appeared to cease likewise increased with temperature. In the limit of temperatures compatible with steady-state growth, the time at which cell division ceased became infinite. These results indicate that the initial accumulation of metaphase cells in the cold is not due to an absolute block of cell division. The accumulation is a consequence of the prolongation of mitosis. Mitosis being more sensitive to temperature than G_1, S, and G_2 (Fig. 2), the increase in the duration of mitosis was significantly greater than that of the other phases of the mitotic cycle. This resulted in more cells entering, but fewer completing mitosis; thus, the mitotic index increased. Mitotic accumulation in the newt, as a result of cold treatment, has previously been observed by Barber and Callan (1943).

Synchronized cells were used to study the nature of the mitotic accumulation. For this purpose cells were synchronized by the use of excess thymidine as described in the next section.

V. INDUCTION OF DIVISION SYNCHRONY BY EXCESS THYMIDINE

An elegant and efficient method of synchronizing tissue culture cells, discovered by Xeros (1962), is based on his findings that DNA synthesis

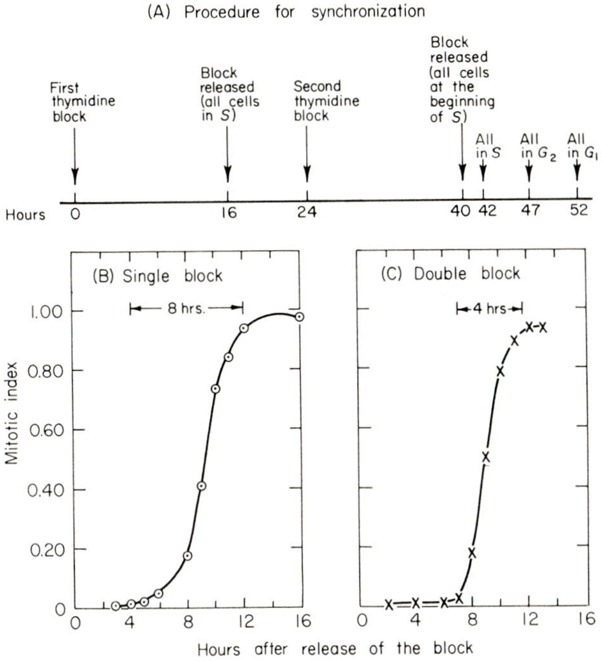

FIG. 6. (A) Time sequence for synchronizing HeLa cells by the excess thymidine double-block technique. (B) Rate of mitotic accumulation in Colcemid treated cultures following the release of the first thymidine block. The accumulation is a measure of the increase in cell number which would occur in the absence of Colcemid. (C) Same as B but following the release of second thymidine block.

is completely inhibited in the presence of an excess of thymidine in the culture medium. When the excess thymidine is removed the cells proceed with DNA synthesis from the point in the *S* period at which they had been halted (Puck, 1964). During the period of inhibition protein synthesis and RNA synthesis continues, and cells in G_1, G_2, and M proceed through the mitotic cycle until they reach the beginning of the *S* period.

Our HeLa cultures were synchronized by the use of excess thymidine as follows (Fig. 6A). After the thymidine was added to a level of 2.5 mM we waited until most of the cells were either in *S* or at the end of G_1. The cells at the end of G_1 were those which had been in G_2, M, and G_1 at the time the thymidine was added. When the thymidine was then removed the cells were released and a burst of mitosis was observed about 8–9 hours later (Fig. 6B). The degree of synchrony was increased by use of a second thymidine block applied after the cells released from the first block had passed through S (Puck, 1964; Bootsma *et al.*, 1964; Petersen and Anderson, 1964). The cells passed around the cycle and accumulated at the end of G_1 where DNA synthesis would normally be initiated. When almost all the cells had accumulated at the beginning of S the thymidine block was released. In the case of our HeLa cells about 90% of the cells then divided between 7–11 hours after the release (Fig. 6C). The synchrony was enhanced by the double-block procedure because, in contrast to the single-block procedure where cells were halted by the block both at the end of G_1 and throughout S, the double-block yielded cells which accumulated only at the end of G_1.

Although 98% of the cells were initially accumulated within a region no larger than 5% of the total life cycle, i.e., at the beginning of S (Puck, 1964), by the time the cells reached mitosis they were no longer in perfect synchrony for 2 reasons. First, intrinsic variability (Sisken, 1963; Engelberg, 1964) led to differences in the rates with which different cells traversed the mitotic cycle. Second, the released cells were not a homogeneous population. Cells which arrived early at the block were in a different state at the time of release than cells which arrived last.

In unsynchronized cells at 37°C, the duration of S plus G_2 was 10.5 hours. In Fig. 6C, however, we see that cells arrive in mitosis already 7 hours after their release (see also Bootsma *et al.*, 1964). It is presumed that cells waiting at the end of G_1 proceeded with the biosynthesis of constituents other than DNA and were able, therefore, to traverse the portion of the mitotic cycle from the initiation of DNA synthesis through mitosis in less time than normal cells.

VI. EFFECT OF TEMPERATURE SHIFT ON THE DURATION OF MITOSIS IN A SYNCHRONIZED CULTURE

Cells grown at 37°C, synchronized by the excess thymidine technique and collected in G_1, S or G_2 were shifted to 29°C with the addition of Colcemid. Rough estimates for the duration of mitosis of cells which were in G_1, S, and G_2 at the time of the temperature shift were obtained from the data shown in Fig. 7. The estimate was taken to be the time elapsed between the appearance of the first mitoses to the point where the curves for the Colcemid treated and control curves diverged. The divergence occurred when the cells in the control culture started dividing. The durations were on the order of 3, 6, and 28 hours for cells which were, respectively, in G_2, S, and G_1 at the time of the temperature shift.

These results indicate that the duration of mitosis (M) depends *both* on the temperature *and* on the time which the cell spends at the new temperature before entering mitosis. The mean duration of mitosis in

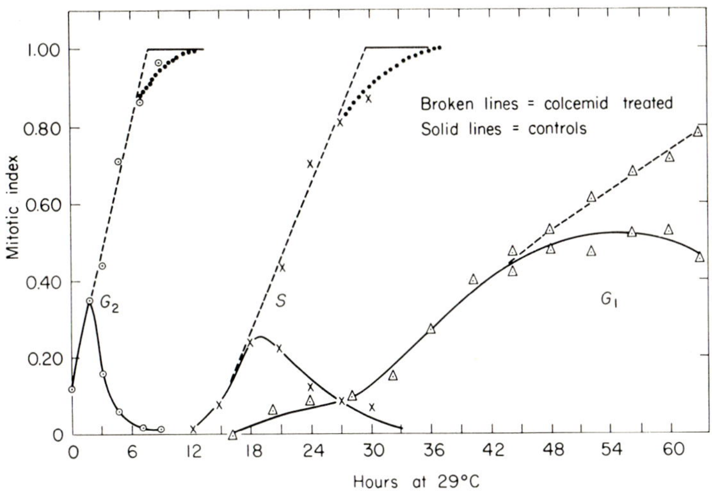

FIG. 7. Effect of temperature shift (from 37°C to 29°C) on cells in G_1, S, and G_2 at the time of the shift. The shift occurred at 0 hours. The difference in the mitotic index curves of the Colcemid treated cells and the controls is a measure of the number of cells which divided by a given time: The greater the difference, the greater the number of cells which divided.

HeLa cells growing in the steady-state at 37°C was 0.9 hours. When the temperature of such cells was dropped to 29°C the duration of mitosis immediately increased from 0.9 hours to 3 hours. This increase could reflect the decrease in reaction rates at the lower temperature, since, presumably, the number and kinds of cell constituents were the same at the 2 temperatures. The cells which were in G_1 at the time of the temperature shift, on the other hand, spent about 28 hours at 29°C before completing division. The 28 hour duration for mitosis of these cells is attributable in part to lower reaction rates at 29°C, though for the most part this large increase in duration must be credited to changes in cellular composition. The effect reflects the previous history of the cell's life cycle at the lower temperature.

HeLa cells in G_2 and S when transferred to 29°C subsequently divided normally. G_1 cells, however, manifested an abnormal anaphase at the time of mitosis. In this anaphase the chromosomes traversed the distance between the metaphase plate and the pole individually or in small groups. A second characteristic of the anaphase was that the sister chromatids of each chromosome remained together: The chromosomes arriving at the poles consisted of sister chromatid pairs of the kind usually seen at metaphase. A further study of this phenomenon is in progress.

VII. SYNCHRONIZATION BY TEMPERATURE SHOCKS

A. Brief Temperature Shocks

Exposure of HeLa cells to a 4°C temperature for 60 minutes yielded little synchrony. When these cells were returned to 37°C there was a lag of 2 hours which was followed by exponential growth. A 45-minute heat shock at 43°C, on the other hand, resulted in a lag of 21 hours after which 50% of the cells divided within 3 hours.

B. Effect of Long Exposure to Cold on the Degree of Synchrony

As an alternative to brief cold shocks cells were maintained at temperatures between 26° and 30°C for durations of 6, 12, 24, and 48 hours and then transferred to 37°C. When cells grown at 29°C for 6 hours were returned to 37°C all the cells divided between 6 and 17 hours after the end of the cold treatment. The degree of synchrony obtained by other temperature-time patterns was much less pronounced. The longer the cells were maintained at 29°C the lower the number of cells that divided when they were brought back to 37°C (Table III).

The pattern of mitotic accumulation during the treatment at 29°C and at 37°C following the treatment is of interest. At 29°C there was no change in mitotic index for about 12 hours followed by a quasilinear increase in mitotic index with time. The mitotic index reached a peak within 4–6 hours after returning to 37°C (Fig. 8). Cells maintained at 29°C and then returned to 37°C exhibited a partial synchrony. This phenomenon may be explained as follows. At 37°C the cells have a G_2 period of 4 hours. At 29°C all the phases of the cell's life cycle (G_1, S, G_2, and M) are slowed down considerably. As discussed above G_2 is least affected during immediately after the temperature shift. The greater sensitivity of S than G_2 to cold could account for the partial synchrony: When the cells were brought back to 37°C DNA synthesis was completed rapidly and the cells accumulated in S at 29°C, taking about 4 hours to complete G_2, reached mitosis in greater numbers.

TABLE III
Effect of Cold Treatment on Cell Division

Duration at 29°C (hours)	% of increase in cell number at 37°C following cold treatment	Period during which increase in cell number occurred following cold treatment (hours)
6.0	100	6.0–17 (11.0 hours)
12.0	70	3.5–13 (9.5 hours)
24.0	35	7.0–12 (5.0 hours)
48.0	0	—

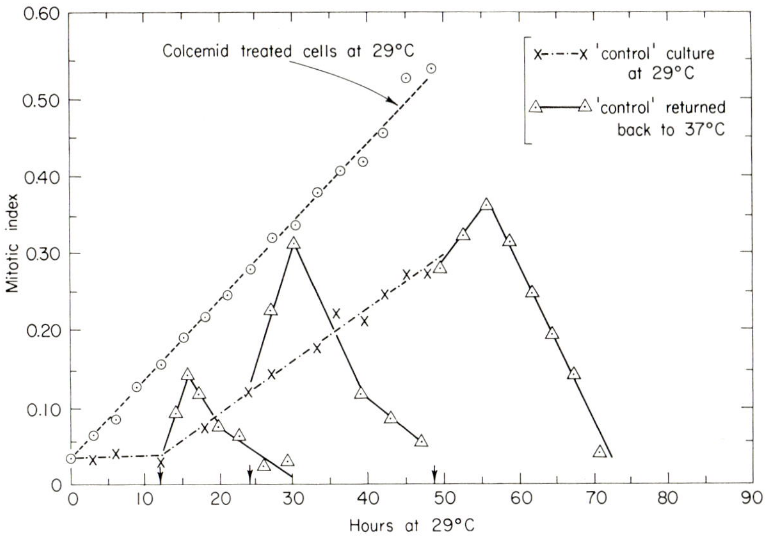

FIG. 8. The temperature of a HeLa cell culture at 37°C was shifted to 29°C. Twelve, 24, or 48 hours later it was shifted back to 37°C. The increase in mitotic index in Colcemid treated cells is compared with the control (no Colcemid added) at 29°C. Note the synchronous burst of cell division (seen as a sharp rise in mitotic index) when cultures are returned to 37°C.

Efforts to synchronize suspension cultures of HeLa cells by subjecting them to brief cold or heat shocks, prolonged cold treatment, gradual cooling as well as various other temperature-time patterns yielded disappointing degrees of synchrony. This is no surprise in the light of the general experience of other workers with mammalian cell lines. Newton and Wildy (1959) for example, obtained good synchrony by temperature shock only when cells in early logarithmic phase of growth were selected for subculture. These cells were in the stationary phase before subculturing (Newton, 1964). From this it is clear that in mammalian cells the nutritional state of the culture is also involved in the induction of division synchrony by temperature shock. The cell line is also of importance (Toolan, 1954; Rapp, 1960). Similarly, Newton and Wildy induced synchrony in monolayer cultures of wild-type HeLa and strain L cells but failed to do so with the S-3 substrain of HeLa cells (Newton, 1964).

VIII. SUMMARY

(*a*) Steady-state exponential growth of HeLa cells was observed only within the range of 33°–40°C. In this range *growth rate* (reciprocal of generation time) manifested a quasilinear relationship with temperature on either side of the growth-rate maximum. Outside this temperature range cells, sooner or later, failed to divide. Growth rate varied inversely with mitotic index, the growth rate being a maximum when the mitotic index was a minimum.

When a HeLa cell culture was transferred from a 37°C to a lower temperature in this range the growth rate dropped immediately to a new value which further decreased with time until a steady-state was reached. However, following a sudden increase in temperature a lag period during which no cell division occurred preceded exponential growth at the higher temperature.

(*b*) Life cycle analyses of HeLa cells in suspension culture at temperatures between 33° and 40°C were carried out both immediately after a shift from 37°C to a new temperature and after a steady-state was reached at the new temperature. Mitosis and particularly metaphase was the most temperature sensitive part of the cell's life cycle. A sudden change in temperature resulted in an immediate, relatively great increase in the duration of G_1 compared to S or G_2. Each of the phases, including mitosis, passed through a transient period following a temperature shift.

(*c*) There was a significant accumulation of cells in metaphase at temperatures ranging from 26° through 32°C. A record high of 44 mitoses per 100 cells was observed after 60 hours at 29°C and pH 7.7.

(*d*) Cells synchronized by the excess thymidine technique were collected in G_1, S or G_2 and transferred from 37° to 29°C. The time required for the first mitosis at 29°C was on the order of 3, 6, and 28 hours for cells which were, respectively, in G_2, S, and G_1 at the time of the temperature shift. Cells in G_2 and S when transferred to 29°C subsequently divided normally. G_1 cells, however, manifested abnormal anaphases involving an asynchronous movement of chromosomes at the time of division.

(*e*) Attempts to synchronize our HeLa cells by brief temperature shocks or prolonged exposures to subnormal temperatures yielded disappointing degrees of synchrony.

Acknowledgments

We thank Mrs. Nora Mitchell, Mrs. Mattie Mitchell, and Mr. Christopher Landrum for their valuable technical assistance. We are indebted to Dr. Jesse E. Sisken for making available to us unpublished data on the effects of temperature on human amnion cells.

This work was supported by Public Health Service Research Grant NIH-CA-06835 from the National Cancer Institute and by Research Grant E-303 from the American Cancer Society.

REFERENCES

Barber, H. N., and Callan, H. G. (1943). *Proc. Roy. Soc.* **B131,** 258.

Blakeslee, A. F., and Cartledge, J. L. (1927). *Mem. Hort. Soc. N.Y.* **3,** 305.

Bootsma, D., Budke, L., and Vos, O. (1964). *Exptl. Cell Res.* **33,** 301.

Brown, R. (1951). *J. Exptl. Botany* **2,** 96.

Brues, A., and Jackson, E. (1937). *Am. J. Cancer* **30,** 504.

Chevremont-Comhaire, S., and Chevremont, M. (1956). *Compt. Rend. Soc. Biol.* **150,** 1046.

de Wildeman, E. (1891). *Ann. Soc. Belge Microscop.* **15,** 5.

Drost-Hansen, W. (1965). *Proc. Symp. Forms of Water in Biol. Systems, Ann. N.Y. Acad. Sci.* **125,** 471.

Engelberg, J. (1964). *Exptl. Cell Res.* **36,** 647.

Ephrussi, B. (1926). *Protoplasma* **1,** 105.

Evans, H. J., and Savage, J. R. K. (1959). *Exptl. Cell Res.* **18,** 51.

Gey, G. O., Bang, F. B., and Gey, M. K. (1954). *Ann. N.Y. Acad. Sci.* **58,** 976.

Glasstone, S. (1946). "Textbook of Physical Chemistry," 2nd ed., p. 1088. Van Nostrand, Princeton, New Jersey.

Hinshelwood, C. N. (1952). "The Chemical Kinetics of the Bacterial Cell." Oxford Univ. Press (Clarendon), London and New York.

Howard, A., and Pelc, S. R. (1953). *Heredity* Suppl. 6, 261.

Johnson, F. H., and Lewin, I. (1946). *J. Cellular Comp. Physiol.* **28,** 47.

Lewis, M. R. (1933). *Arch. Exptl. Zellforsch.* **14,** 464.

Makino, S., and Nakahara, H. (1953). *Cytologia (Tokyo)* **18,** 128.

Mitchell, W. H. (1929). *J. Exptl. Zool.* **54,** 383.

Miura, T., and Utakoji, T. (1961). *Exptl. Cell Res.* **23,** 452.

Mucibabic, S. (1956). *J. Exptl. Biol.* **33,** 627.

Newton, A. A. (1964). *In* "Synchrony in Cell Division and Growth" (E. Zeuthen, ed.), pp. 441–466. Academic Press, New York.

Newton, A. A., and Wildy, P. (1959). *Exptl. Cell Res.* **16,** 624.

Petersen, D. F., and Anderson, E. C. (1964). *Nature* **203,** 642.

Phelps, A. (1946). *J. Exptl. Zool.* **102,** 277.

Prescott, D. M. (1957). *J. Protozool.* **4,** 252.

Prescott, D. M. (1964). *Methods Cell Physiol.* **1,** 365–370.

Puck, T. T. (1964). *Science* **144,** 566.

Puck, T. T., and Steffen, J. (1963). *Biophys. J.* **3,** 379.

Puck, T. T., Sanders, P., and Petersen, D. F. (1964). *Biophys. J.* **4,** 441.

Rao, P. N., and Engelberg, J. (1965). *Science* **148,** 1092.

Rapp, F. (1960). *Virology* **10,** 86.

Sanford, K. K., Earle, W. R., Evans, V. J., Waltz, H. K., and Shannon, J. E. (1951). *J. Natl. Cancer Inst.* **11,** 773.

Savage, J. R. K., and Evans, H. J. (1959). *Exptl. Cell Res.* **16,** 364.

Scherbaum, O. H. (1957). *Exptl. Cell Res.* **13,** 11.

Sisken, J. E. (1963). *In* "Cinemicrography in Cell Biology" (G. Rose, ed.), pp. 143–168. Academic Press, New York.

Smith, J. D. (1940). *Biol. Bull.* **79,** 379.

Spear, F. G. (1929). *Arch. Exptl. Zellforsch.* **7,** 484.

Stubblefield, E. (1964). *In* "Cytogenetics of Cells in Culture" (R. J. C. Harris, ed.), pp. 223–248. Academic Press, New York.

Thompson, T. E. (1964). *In* "Cellular Membranes in Development" (R. Locke, ed.), pp. 83–96. Academic Press, New York.

Thormar, H. (1962). *Exptl. Cell Res.* **28,** 269.

Toolan, H. W. (1954). *Cancer Res.* **14,** 660.

Wildy, P., and Newton, A. A. (1958). *Biochem. J.* **68,** 14P.

Woodruff, L. L., and Baitsell, G. A. (1911). *Am. J. Physiol.* **29,** 147.

Xeros, N. (1962). *Nature* **194,** 682.

Zeuthen, E. (1958). *Advan. Biol. Med. Phys.* **6,** 37.

Zeuthen, E. (1964). *In* "Synchrony in Cell Division and Growth" (E. Zeuthen, ed.), pp. 99–158. Wiley (Interscience), New York.

CHAPTER 18

Some Macromolecular Characteristics of Synchronized HeLa Cells

Elliott Robbins and Matthew Scharff

I. INTRODUCTION

The life cycle of the animal cell is characterized by a well-coordinated synthesis of various macromolecules leading to an increase in cell mass, replication of genetic material, and ultimately a sharing of macromolecules as a result of mitosis (Prescott, 1964). Until recently, most of the information concerning these events was obtained by observing single cells in randomly dividing cultures (Prescott, 1964; Taylor, 1960; Prescott and Bender, 1962; Seed, 1963; Konrad, 1963; Feinendegen and Bond, 1963; Killander and Zetterberg, 1965), but for detailed biochemical investigations large numbers of synchronized cells are obviously required. In general, attempts at synchronization have employed either temperature shock (Newton and Wildy, 1959), or reversible chemical inhibition of DNA synthesis, the subsequent release of this inhibition resulting in a burst of mitotic activity several hours later (Rueckert and Mueller, 1960; Xeros, 1962; Littlefield, 1962). A more satisfactory approach, which avoids the ambiguities introduced by chemical and physical agents, is synchronization by selection of cells at a particular stage in the life cycle. Terasima and Tolmach, taking advantage of the relatively tenuous attachment of dividing cells to glass, detached mitotic cells by gently passing a stream of medium over monolayer cultures (Terasima and Tolmach, 1963). Although the number of cells collected in this fashion was small, they were able to investigate the chronology of macromolecular synthesis in HeLa cells. Their technique, with some modifications, has been used in combination with cold shock by Sinclair with the goal of increasing the yield of mitotic cells (W. K. Sinclair and Morton, 1963). This investigator also used gradient centrifugation to select cells based on their size (R. Sinclair and Bishop, 1965). Both of these approaches detract significantly from the salient advantage of the Terasima and Tolmach technique—the avoidance of extraneous physical and chemical manipulation. However, these and other modifications may be useful in certain instances as will be pointed out below. Robbins and Marcus have shown that when the growth medium contains no calcium other than that present in the serum supplement, the attachment of mitotic cells is so tenuous that they may be preferentially detached by mild agitation of the overlying medium (Robbins and Marcus, 1964c). This modification makes it possible to use mild shearing forces for rapidly collecting populations of virtually pure, untreated, dividing cells (Fig. 1), in numbers large enough for any biochemical analysis.

In the sections which follow we will describe the technique of synchronization by selective detachment as it is currently used in our laboratories and also give some of our preliminary results using cells synchronized by this technique. We will not attempt to review the literature systematically since this has been done recently (Prescott, 1964; Newton, 1964; Burns, 1962; Symposium on Macromolecular Aspects of the Cell Cycle, 1963), nor will we refer to the many pertinent investigations on *in vivo* systems, such as regenerating liver.

II. METHOD

A. Cell Synchronization

Monodisperse suspensions obtained from continuous suspension cultures or by enzymatic detachment of monolayer cultures are planted on

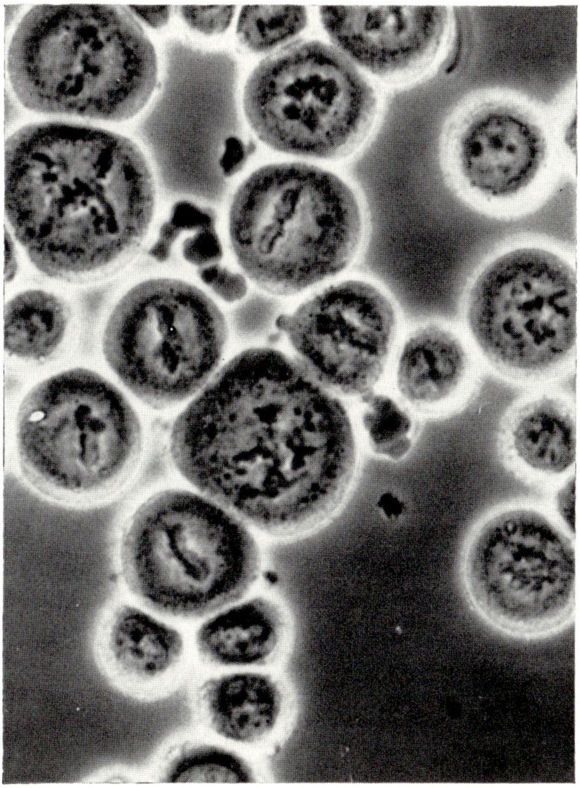

FIG. 1. These are synchronized mitotic HeLa cells obtained by selective detachment.

Blake (or other large surface area bottles) 1–2 days prior to synchronization. The medium presently used for HeLa cells is that described by Eagle for suspension cultures (Eagle, 1959) supplemented with 3.5% each of calf and fetal calf serum and nonessential amino acids. While this medium contains only the calcium in the serum, all of the cells we have employed will attach to glass and propagate in it with a normal generation time. The optimum number of cells per bottle varies with different cell types, but in all cases it is desirable to have only partially confluent cultures at the time of synchronization.

Before collection of synchronized cells, the bottles are vigorously shaken so as to detach any loosely adherent cells or debris, and the spent medium decanted. After rinsing the monolayer with an additional 10 ml of medium, 30 ml of fresh medium is added to each bottle to cover the cell sheet. Thirty minutes to 3 hours later mitotic cells are obtained in the desired population densities by serially transferring the collecting medium through approximately 10 bottles as follows: after decanting the first bottle and adding 20 ml of fresh medium, the bottle is gently rocked from side to side 20 times; the medium containing the harvested mitotic cells is then transferred to a second previously decanted Blake bottle which is similarly harvested, and so on. In this manner, 2 workers can collect the desired cells from 40 Blake bottles in 10–15 minutes and the potential trauma of concentration by centrifugation is avoided. HeLa S-3 cells thus treated yield about 3×10^5 synchronized cells per Blake bottle. If medium is returned to the harvested bottles they can be used again within 30–60 minutes.

While we stress the avoidance of centrifugation as a precautionary measure, it should be mentioned that at present we have no evidence that gentle packing has any adverse effect on subsequent events in HeLa. However, special care must be taken with certain other aspects of the procedure to obtain consistent results. All operations must be carried out at 37°C; medium and glassware must be prewarmed; and pH must be maintained at physiological levels. A common pitfall in our experience is the use of overcrowded monolayers which leads to both interphase cell detachment at the time of collection, and degeneration of a significant fraction of cells in the course of the subsequent experiment. Finally, synchronized cells should be grown at optimum population densities. While HeLa S-3 cells will often initiate DNA synthesis at the predicted time even when diluted to cell concentrations which normally lead to suboptimal growth in random cultures, other cell types more frequently show a delay in the onset of this synthesis. In general, the use of overcrowded stock cultures for harvesting or low population densities after harvest, cause an artifactual plateau in the rate of RNA and protein synthesis per cell during late G_1 (the period prior to DNA synthesis), S (period of DNA synthesis), and G_2 (period between DNA synthesis and mitosis) associated with morphological abnormalities, i.e., cell aggregation and unusually large numbers of cytoplasmic granules. While the mechanism of this granulation is unknown it is associated with the accumulation of membrane bound organelles which themselves contain aggregates and stacks of electron dense membranes (Fig. 2). These frequently resemble myelin-like figures and are most probably a manifestation of lysosomal transformation (residual bodies) similar to that described by numerous investigators under a variety of conditions (Novikoff *et al.*, 1964; de Duve, 1964; Robbins *et al.*, 1964).

In the cell lines with which we have been concerned, addition of various nutritional supplements including insulin, lactalbumin hydrolysate, extra serum, "spent" medium, etc., at the time of collection, have not influenced either the subsequent degree of synchrony or the synthetic activity of the cells.

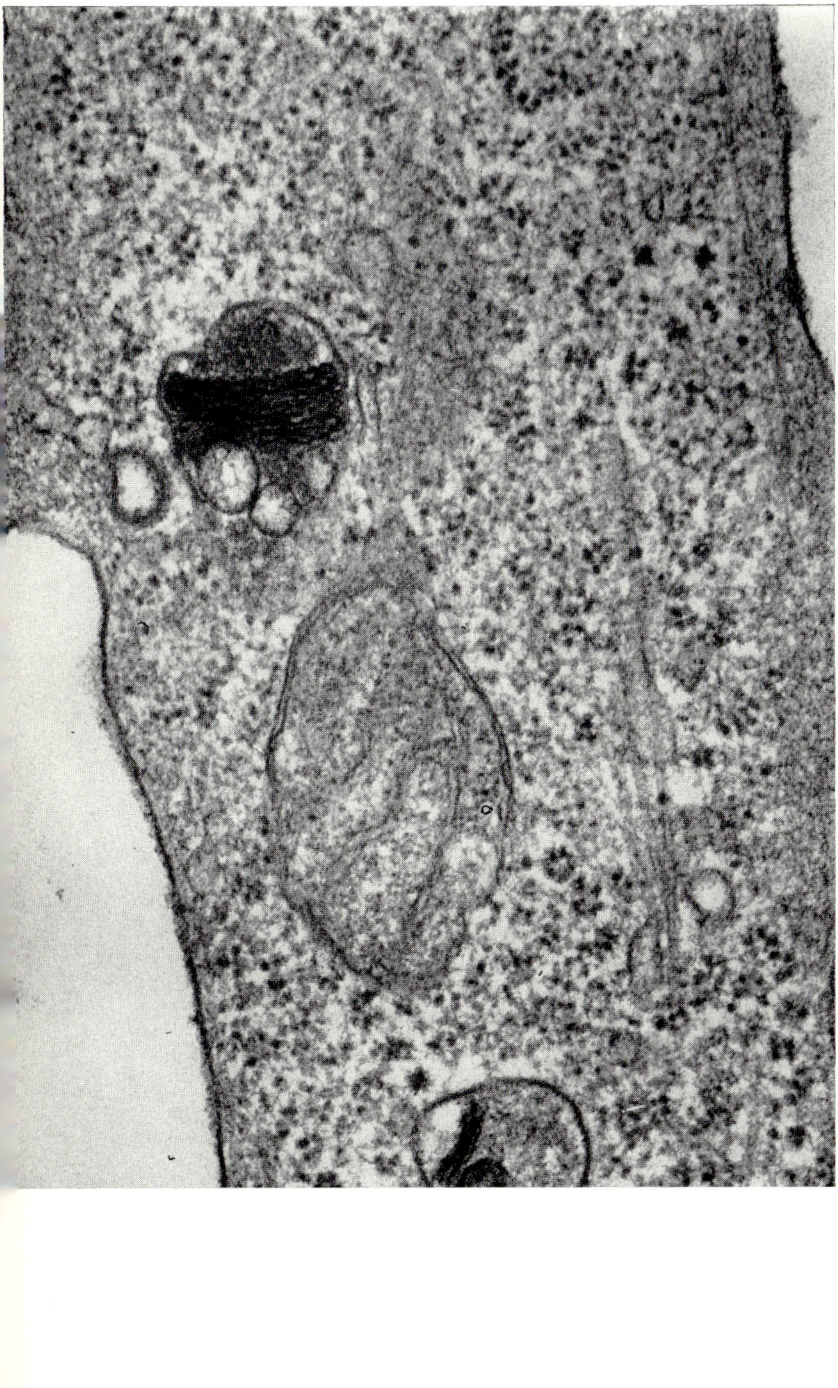

Fig. 2. Electron micrograph of HeLa S-3 cell in late *S*. Cell was derived from overcrowded monolayer. Note membrane bound bodies containing electron-dense membrane aggregates. × 88,000.

We have used this synchronization procedure with several cell types. Preparations consisting of 95–100% of cells in various phases of mitosis are consistently obtained from cultures of Chinese hamster lung, HeLa, and KB cells. Earle's L cells which in general have a tenuous attachment to glass have given unsatisfactory preparations, with 40% interphase contamination even upon gentle agitation. Cultures of some human diploid cells and of chick embryo heart have been successfully synchronized, but the yields are smaller than those from established cell lines.

The yield of synchronized cells can be increased by combining chemical inhibitors with the preferential selection technique. However, the drawbacks of the additional variable should be recognized. Two m*M* thymidine arrests cells in the *S* phase of the growth cycle and is immediately reversible without any observable residual effects (Xeros, 1962; Bootsma *et al.*, 1964). When monolayers are exposed for 15 hours, decanted, and fresh thymidine-free medium added for 7–10 hours before collecting, the yield of synchronized cells is increased tenfold. We have confirmed the finding that these cells complete mitosis in the usual time and clone with a normal plating efficiency (Bootsma *et al.*, 1964).

Selective detachment, combined with the use of spindle inhibitors, such as colchicine, is also a useful tool for biochemical characterization of cells in metaphase. Before, cells had to be exposed to the drug for three quarters of a growth cycle or more to obtain populations predominantly in mitosis. In such circumstances it was difficult to separate the effects of the drug from those attributable to metaphase per se, since adequate controls were not available. By treating the monolayer cultures with colchicine and then preferentially detaching the mitotic cells, it is possible to acquire pure populations of metaphase-arrested cells which have been exposed to the drug for only a few hours. Perhaps more important, the metaphase-arrested cells may be compared with interphase cells incubated with the drug for the same length of time. Such colchicine-treated interphase cells may be obtained by trypsinization of the population remaining on the glass substrate after detachment of metaphase cells or simply by exposing suspension cultures to colchicine. In the latter case only small numbers of metaphase cells will be present in the interphase control if the time of drug exposure is brief.

The comparison of colchicine-treated metaphase and interphase populations removes some of the ambiguities involved in the use of the drug. It is, however, important to remember that while treated interphase cells are in a dynamic state, progressing toward mitosis in a fashion not unlike their untreated counterparts, metaphase-arrested cells do not complete division and thus are not strictly comparable to normal mitotic cells. In addition, although no significant biochemical differences have been found between normal and colchicine-treated interphase cells, colchicine does induce characteristic morphological changes in interphase cells (Robbins and Gonatas, 1964a). In spite of these shortcomings, the system has proved useful in certain studies.

B. Isotopic Techniques

Since isotopic techniques are particularly applicable to the study of macromolecular synthesis in synchronized cells, we will briefly describe the 2 methods of labeling with radioactive precursors which we have found most expedient. In the first, the complete harvest of synchronized cells is placed in suspension culture at a concentration of 1–4 $\times$ 10^5 cells/ml and stirred by means of a suspended magnetic bar (McLimans *et al.*, 1957). As the cells progress through the growth cycle, measured aliquots are removed and added to test tubes which contain appropriate radioactive precursors such as thymidine-C^{14}, amino acids-C^{14}, uridine-C^{14} or $P^{32}O_4$. The tubes are placed in an atmosphere of 5% CO_2 and the cells are incubated in suspension for a prescribed period. The incorporation of radioactive precursor is stopped by cooling in ice, following by addition of cold balanced salt solution. The cells are washed, pelleted, and treated with 5% TCA. The acid-precipitable material is collected on Millipore filters which are then glued to planchets and counted in a low background counter. Replication error with this technique is less than 10%.

The second procedure is an adaptation of one already described (Baltimore and Franklin, 1962) and is useful for cells which cannot be maintained in suspension culture, for experiments in which rapid environmental alterations are to be carried out without centrifugation, and for histochemical, morphological, or autoradiographic studies. Equal volumes of synchronized cells are placed on individual ⅝ inch coverslips supported on small corks in petri dishes and maintained in a CO_2 incubator. Each coverslip is a replicate of all others and once the cells have attached to the glass substrate, environmental manipulation or isotopic labeling may be conveniently carried out simply by removing existing medium with suction and replacing it with the test solution. Isotope incorporation is stopped by rinsing the coverslip in cold balanced salt solution followed by fixation in 5% TCA, and washing in distilled water. The coverslip is then dried and glued to a planchet for counting in the usual way.

III. MACROMOLECULAR SYNTHESIS

RNA, DNA, and Protein

The availability of cells synchronized by selection provides a unique tool with which to relate macromolecular synthesis to the various stages of the life cycle. Terasima and Tolmach (1963), using this technique and autoradiography, found G_1 in their HeLa cells to be about 8 hours. Following initiation of DNA synthesis there was a rapid increase in both the number of cells making DNA and the rate of synthesis per cell. Fifteen hours after synchronization over 90% of the population was incorporating thymidine and a maximum rate of DNA synthesis was recorded. In contrast, RNA synthesis per cell increased progressively throughout interphase. We have confirmed and extended these findings using large populations of HeLa S-3 cells synchronized by selective detachment, and examining the incorporation of radioactivity into acid-precipitable material (Scharff and Robbins, 1965a). In Fig. 3 the rates of total RNA, protein, and DNA synthesis in relation to cell division are illustrated. The initial population consisted of 55% metaphase cells, 43% anaphase or telophase cells, and 2% early interphase cells. Within 1 hour after collection, more than 90% of the cells were in G_1, nuclear membranes and nucleoli were readily recognized, and in many instances the daughter cells had separated. By 15 hours significant numbers of mitotic cells began to appear and by 17 hours the mitotic index was 24%. Since mitosis lasts 30–45 minutes, an appreciable mitotic index persisting for several hours indicates the loss of synchrony which characteristically occurs during one division cycle. This loss of synchrony makes

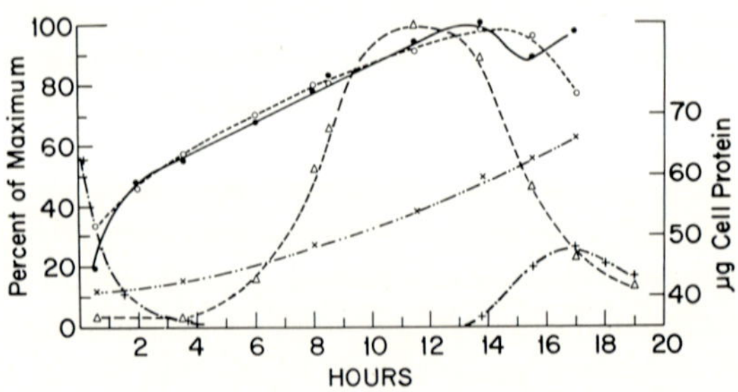

FIG. 3. The rate of synthesis of RNA, DNA, and protein in synchronized HeLa S-3 cells is determined by a procedure using radioactive isotopes. Cells were synchronized as described and maintained in suspension culture. At the indicated time 10^5 cells in 0.5 ml aliquots were incubated in duplicate for 30 minutes with either 0.2 μc uridine-C^{14} (30 mc/mmole), 0.5 μc thymidine-C^{14} (30 mc/mmole), or 1 μc of a mixture of threonine-C^{14}, valine-C^{14}, and phenylalanine-C^{14} (100–300 mc/mmole). The cells were then washed and analyzed for acid-precipitable radioactivity and the results plotted as the percent of maximum incorporation per milliliter of cells. Cell protein was determined by the method of Lowry. The data is not corrected for the 8–10% of the acid-precipitable uridine-C^{14} counts which are found in DNA during *S*. Metaphase index, ×–·–·×; thymidine-C^{14} incorporation, △– – –△; uridine-C^{14} incorporation, ●——●; amino acid-C^{14} incorporation, ○····○; cell protein, +–··–+.

it impossible to obtain a uniform population of cells which have completed DNA synthesis but have not yet entered mitosis (G_2). Cell populations which we will call "G_2" are therefore in fact contaminated with significant numbers of cells which are still in mitosis or synthesizing DNA, as well as a few which have begun a second interphase.

DNA replication occurred between 6 and 16 hours with a peak incorporation of labeled thymidine at 11 hours. The onset of DNA synthesis at 6 hours is significantly earlier than that reported for the HeLa strain used by Terasima and Tolmach (1963). The length of the cell cycle is in fact highly variable, depending upon environmental conditions as well as the cell strain being examined. We have found, for instance, that one strain of Chinese Hamster lung exhibits virtually no detectable G_1, thymidine incorporation starting within ½ hour after the completion of mitosis and reaching a maximum as early as 3½–4 hours later.

The rate of protein synthesis per milliliter of cells increased throughout interphase and, consistent with the findings of others (Prescott and Bender, 1962; Konrad, 1963), was depressed during cell division. Since there is virtually no RNA synthesis during metaphase (Taylor, 1960; Prescott and Bender, 1962; Konrad, 1963; Feinendegen and Bond, 1963), the initial rate of uridine-C^{14} incorporation was very low, but rose rapidly as the cells progressed into G_1, and increased steadily throughout interphase. At 15 hours a moderate drop in rate occurred, concurrent with a rise in motitic index to 22%. As more cells completed the second mitosis and entered interphase, the rate of RNA synthesis again rose.

In a number of experiments we have been unable to detect either a decrease or increase in RNA synthesis during DNA replication. Similar findings have been reported by Terasima and Tolmach (1963), using synchronized cells and autoradiography, by Seed (1963), employing autoradiography of single cells the age of which had been determined by time lapse photography, and by Killander and Zetterberg (1965), using time lapse photography and microspectrophotometry. Although Reiter and Littlefield (1964) have reported a decrease in the rate of nuclear RNA synthesis during early S in cells which have been synchronized with FUDR, it is difficult to rule out nonspecific effects of the drug unrelated to the stage of the cell cycle.

We have also related the synthesis of the various species of RNA which can be separated on sucrose gradients to the different stages of the division cycle. In randomly dividing, logarithmically growing, HeLa cells it has been shown by a number of investigators that most of the RNA synthesized is ribosomal RNA (Scherrer *et al.*, 1963; Perry, 1962; Girard *et al.*, 1964; Girard *et al.*, 1965; Rake and Graham, 1964). Using short pulses of uridine-C^{14}, these workers found that the first species of

RNA to be made has a value of approximately 45 S and is found only in the nucleus. After 20–30 minutes exposure to uridine-C^{14}, small amounts of newly synthesized 28 S and 16 S RNA are detectable and with longer labeling periods begin to appear in the cytoplasm in association with the 60 S and 40 S ribosomal subunits, respectively. The experiments depicted in Figs. 4 and 5 indicate that the same sequence of events occurs throughout all of interphase. In the experiment depicted in Fig. 4 cells were exposed to uridine-C^{14} for 20 minutes during G_1, S, and G_2; then the RNA was extracted and analyzed on sucrose gradients (Scharff and Robbins, 1965a). Most of the newly synthesized RNA was found in the 45 S area of the gradient while small amounts of 28 S and 16 S RNA were detectable. For convenient comparison of the relative quantities of the different RNA species the curves have been "normalized" so that there is the same number of total counts in each gradient. Thus, while there is less total RNA synthesis per cell during G_1 than during S, this is not

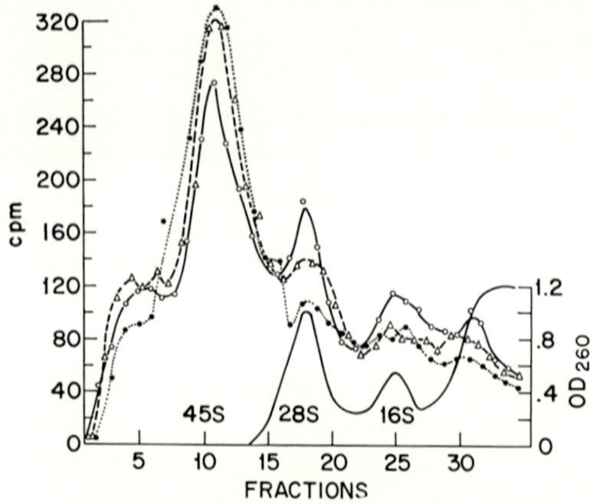

FIG. 4. The synthesis of high molecular weight RNA in G_1, S, and G_2 is shown. Separate preparations of synchronized HeLa cells were collected at 2, 9, and 15 hours prior to labeling so as to allow simultaneous processing of G_1, S, and G_2 populations. Four ml of cells at 2×10^5 cells/ml were incubated with 5 μc of uridine-C^{14} for 20 minutes. Following washing in Earle's salts solution the cells were resuspended in a hypotonic buffer (10^{-2} M NaCl, 10^{-2} tris-HCl, pH 7.4; 1.5×10^{-3} M $MgCl_2$), mixed with 1 ml of unlabeled HeLa cell cytoplasm from 2×10^7 cells and extracted with 1% sodium dodecyl sulfate (SDS). (Similar results were obtained by extraction with hot phenol and SDS.) The samples were then layered on a linear 15–30% sucrose gradient containing 0.1 M NaCl, 5×10^{-3} tris-HCl, pH 7.4; and 0.5% SDS and centrifuged at 20°C for 14 hours at 19,000 rpm in a "SW 25" swinging bucket rotor (Spinco). OD_{260} was monitored with a recording spectrophotometer (Gilford) and acid-precipitable radioactivity determined by making the samples 12% with TCA, collecting the samples on Millipore filters, and counting them in a low background gas flow counter. G_1, ○——○; S, △- - -△; G_2, ●····●.

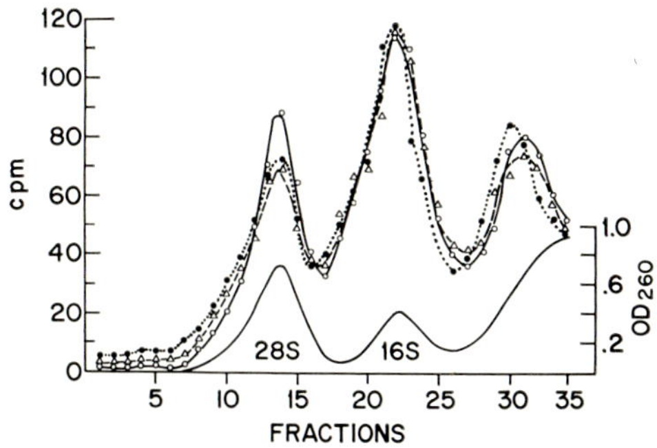

FIG. 5. The synthesis of cytoplasmic RNA: Cells from G_1, S, and G_2 were prepared as described in Fig. 4. Two ml of cells was exposed to 2.5 μc uridine-C^{14} for 2 hours, washed, and mixed with 2×10^7 unlabeled HeLa cells in hypotonic buffer. Cytoplasm was prepared by Dounce homogenization and after extraction with 1% SDS analyzed on sucrose gradients centrifuged at 23,000 rpm for 16 hours. G_1, ○——○; S, △---△; G_2, ●····●.

apparent from the curves as drawn. When cells at different stages of the cycle were exposed to uridine-C^{14} for 2 hours and the cytoplasm alone examined (Fig. 5), 28 S and 16 S ribosomal RNA as well as 4 S transfer RNA were found in almost the same relative amounts at all times. As with randomly growing cultures, newly synthesized 16 S RNA were present in larger amounts than 28 S RNA due to its more rapid entry into the cytoplasm. Similar studies have been carried out on cells at several other time points in G_1 and S with the same results, indicating that, as with total RNA synthesis, the synthesis of 45 S, 28 S, and 16 S RNA occurs at essentially the same relative rates throughout interphase.

Since Prescott and Kimball (1961), Taylor (1960), and others have shown that RNA is not synthesized on replicating DNA in various cell systems, it is important to consider whether a decrease in RNA synthesis should be detectable during S in synchronized HeLa cells. Between 11½ and 12 hours, the population of synchronized cells described in Fig. 2 synthesized about 5% of its total DNA. If this 5% represents a random sample of the cistrons responsible for RNA synthesis, then a parallel 5% decrease in RNA synthesis would be the maximum that could be expected, and this is within our experimental error. Recent evidence has implicated about 400 cistrons in the synthesis of 28 S ribosomal RNA (McConkey and Hopkins, 1964) and has suggested that these are in close spatial relationship (McConkey and Hopkins, 1964; Ritossa and Spiegelman, 1965). It is therefore possible, although not supported by any direct evidence, that the DNA involved in ribosomal RNA synthesis is not randomly replicated but may all be synthesized over a short period of time in any one cell. Since ribosomal RNA constitutes a major fraction of the total cellular RNA and is composed of easily identifiable components, a decrease in uridine incorporation into 28 S and 16 S RNA might be expected in a perfectly synchronized system. The loss of synchrony in our cells however, precludes such an observation. This question might be resolved, at least with regard to total RNA, by autoradiographic examination of synchronized cells exposed to label for very short periods. Under these conditions one might find some cells in *S* incorporating very little label.

Of great interest is the lack of any detectable increase in RNA synthesis during *S*. Since the amount of cell DNA doubles during this period, a greatly increased rate of RNA synthesis might be expected. That such an increase does not occur implies that either all of the newly made DNA is repressed or that large portions of the previously derepressed DNA are no longer able to prime for RNA synthesis.

Approaches similar to those described for ribosomal RNA, applied to the synthesis of various other macromolecules such as histones, phospholipids, gangliosides, polymerases, etc., should provide important information on the sequence of events which ultimately leads to mitosis.

IV. MORPHOLOGICAL STUDIES

While the extensive morphological changes which occur during mitosis have received well deserved attention in both the light microscopic and electron microscopic literature (Porter and Machado, 1960; Harris, 1961; Robbins and Gonatas, 1964b; Hughes, 1952; Schrader, 1953) there are no data indicating that significant organelle transitions characterize the stages of interphase. This lack of information is partly due to the fact that there has been no satisfactory way to distinguish between interphase cells in specific phases of the growth cycle. In a preliminary report on synchronized cells Blondell and Tolmach (1965) were unable to see any ultrastructural alterations of the nucleus which were specific for particular stages in the cycle. In comparable investigations our initial impression is that, aside from the changes in total mass, there is very little visible transformation in the Golgi apparatus, lysosomes, endoplasmic reticulum, mitochondria, or intranuclear structures in HeLa cells as they pass from one mitosis to the next.

V. BIOCHEMICAL CHARACTERISTICS OF METAPHASE-ARRESTED CELLS

A. Protein Synthesis

As already described, it is possible to collect large populations of metaphase-arrested cells following short exposures to colchicine, for compar-

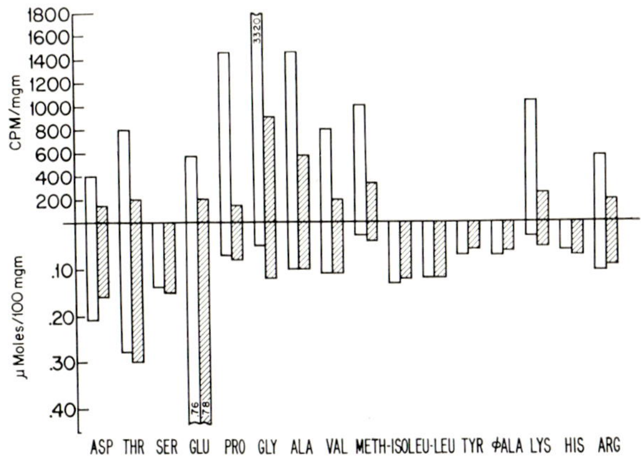

FIG. 6. The top half of the figure shows the incorporation of several individual amino acids into acid precipitable material. Aliquots of 2×10^5 cell/ml were exposed to 0.2 mc/ml of various amino acids for 30 minutes, washed, precipitated with 5% TCA, and plated on Millipore filters for determination of isotope incorporation. Uptake of all amino acids into acid-precipitable material was depressed during metaphase. The bottom half of the figure shows the relative size of individual amino acid pools in metaphase and interphase. One-hundred mg of cells were precipitated with 3 ml of 1% picric acid. The supernatant fluid was passed through Dowex 2-X8 ion exchange resin to remove picrate and the amino acids were eluted with 0.02 *N* HCl. The effluent, after concentration by lyophilization, was chromatographed on a Phoenix Amino Acid Analyzer. Cross-hatched bar—metaphase; clear bar—interphase.

ison with interphase cells which have been exposed to the drug for the same time. Several investigators, using autoradiography, have demonstrated that the rate of protein synthesis is depressed in metaphase (Prescott and Bender, 1962; Konrad, 1963). We have used the colchicine, metaphase-colchicine, interphase system to extend these observations (Robbins and Scharff, 1965). Figure 6 (upper part) shows that during mitosis the rates of incorporation of several C^{14}-labeled amino acids were one-third to one-fifth that seen in interphase cells, and that proline was even lower. The size of the individual amino acid pools was measured on the supernatant of picric acid precipitated cells according to the method of Stein and Moore (1954) (Fig. 6). Although the absolute size of each pool varied somewhat in individual experiments there was no significant difference between the pool sizes in interphase and metaphase cells with the possible and unexplained exception of glycine.

Two other possible mechanisms for the lower rate of protein synthesis during mitosis are: (*a*) depressed transport of amino acids across the

TABLE I

UPTAKE OF C^{14}-LABELED AMINO ACIDS IN COLCHICINE-TREATED INTERPHASE AND METAPHASE-ARRESTED CELLS[a]

Fraction	Colchicine-interphase			Metaphase-arrested		
	2 min	6 min	12 min	2 min	6 min	12 min
Unwashed packed cells	350	461	746	267	395	616
Extracellular fluid	22	22	22	24	24	24
TCA-precipitable counts	88	128	242	24	52	108
Acid-soluble pool	240	311	482	219	319	484

[a] Metaphase-arrested cells were obtained by selective detachment after 2 hours in colchicine. After colchicine-treated interphase cells were exposed to the drug in spinner culture for 2 hours, 6% were in mitosis. Four × 10^7 cells were washed once and resuspended in 35 ml of medium containing $\frac{1}{20}$ the normal amount of amino acids. At 0 time C^{14}-amino acids (algal hydrolyzate) was added to a final concentration of 0.01 μc/ml. Two 5 ml aliquots were withdrawn from each sample and immediately centrifuged at 6500 g in thrombocytocrits (A. H. Thomas). All supernatant fluid was removed by suction and the part of the tubes above the packed cells were thoroughly rinsed with running water. The cells were then either resuspended in 50% ethanol and the entire sample plated on ridged planchets for counting or they were first resuspended in 5% TCA, washed twice and finally dispersed in 50% ethanol for plating. A second and third sample was taken at 6 and 12 minutes. Since all samples contained the same amount of material no correction was necessary for self adsorption, but it should be noted that the results are relative. Using C^{14}-inulin, the trapped extracellular fluid of the unwashed sample was determined to be 20% of the total packed cell volume and the radioactivity in an equivalent volume of medium was subtracted from the total counts. Counts in the acid-soluble pool are recorded as the difference between the counts in the unwashed sample (corrected for the trapped extracellular fluid) and those in the TCA precipitable material.

plasma membrane and (*b*) altered functional capacity of the sites of nascent protein synthesis, namely the polyribosomes (Warner *et al.*, 1962; Gierer, 1963; Wettstein *et al.*, 1963; Penman *et al.*, 1963; Marks *et al.*, 1962). Since there is no shrinkage in pool size during metaphase, transport of amino acids into the cell probably functions normally. Direct evidence for this has been obtained by exposing amino acid-depleted metaphase-arrested or interphase cells to labeled amino acids. Both accumulate acid-soluble material at exactly the same rate (Table I).

There were, however, striking differences in the ribosomal organization and activity of cells in metaphase and interphase. When colchicine-treated metaphase and interphase cells were exposed to C^{14}-labeled amino acids for $1\frac{1}{2}$ minutes, disrupted by Dounce homogenization, and analyzed on sucrose gradients for 260 mμ absorbing material and acid-precipitable radioactivity (Scharff *et al.*, 1963), curves such as those shown in Fig. 7 were obtained. The ultraviolet-absorbing material with S

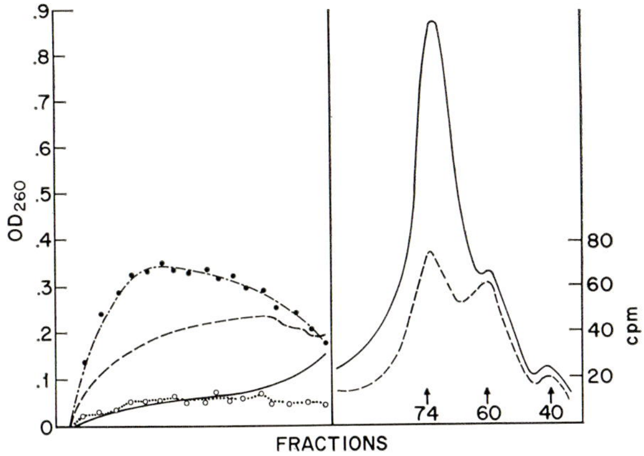

FIG. 7. Polyribosomes and polyribosome associated protein synthesis in colchicine metaphase and colchicine interphase cells: Monolayers of HeLa S-3 cells were treated with 0.1 μg of colchicine/ml of media for 3 hours and then shaken so as to harvest metaphase cells. These cells and a random population of logarithmically growing cells which had been treated with colchicine for the same amount of time were adjusted to the same mass of cells/ml of media, allowed to incubate with 10 μc of C^{14}-amino acids/10^7 cells (algal hydrolyzate) for 1½ minutes, and then processed as described in a previous publication (Scharff *et al.*, 1963). Half of each sample was centrifuged for 130 minutes at 24,000 rpm (left panel) and half for 8 hours at 25,000 rpm (right panel) in 15-30% linear sucrose gradients. Ultraviolet absorbancy at 260 mμ and acid-precipitable radioactivity were determined. Interphase OD_{260}----; metaphase OD_{260}——; metaphase radioactivity ○····○, interphase radioactivity ●---●.

values between 100 and 350 in the left panel of Fig. 7 consists of aggregates of single ribosomes connected by RNA (i.e., polyribosomes) (Warner *et al.*, 1962; Gierer, 1963; Wettstein *et al.*, 1963; Penman *et al.*, 1963; Marks *et al.*, 1962); while single ribosomes (74 S) and ribosomal subunits (60 S and 40 S) appear in the right panel. In the metaphase cells polyribosomes have been degraded to smaller aggregates and to single ribosomes as compared to interphase controls (Robbins and Scharff, 1965). The loss of ultraviolet absorbing material from the polyribosome area is completely accounted for by the increase in single ribosomes (74 S) with no change in the amounts of ribosomal subunits (40 S and 60 S). Polyribosome associated protein synthesis is similarly decreased in the metaphase as compared to the interphase cells.

These findings are confirmed by electron micrographs of intact cells. While cells in interphase (Fig. 8) display typical helical, spiral, and amorphous aggregates of ribosomes scattered throughout the cytoplasm, these are greatly diminished in number during metaphase (Fig. 9). The cytoplasm instead is almost homogenously punctuated with single ribo-

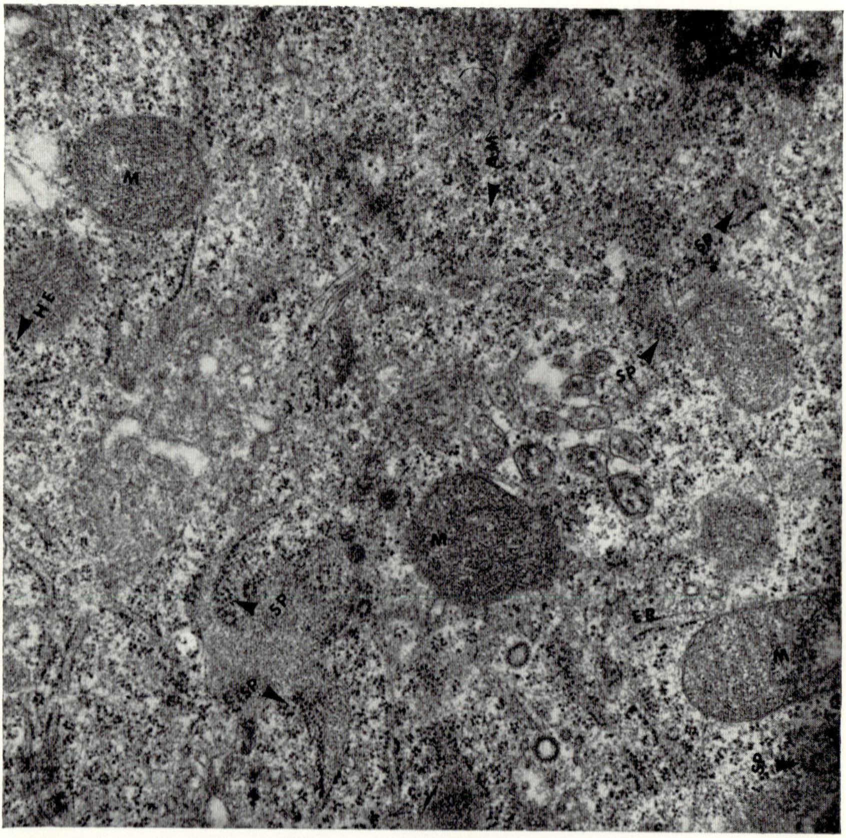

FIG. 8. An electron micrograph showing HeLa S-3 cell in late G_2. Cytoplasmic ribosomal aggregates of varying helical, spiral and amorphous patterns (arrows) are scattered between mitochondria and other organelles. Spiral forms are characteristically embedded in a more dense cytoplasmic matrix than the other aggregates. Cells fixed in 5% glutaraldehyde, postfixed with 2% OsO_4; stained with lead hydroxide and uranyl acetate. ER, endoplasmic reticulum; M, mitochondria; N, nucleus; SP, spiral aggregated; AM, amorphous aggregated; HE, helical aggregates. × 27,000, original × 42,000.

somes. The occasional polysome which is seen is not localized in any particular part of the cell although we have noted their consistent absence in the pericentriolar region.

While the cause of polyribosome breakdown is not known, preliminary data suggest that mRNA remains intact implying that only the *association* of ribosomes and mRNA has been disrupted. This would be consistent with our finding that polyribosomes break down when *untreated*

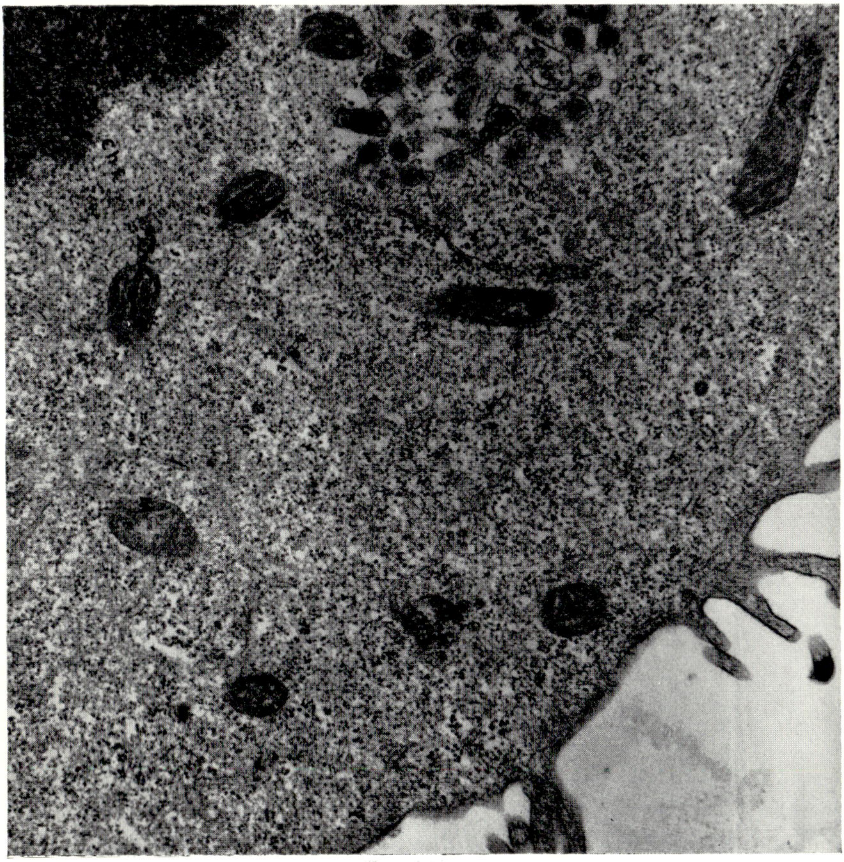

FIG. 9. An electron micrograph showing HeLa S-3 cell in metaphase. Condensed chromosomal masses are visible in upper left hand corner. Recognizable patterns of ribosomal aggregates are sparse in striking contrast to those seen in the previous figure. Chr, chromosomes; M, mitochondria. × 27,000, original × 42,000.

cells enter metaphase, and have reformed by the time they reenter interphase 30–45 minutes later (Scharff and Robbins, 1965b).

B. PHOSPHOLIPIDS IN METAPHASE-ARRESTED AND COLCHICINE-TREATED INTERPHASE CELLS

Mitosis in HeLa cells is accompanied by extensive membrane transformations. Along with the breakdown of the nuclear membrane, we have described the apparent disappearance of the Golgi apparatus (in HeLa cells) and profound morphological changes in the lysosomes (Rob-

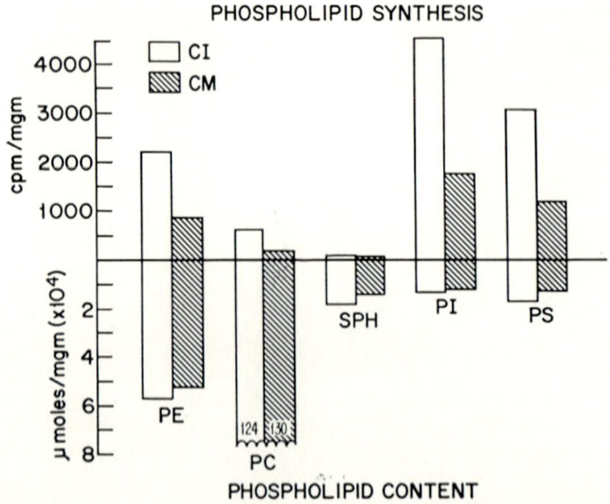

FIG. 10. The top half of the figure shows the incorporation of $P^{32}O_4$ into individual phospholipids. Four-hundred μc of $P^{32}O_4$ were added to 5 ml of HeLa cell suspension containing 5×10^5 cells/ml of PO_4 free medium. After 30 minutes the cells were packed by centrifugation and extracted with chloroform, methanol 2 to 1. The extract was separated into 2 phases by addition of ⅕ volume of water, the lower phase recovered and evaporated to dryness under a continuous stream of nitrogen. The residue was dissolved in chloroform, spotted on thin layer chromatographic plates (Silica gel H), and run at room temperature in the solvent system of Muldner. The spots were developed in iodine vapor, individually scraped off the plates into ridged counting planchets, covered with Parafilm, and counted in a low background gas flow counter. The lower half of the figure shows phospholipid content in colchicine-treated interphase and metaphase-arrested cells. Three-hundred mg of cells were packed by centrifugation and phospholipids were separated on thin layer chromatographic plates as above. The individual spots were scraped off and PO_4 content per spot determined on the scrapings. Striped bar, metaphase; clear bar, interphase; PE, phosphatidyl ethanolamine; PC, phosphatidyl choline; SPH, sphingomyelin; PI, phosphatidyl inositol; PS, phosphatidyl serine.

bins and Gonatas, 1964b). These transitions of phospholipid-containing moieties, however, are not accompanied by any parallel changes in the levels of either total cellular phospholipid or of individual fractions prepared according to Muldner *et al.* (1962) and separated by thin layer chromatography (Fig. 10 bottom portion). The top portion of Fig. 10 shows that as with labeled amino acids, the level of incorporation of $P^{32}O_4$ into individual phospholipids is reduced during metaphase. The reduction in isotopic incorporation ranged between 37% and 47% for the different fractions. No fraction increased its specific activity in metaphase-arrest and, in fact, the decrease in the rate of synthesis is surprisingly similar for all fractions.

It is noteworthy that phosphatidyl inositol has a rate of incorporation 29 times that of phosphatidyl choline in both groups of cells.

Phosphatidyl inositol which accounts for only about 4% of the total phospholipid has 46% of the radioactivity, while phosphatidyl choline which makes up about 50% of the total content accounts for 4% of the counts.

C. Energy Metabolism

All of our data, as well as that of others (Mazia, 1961), indicate that mitosis is a stage of decreased metabolic activity. One would assume that this decrease would be reflected by diminished utilization of oxygen as well as diminished lactic acid production. The classic studies of Zeuthen (1955) on single *Tetrahymena* did in fact show a cyclic drop in respiration, concomitant with division, but this decrease was very small and its existence has been disputed by Scholander *et al.* (1958). Recent work by Gershenson *et al.* (1965) on HeLa cells synchronized with aminopterin gives new support to the original work of Zeuthen (1955). As part of our overall investigation of the mitotic process we have used standard techniques to measure oxygen utilization (Rikmenspoel, 1965) lactic acid production (Barker and Summerson, 1941) and ATP concentration (McElroy *et al.*, 1953) in colchicine-treated metaphase and interphase cells. It is clear from Table II that in col-

TABLE II

Relation of ATP, Oxygen Uptake, and Lactic Acid Production in Colchicine-Treated Interphase and Metaphase-Arrested Cells[a]

	Colchicine-interphase	Metaphase-arrested
ATP (μg/100 mg)	1.24	1.43
O_2 uptake (μl/100 mg/hr)	7.7	4.6
Lactic acid production (μg/100 mg/hr)	3.8	2.3

[a] ATP was measured using the firefly luminescence technique (McElroy *et al.*, 1953). Fifty mg of cells were packed in thrombocytocrit tubes at 6500 *g* for 1 minute at 4°C. The supernatant fluid was quickly removed with suction and the packed cells rapidly extruded into 1 ml of boiling water. Following centrifugation, 0.1 ml of supernatant fluid was added to spectrophotometer cuvettes containing the buffer system and enzyme specified by McElroy *et al.* (1953). Readings of luminescence were taken at exactly 10 seconds following addition of the sample and compared to standards.

Oxygen uptake was measured with the Clark oxygen electrode. Ten mg of cells/ml of medium were added to the measuring cuvette and the oxygen tension (in microvolts) was measured as a function of time.

Lactic acid production was measured by resuspending cells in fresh medium at a population density of 5 mg of cells/ml and taking timed aliquots of suspension for analysis. The cells were removed by centrifugation and the method of Barker and Summerson (1941) was applied to the supernatant medium. Results are the average of 3 experiments.

chicine-treated mitotic cells both oxygen utilization and lactic acid production follow the expected pattern, dropping to about 60% of control levels. ATP concentrations are the same in metaphase and interphase possibly reflecting both decreased formation and decreased breakdown, a pattern which would be consistent with the lowered levels of macromolecular synthesis during mitosis. Plesner has reported detailed studies of nucleotide triphosphate levels in heat synchronized *Tetrahymena* (Plesner, 1964). Although we cannot extrapolate this system to the mammalian cell, it is interesting to note that there is a sharp rise, in both ATP and GTP, 10–20 minutes prior to division with an equally sharp drop beginning a few minutes before cytoplasmic fission. It is unfortunate that the significant loss of synchrony which occurs in our system by G_2 prevents an analysis of the crucial events occurring at this time. Studies such as Plesner's, as well as others, might well be expected to provide useful information on the actual initiation of the mitotic process.

VI. CONCLUSION

We have attempted to establish the usefulness of synchronization by selective detachment for investigating macromolecular characteristics of mammalian cells at different stages of the life cycle. We have extended results obtained from studies of single cells by characterizing and quantitating individual molecules and subcellular structures. With the general approach described it should be possible to elucidate the factors responsible for initiation of DNA synthesis and mitosis in the not too distant future.

ACKNOWLEDGMENTS

Supported by grants from the NIH (GM 11558, GM 12182, AI 05231, AI 104153), NSF (GB 984) and The American Cancer Society (E-379). Both authors are recipients of Career Development Awards from the NIH.

REFERENCES

Baltimore, D., and Franklin, R. M. (1962). *Proc. Natl. Acad. Sci. U.S.* **48,** 1383.

Barker, S. B., and Summerson, W. H. (1941). *J. Biol. Chem.* **138,** 535.

Bloch, D. P. and Godman, G. C. (1955). *J. Biophys. Biochem. Cytol.* **1,** 17.

Blondell, J., and Tolmach, L. J. (1965). *Exptl. Cell Res.* **37,** 497.

Bootsma, D., Budke, L., and Vos, O. (1964). *Exptl. Cell Res.* **33,** 301.

Burns, V. (1962). *Progr. Biophys. Biophys. Chem.* **12,** 1.

de Duve, C. (1964). *Federation Proc.* **23,** 1045.

Eagle, H. (1959). *Science* **130,** 432.

Feinendegen, L. E., and Bond, V. P. (1963). *Exptl. Cell Res.* **30,** 393.

Gershenson, L. E., Strasser, F. F., and Rounds, D. E. (1965). *Life Sciences* **4,** 927.

Gierer, A. (1963). *J. Mol. Biol.* **6,** 148.

Girard, M., Penman, S., and Darnell, J. E. (1964). *Proc. Natl. Acad. Sci. U.S.* **51,** 205.

Girard, M., Latham, H., Penman, S., and Darnell, J. E. (1965). *J. Mol. Biol.* **11,** 187.

Harris, P. (1961). *J. Biophys. Biochem. Cytol.* **11,** 419.
Hughes, A. W. (1952). "The Mitotic Cycle," p. 145. Academic Press, New York.
Killander, D., and Zetterberg, A. (1965). *Exptl. Cell Res.* **38,** 272.
Konrad, C. G. (1963). *J. Cell Biol.* **19,** 267.
Littlefield, J. W. (1962). *Exptl. Cell Res.* **26,** 318.
McConkey, E. H., and Hopkins, J. W. (1964). *Proc. Natl. Acad. Sci. U.S.* **51,** 1197.
McElroy, W. D., Hastings, J. W., Coulombre, J., and Sonnenfeld, V. (1953). *Arch. Biochem. Biophys.* **46,** 399.
McLimans, W. F., Davis, E. V., Glover, F. L., and Rake, G. W. (1957). *J. Immunol.* **79,** 428.
Maizel, J. V. (1966). *Science* (in press).
Marks, P., Burka, E., and Schlessinger, D. (1962). *Proc. Natl. Acad. Sci. U.S.* **48,** 2163.
Mazia, D. (1961). *In* "The Cell" (J. Brachet and A. E. Mirsky, eds.), Vol. 3, pp. 77–412. Academic Press, New York.
Muldner, H. G., Wherret, J. R., Cumings, J. N. (1962). *J. Neurochem.* **9,** 607.
Newton, A. A. (1964). *In* "Synchrony in Cell Division and Growth" (E. Zeuthen, ed.), pp. 441–466. Wiley (Interscience), New York.
Newton, A. A., and Wildy, P. (1959). *Exptl. Cell Res.* **16,** 624.
Novikoff, A. B., Essner, E., Quintana, N. (1964). *Federation Proc.* **23,** 1010.
Penman, S., Scherrer, K., Becker, Y., and Darnell, J. E. (1963). *Proc. Natl. Acad. Sci. U.S.* **49,** 654.
Perry, R. P. (1962). *Proc. Natl. Acad. Sci. U.S.* **48,** 2179.
Plesner, P. (1964). *Compt. Rend. Trav. Lab. Carlsberg* **34,** 1.
Porter, K. R., and Machado, R. D. (1960). *J. Biophys. Biochem. Cytol.* **7,** 167.
Prescott, D. M. (1964). *In* "Synchrony in Cell Division and Growth" (E. Zeuthen, ed.), p. 71. Wiley (Interscience), New York.
Prescott, D. M., and Bender, M. A. (1962). *Exptl. Cell Res.* **26,** 260.
Prescott, D. M., and Kimball, R. F. (1961). *Proc. Natl. Acad. Sci. U.S.* **47,** 686.
Rake, A. V., and Graham, A. F. (1964). *Biophys. J.* **4,** 267.
Reiter, J. M., and Littlefield, J. W. (1964). *Biochim. Biophys. Acta* **80,** 562.
Rikmenspoel, R. (1965). *Exptl. Cell Res.* **37,** 2.
Ritossa, F. M., and Spiegelman, S. (1965). *Proc. Natl. Acad. Sci. U.S.* **53,** 737.
Robbins, E., and Gonatas, N. K. (1964a). *J. Histochem. Cytochem.* **12,** 204.
Robbins, E., and Gonatas, N. K. (1964b). *J. Cell Biol.* **21,** 429.
Robbins, E., and Marcus, P. I. (1964c). *Science* **144,** 1152.
Robbins, E., and Scharff, M. D. (1965). *Federation Proc.* **24,** 445.
Robbins, E., Marcus, P. I., and Gonatas, N. K. (1964). *J. Cell Biol.* **21,** 49.
Rueckert, R. R., and Mueller, G. C. (1960). *Cancer Res.* **20,** 1584.
Scharff, M. D., and Robbins, E. (1965). *Nature* **208,** 464.
Scharff, M. D., and Robbins, E. (1966). *Science* (in press).
Scharff, M. D., Shatkin, A. J., and Levintow, L. (1963). *Proc. Natl. Acad. Sci. U.S.* **50,** 686.
Scherrer, K., Latham, H., and Darnell, J. E. (1963). *Proc. Natl. Acad. Sci. U.S.* **49,** 240.
Scholander, P. F., Leivestad, H., and Sundnes, G. (1958). *Exptl. Cell Res.* **15,** 505.
Schrader, H. W. (1953). "Mitosis." Columbia Univ. Press, New York.
Seed, J. (1963). *Nature* **198,** 147.
Sinclair, R., and Bishop, D. H. L. (1965). *Nature* **205,** 1272.
Sinclair, W. K., and Morton, A. A. (1963). *Nature* **199,** 1158.

Stein, W. H., and Moore, S. (1954). *J. Biol. Chem.* **211,** 915.
Symposium on Macromolecular Aspects of the Cell Cycle. (1963). *J. Cellular Comp. Physiol.* **62,** Suppl. 1, 1.
Taylor, J. H. (1960). *Ann. N.Y. Acad. Sci.* **90,** 409.
Terasima, T., and Tolmach, L. J. (1963). *Exptl. Cell Res.* **30,** 344.
Umana, R., Updike, S., Randall, J., and Dounce, A. (1964). *In* "The Nucleohistones" (J. Bonner, P. T'so, eds.), Vol. 1, pp. 200–229. Holden-Day Inc., San Francisco.
Warner, J. R., Rich, A., and Hall, C. E. (1962). *Science* **138,** 1399.
Wettstein, F. O., Staehelin, T., and Noll, H. (1963). *Nature* **197,** 430.
Xeros, N. (1962). *Nature* **194,** 682.
Zeuthen, E. (1955). *Biol. Bull.* **108,** 366.

Author Index

Numbers in italics refer to pages on which the complete references are listed.

Subject Index

Most of the headings and subheadings found under specific organisms or cell types in this index deal with synchronized cells.